The IMA Volumes in Mathematics and its Applications

Volume 145

Series Editors
Douglas N. Arnold   Arnd Scheel

Institute for Mathematics and its Applications (IMA)

The Institute for Mathematics and its Applications was established by a grant from the National Science Foundation to the University of Minnesota in 1982. The primary mission of the IMA is to foster research of a truly interdisciplinary nature, establishing links between mathematics of the highest caliber and important scientific and technological problems from other disciplines and industries. To this end, the IMA organizes a wide variety of programs, ranging from short intense workshops in areas of exceptional interest and opportunity to extensive thematic programs lasting a year. IMA Volumes are used to communicate results of these programs that we believe are of particular value to the broader scientific community.

The full list of IMA books can be found at the Web site of the Institute for Mathematics and its Applications:

http://www.ima.umn.edu/springer/volumes.html

Presentation materials from the IMA talks are available at

http://www.ima.umn.edu/talks/

Douglas N. Arnold, Director of the IMA

* * * * * * * * * *

IMA ANNUAL PROGRAMS

1982–1983	Statistical and Continuum Approaches to Phase Transition
1983–1984	Mathematical Models for the Economics of Decentralized Resource Allocation
1984–1985	Continuum Physics and Partial Differential Equations
1985–1986	Stochastic Differential Equations and Their Applications
1986–1987	Scientific Computation
1987–1988	Applied Combinatorics
1988–1989	Nonlinear Waves
1989–1990	Dynamical Systems and Their Applications
1990–1991	Phase Transitions and Free Boundaries
1991–1992	Applied Linear Algebra
1992–1993	Control Theory and its Applications
1993–1994	Emerging Applications of Probability
1994–1995	Waves and Scattering
1995–1996	Mathematical Methods in Material Science
1996–1997	Mathematics of High Performance Computing

(Continued at the back)

Pao-Liu Chow Boris Mordukhovich
George Yin
Editors

Topics in Stochastic Analysis and Nonparametric Estimation

Springer

Pao-Liu Chow
Department of Mathematics
Wayne State University
Detroit, MI 48202

Boris Mordukhovich
Department of Mathematics
Wayne State University
Detroit, MI 48202
http://www.math.wayne.edu/~boris/

George Yin
Department of Mathematics
Wayne State University
Detroit, MI 48202
http://www.math.wayne.edu/~gyin/

ISBN-13: 978-0-387-75110-8 e-ISBN-13: 978-0-387-75111-5

Library of Congress Control Number: 2007937098

Mathematics Subject Classification (2000): 60F99, 62G05, 62G20, 60H10, 60H15, 60J60

Camera-ready copy provided by the IMA.

9 8 7 6 5 4 3 2 1

springer.com

FOREWORD

This IMA Volume in Mathematics and its Applications

TOPICS IN STOCHASTIC ANALYSIS AND NONPARAMETRIC ESTIMATION

contains papers that were presented at the IMA Participating Institution conference on "Asymptotic Analysis in Stochastic Processes, Nonparametric Estimation, and Related Problems" held on September 15–17, 2006 at Wayne State University. The conference, which was one of approximately ten selected each year for partial support by the IMA through its affiliates program, was dedicated to Professor Rafail Z. Khasminskii on the occasion of his 75th birthday, in recognition of his profound contributions to the field of stochastic processes and nonparametric estimation theory.

We are grateful to the participants and, especially, to the conference organizers, for making the event so successful. Pao-Liu Chow, Boris Mordukhovich, and George Yin of the Department of Mathematics at Wayne State University did a superb job organizing this first-rate event and in editing these proceedings.

We take this opportunity to thank the National Science Foundation for its support of the IMA.

Series Editors

Douglas N. Arnold, Director of the IMA

Arnd Scheel, Deputy Director of the IMA

DEDICATED TO

PROFESSOR RAFAIL Z. KHASMINSKII ON THE OCCASION OF HIS SEVENTY-FIFTH BIRTHDAY

PREFACE

Research on stochastic analysis and nonparametric inference has witnessed tremendous progress in recent years resulting in significant impact on mathematics and its applications to diverse fields in science and technology. The progress in technology and the emerging applications have presented new challenges to the study of stochastic processes and the statistical estimation theory. To assess the past achievement and to provide a road map for future research, an IMA participating institution conference entitled "Conference on Asymptotic Analysis in Stochastic Processes, Nonparametric Estimation, and Related Problems" was held at Wayne State University, September 15–17, 2006. On the occasion of his seventy-fifth birthday, this conference was also held to honor Professor Rafail Z. Khasminskii for his fundamental contributions to many aspects of stochastic processes and the nonparametric estimation theory. Presenting a concerted effort of researchers from multiple institutions, it assembled an impressive list of invited speakers, who are renowned leaders in the fields of probability theory, stochastic processes, stochastic differential equations, as well as in the nonparametric estimation theory, and related fields. A number of invited speakers were early developers of the fields of probability and stochastic processes, establishing the foundation of the Modern probability theory. Their earlier works have stimulated much of subsequent progress in these fields. It is conceivable that the conference will have a lasting impact on the further development of the conference subjects.

After the conference, to commemorate this special event, we felt it was fitting to put together an IMA volume dedicated to Professor Rafail Z. Khasminskii for his outstanding scientific achievement. So we invited the distinguished speakers of the conference to contribute to this commemorative volume and most of them graciously agreed. It consists of nine papers on various topics in probability and statistics. They include authoritative expositions as well significant research papers of current interest. Roughly, they are grouped in three major areas: four papers in asymptotic analysis of stochastic differential equations; three papers on nonparametric estimation problems, and the remaining two concerning stochastic analysis of partial differential equations. The papers will be sequentially arranged accordingly.

Without the help, assistance, support, and encouragement of many people, this volume could not have come into being. First of all, we would like to express our deep gratitude to the authors for their valuable contributions. Also we would like to take this opportunity to thank IMA, NSF, and Wayne State University for their financial support of the conference. We thank all of the conference participants, including the invited speakers, the poster presenters, and all attendees for making it a successful event.

Our thanks go to Douglas N. Arnold and Arnd Scheel for helping us with the preparations of the conference and this volume. We are grateful to the IMA staff, in particular to Patricia V. Brick and Dzung N. Nguyen for their help and assistance in putting the final product together as an IMA volume.

Pao-Liu Chow Boris Mordukhovich George Yin
Department of Mathematics, Wayne State University

CONTENTS

Part I: Asymptotic Analysis Involving Stochastic Differential Equations

SOME RECENT RESULTS ON AVERAGING PRINCIPLE

MARK FREIDLIN* AND ALEXANDER WENTZELL†

1. Non-perturbed system. Averaging principle is one of the main methods in perturbation theory. It came into being more than two centuries ago in celestial mechanics, but even now there are many open questions having to do with applications of this principle. And these questions are not just about rigorous justification of procedures applied, but even it is unclear sometimes how to apply averaging procedures.

One can see some of the difficulties in a relatively simple case of an oscillator with one degree of freedom. The non-perturbed system in this case is described by the equation

$$\ddot{q}_t = -F'(q_t). \tag{1.1}$$

If we introduce the Hamiltonian $H(q,p) = \frac{1}{2}p^2 + F(q)$, $(q,p) = x \in \mathbb{R}^2$, the equation (1.1) can be written as a system:

$$\dot{X}_t = \overline{\nabla} H(X_t), \qquad \overline{\nabla} H(x) = \left(\frac{\partial H}{\partial p}, -\frac{\partial H}{\partial q}\right). \tag{1.2}$$

It is well known that the energy $H(X_t)$ is preserved: $H(X_t) \equiv H(X_0)$, and the flow X_t preserves the Lebesgue measure in $\mathbb{R}^2$. Let us assume, at first, that the Hamiltonian has just one well (the potential $F(x)$ has just one minimum, and $\lim_{|q|\to\infty} F(q) = \infty$). The trajectories X_t are periodic, and each one moves along the corresponding level curve $C(z) = \{x \in \mathbb{R}^2 : H(x) = z\}$ of the Hamiltonian (see Figure 1).
The period of rotation along the curve $C(z)$ is equal to $T(z) = \oint_{C(z)} \frac{d\ell}{|\nabla H(x)|}$. It is well known that if $S(z)$ is the area of the region $G(z) \subset \mathbb{R}^2$ bounded by $C(z)$, then $T(z) = \frac{dS(z)}{dz}$. The flow X_t can be considered on the curve $C(z)$, $z > \min H(x)$, and it has a unique normalized invariant measure μ_z on this curve with density $m_z(x) = \big(T(z)|\nabla H(x)|\big)^{-1}$ with respect to the curve length.

If the Hamiltonian $H(x)$ has more than one well, as in Figure 2, then the level set $C(z) = \{x \in \mathbb{R}^2 : H(x) = z\}$ can consist of several connected components $C(z,k)$: $C(z) = \bigcup_{k=1}^{n(z)} C(z,k)$. In this case, an invariant measure with density $m_{z,k}(x) = \big(T(z,k)|\nabla H(x)|\big)^{-1}$ is concentrated on each $C(z,k)$, where $T(z,k) = \oint_{C(z,k)} |\nabla H(x)|^{-1}\, d\ell$ is the period of rotation

*Department of Mathematics, University of Maryland, College Park, MD 20742.
†Department of Mathematics, Tulane University, New Orleans, LA 70118.

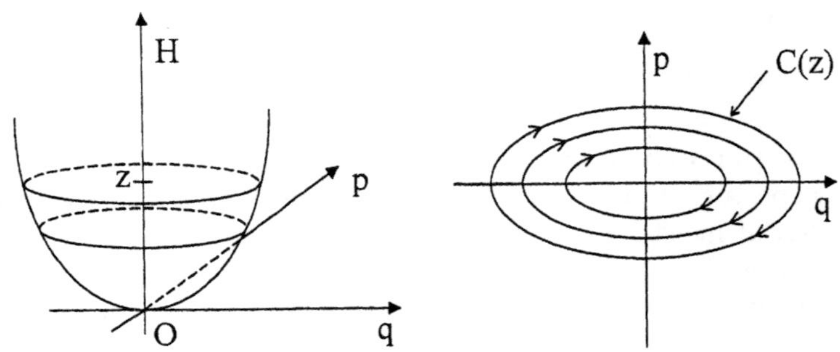

FIG. 1.

along $C(z,k)$. We'll see that non-uniqueness of the invariant measure on the level set $C(z) = \bigcup_k C(z,k)$ leads to essential changes in the averaging principle. If z is the value of the Hamiltonian $H(x)$ at a saddle point, the connected component of the level set $C(z)$ containing this saddle point consists of three trajectories (one is the equilibrium point).

Let us denote with Γ (Fig. 2) the graph that is homeomorphic to the set of connected components of all level sets (provided with the natural topology). In the case of a one-well potential, Γ consists of just one edge. Let us number the edges I_1, ..., $I_n \in \Gamma$. Each point of the graph Γ can be characterized by the value z of the Hamiltonian on the level set component corresponding to this point and the number k of the edge containing this point. The pair (z,k) forms a global coordinate system in Γ.

Let $Y : \mathbb{R}^2 \mapsto \Gamma$ be the projection of $\mathbb{R}^2$: for $x \in \mathbb{R}^2$, $Y(x)$ is the point of Γ with coordinates $\big(H(x),\ k(x)\big)$, where $k(x)$ is the number of the edge in Γ containing the point corresponding to the level set component containing x. It is clear that not just $H(x)$ is preserved by the flow X_t, but $k(x)$ as well. So $k(x)$ is a kind of additional first integral for our system (a discrete one). One can say that the whole $Y(x)$, with its two coordinates $H(x)$, $k(x)$ is a first integral.

Let us denote with $G(z,k)$ the region in $\mathbb{R}^2$ bounded by the curve $C(z,k)$; let $S(z,k)$ be the area of the region $G(z,k)$. An eight-shaped curve (see Fig. 2) is associated to each saddle point of the Hamiltonian $H(x)$. (Note that the Hamiltonian $H(q,p) = \dfrac{1}{2}p^2 + F(q)$ has no local maxima in our case; for general Hamiltonian systems, $H(x)$ can have local maxima as well as minima.)

Now let us consider a Hamiltonian system with many degrees of freedom

$$\dot{X}_t = \overline{\nabla} H(X_t),$$
$$\overline{\nabla} H(x) = \overline{\nabla} H(q,p) = \big(\nabla_p H(q,p),\ -\nabla_q H(q,p)\big), \quad q,\, p \in \mathbb{R}^n.$$

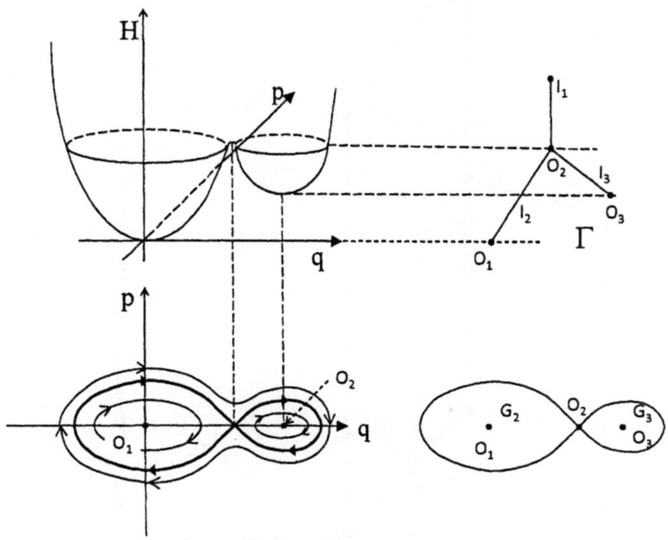

FIG. 2.

This system, in the case of $n > 1$, may have other smooth first integrals besides $H(x)$: $H_1(x) = H(x)$, ..., $H_l(x)$. If $l = n$, and the integrals are independent, the system is called completely integrable (some mild additional assumptions are usually made). Assume that all $H_k(x)$ are smooth enough generic functions. Let us take $C(z) = \{x \in \mathbb{R}^{2n} : H_1(x) = z_1, ..., H_l(x) = z_l\}$, $z = (z_1, ..., z_l) \in \mathbb{R}^l$, and assume that for every $z \in \mathbb{R}^l$ the set $C(z)$ is compact. Some of the sets $C(z)$ may consist of several connected components. After identifying all points of each connected component, we get a topological space which is called an open book. It consists of l-dimensional manifolds which can be glued together at some manifolds of smaller dimension. The points at which the matrix $(\nabla \mathbf{H}(x)) = (\nabla H_1(x), ..., \nabla H_l(x))$ has maximum rank l correspond to interior points of the l-dimensional "pages" of this open book. The "binding" at which the pages are glued together correspond to the points where the rank is less than l. In particular, the graph Γ considered in the case of one degree of freedom is an example of such an open book: its pages are edges, and the binding consists of the vertices.

Another example: Consider n independent one-degree-of-freedom oscillators

$$\dot{X}_k(t) = \overline{\nabla} H_k(X_k(t)), \qquad X_k(t) \in \mathbb{R}^2, \quad k = 1, ..., n. \tag{1.3}$$

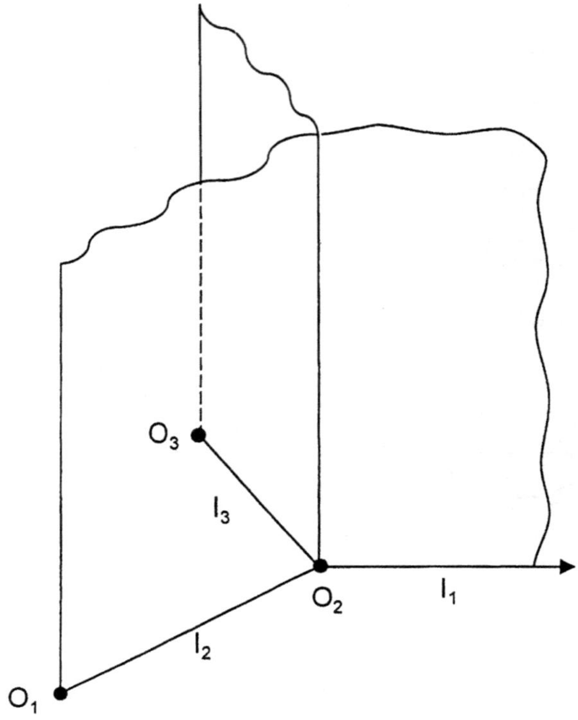

FIG. 3.

Assume that all $H_k(x_k)$ are smooth generic functions, $\lim_{|x_k|\to\infty} H_k(x_k) = \infty$. Then (1.3) considered as a system in $\mathbb{R}^{2n}$ is completely integrable: its first integrals are $H_1(x_1)$, ..., $H_n(x_n)$. It is easy to see that the corresponding open book Π is the product of the graphs Γ_k associated with $H_k(x_k)$, $k = 1$, ..., n: $\Pi = \Gamma_1 \times ... \times \Gamma_n$. For instance, if $n = 2$, $H_1(x_1)$ has just one well as in Figure 1, and $H_2(x_2)$ has two as in Figure 2, then $\Pi = \Gamma_1 \times \Gamma_2$ is shown in Figure 3: Π has three two-dimensional pages and the binding $\Gamma_1 \times \{O_2\}$.

Each page of the open book corresponding to system (1.3) is the product of edges of graphs Γ_k associated with $H_k(x_k)$, one edge from each graph. The edges of each Γ_k are numbered, so a page of the open book Π can be identified by the set of integers i_1, ..., i_n, where i_k is the number of the edge taken from Γ_k. Since in the example corresponding to Figure 3 one of the two graphs has just one edge, the pages are characterized just by the numbers 1, 2, 3.

We can consider the mapping $Y\colon \mathbb{R}^{2n} \mapsto \Pi$ such that $Y^{-1}(y)$, $y \in \Pi$, is the connected component of the level set $C(z)$, $z \in \mathbb{R}^l$, corresponding to y. In the case of system (1.3), the mapping Y is the product of mappings $Y_k : \mathbb{R}^2 \mapsto \Gamma_k$ associated to $H_k(x_k)$: $Y = Y_1 \times ... \times Y_n$. If $y \in \Pi$ is an interior point of an n-dimensional page $\pi_{i_1,...,i_n}$, then $Y^{-1}(y) = C(z_1, i_1; ...; z_n, i_n)$ is an n-dimensional manifold. The density const$\cdot \left(\prod_{k=1}^n |\nabla H_k(x_k)|\right)^{-1}$ on $C(z_1, i_1; ...; z_n, i_n)$ defines an invariant measure for the system (1.3) on this n-dimensional surface. If the frequencies $\omega_k(z_k, i_k)$, $1 \le k \le n$, of rotations along the curves $C_k(z_k, i_k)$ (the i_k-th component of the level set $\{x\colon H_k(x) = z_k\}$) are rationally independent, the normalized invariant measure on $C(z_1, i_1; ...; z_n, i_n)$ is unique. If they are rationally dependent, an infinite set of invariant measures concentrated on tori of dimension smaller than n exists. In the generic case, the set of $z \in \mathbb{R}^n$ for which the frequencies are rationally dependent is dense in $\mathbb{R}^n$. Existence of many normalized invariant measures leads to some difficulties in justification of the averaging principle.

Most of our results can, actually, be generalized to dynamical systems with conservation laws, not necessarily Hamiltonian; but the Hamiltonian structure allows to simplify the results.

Up to the present, we haven't spoken of any perturbations. Let us continue this for a little while.

As the non-perturbed system we can consider *a diffusion process.* Let a process X_t be defined by the stochastic differential equation in $\mathbb{R}^{2n}$:

$$\dot{X}_t = \overline{\nabla} H(X_t) + \sigma(X_t) * \dot{W}_t. \tag{1.4}$$

Here W_t is the Wiener process in $\mathbb{R}^{2n}$, $\dot{W}_t$ the white noise; the $(2n \times 2n)$-matrix $\sigma(x)$ is assumed to have entries with bounded derivatives, and the stochastic term $\sigma(X_t) * \dot{W}_t$ is understood in the sense that the process X_t is governed by the generator L:

$$Lu(x) = \left(\overline{\nabla} H(x), \nabla u(x)\right) + \frac{1}{2} \operatorname{div}\left(a(x) \nabla u(x)\right),$$

where $a(x) = \sigma(x)\, \sigma^*(x)$. It is easy to see that the Lebesgue measure in $\mathbb{R}^{2n}$ is invariant for the process X_t.

Assume that $a(x) \nabla H(x) \equiv 0$. Then $H(x)$ is a first integral for the process X_t: $H(X_t) \equiv H(X_0)$ (this follows from the Itô formula). Assume additionally that for every connected component $C(z, k)$ of the level set $C(z) = \{x \in \mathbb{R}^{2n} : H(x) = z\}$ such that $C(z, k)$ does not contain critical points of the Hamiltonian there exists a positive constant a_0 such that $e \cdot a(x)\, e \ge a_0 (e \cdot e)$ for any vector $e \in \mathbb{R}^{2n}$ such that $e \cdot \nabla H(x) = 0$, and $x \in C(z, k)$. Then on every connected component of this kind the process X_t has a unique normalized invariant measure. This measure has a density $m_{z,k}(x)$. It follows form the fact that the Lebesgue measure is invariant for X_t that $m_{z,k}(x) = \left(|\nabla H(x)| \oint_{C(z,k)} |\nabla H(x)|^{-1}\, ds\right)^{-1}$. If $C(z, k)$ does contain a critical point of $H(x)$ (we assume that it is at most one critical point), then the invariant measure is concentrated at this critical point.

2. Perturbations. Perturbations of a dynamical system can have various nature. They can be deterministic or random, and even their smallness can be understood in different ways. Actually, the perturbed system is the primary object. If the perturbed system depends on a small parameter $\varepsilon > 0$, the non-perturbed system can be constructed as a simpler one that approximates the original system for small ε. Sometimes we may succeed in finding such an approximation, sometimes not.

For a smooth vector field $b(x,\varepsilon)$, $x \in \mathbb{R}^{2n}$, $0 \le \varepsilon \ll 1$, we have: $b(x,\varepsilon) = b(x,0) + \varepsilon\beta(x) + o(\varepsilon)$. Therefore the classical perturbation theory, as a rule, is dealing with additive small perturbations.

Consider at first autonomous deterministic perturbations of the one-degree-of-freedom Hamiltonian system (1.2):

$$\dot{\tilde{X}}_t^\varepsilon = \overline{\nabla} H(\tilde{X}_t^\varepsilon) + \varepsilon\beta(\tilde{X}_t^\varepsilon).$$

Here $\beta(x)$ is a smooth vector field, $0 \le \varepsilon \ll 1$. Under some mild additional assumptions, $\tilde{X}_t^\varepsilon$ is uniformly close to the non-perturbed trajectory X_t on any finite time interval as $\varepsilon \downarrow 0$. However often we are interested in the behavior of $\tilde{X}_t^\varepsilon$ on long, of order ε^{-1}, time intervals. To study $\tilde{X}_t^\varepsilon$ on long time intervals, it is convenient to rescale the time taking $X_t^\varepsilon = \tilde{X}_{t/\varepsilon}^\varepsilon$. then X_t^ε satisfies the equation

$$\dot{X}_t^\varepsilon = \frac{1}{\varepsilon}\overline{\nabla} H(X_t^\varepsilon) + \beta(X_t^\varepsilon). \tag{2.1}$$

Displacements of X_t^ε, for $0 < \varepsilon \ll 1$, have two components: the fast one, which is, actually, the motion along the non-perturbed trajectory at a speed of order ε^{-1}, and the slow component that describes the displacement in the direction transversal to the non-perturbed trajectories. In the case of a one-well Hamiltonian with one degree of freedom (see Fig. 1) the slow motion can be characterized by the evolution of $H(X_t^\varepsilon)$. Taking into account the fact that $\nabla H(x) \cdot \overline{\nabla} H(x) \equiv 0$, we can write:

$$H(X_t^\varepsilon) - H(X_0^\varepsilon) = \int_0^t \nabla H(X_s^\varepsilon) \cdot \beta(X_s^\varepsilon)\, ds. \tag{2.2}$$

Before $H(X_t^\varepsilon)$ changes a little from its initial value $z = H(X_0)$, the trajectory X_t^ε makes many rotations along the loop $C(z) = \{x \in \mathbb{R}^2 \colon H(x) = z\}$. This implies that

$$\int_0^t \nabla H(X_s^\varepsilon) \cdot \beta(X_s^\varepsilon)\, ds = t\Big[\frac{1}{T(z)} \oint_{C(z)} \frac{\nabla H(x) \cdot \beta(x)\, d\ell}{|\nabla H(x)|} + o_\varepsilon(1)\Big] + o(t)$$

(we remember that $\big(T(z)|\nabla H(x)|\big)^{-1}$ is the invariant density on $C(z)$). We conclude from the divergence theorem that

$$\oint_{C(z)} \frac{\nabla H(x) \cdot \beta(x)\, d\ell}{|\nabla H(x)|} = \int_{G(z)} \operatorname{div}\beta(x)\, dx,$$

where $G(z)$ is the region in $\mathbb{R}^2$ bounded by the curve $C(z)$. One can expect that in the one-well case

$$\lim_{\varepsilon\downarrow 0} |H(X_t^\varepsilon) - Z_t| = 0 \tag{2.3}$$

uniformly on any finite time interval, where Z_t is the solution of the equation

$$\dot{Z}_t = \overline{\beta}(Z_t), \qquad \overline{\beta}(z) = \frac{1}{T(z)} \int_{G(z)} \operatorname{div}\beta(x)\, dx, \tag{2.4}$$

with the initial condition $Z_0 = H(X_0)$. The statement (2.3) is a classical manifestation of the averaging principle (see, e. g., [2], Section 52).

Along with deterministic perturbations of the equation (1.2) of the form $\varepsilon\beta(x)$, one can consider random perturbations leading, in the rescaled time, to the perturbed equation of the form

$$\dot{X}_t^\varepsilon = \frac{1}{\varepsilon} \overline{\nabla} H(X_t^\varepsilon) + \beta(X_t^\varepsilon) + \sigma(X_t^\varepsilon)\dot{W}_t. \tag{2.5}$$

Here $\dot{W}_t$ is the Gaussian white noise, and $\sigma(x)$ a matrix with smooth entries. As before, the fast component of X_t^ε is the motion along the non-perturbed trajectories, and the slow component is described by $H(X_t^\varepsilon)$. Applying Itô's formula, we obtain:

$$\begin{aligned} H(X_t^\varepsilon) - H(X_0^\varepsilon) = &\int_0^t \nabla H(X_s^\varepsilon) \cdot \beta(X_s^\varepsilon)\, ds \\ &+ \int_0^t \nabla H(X_s^\varepsilon) \cdot \sigma(X_s^\varepsilon)\, dW_s \\ &+ \frac{1}{2} \int_0^t \sum_{i,j=1}^2 a_{ij}(X_s^\varepsilon) \frac{\partial^2 H}{\partial x^i \partial x^j}\, ds, \end{aligned} \tag{2.6}$$

where $a(x) = (a_{ij}(x)) = \sigma(x)\sigma^*(x)$. A one-dimensional Wiener process $\tilde{W}(t)$ exists such that the stochastic term in (2.6) can be written as $\tilde{W}\Big(\int_0^t a(X_s^\varepsilon)\nabla H(X_s^\varepsilon) \cdot \nabla H(X_s^\varepsilon)\, ds\Big)$. Using this fact and the arguments leading to (2.3)–(2.4) (see below), one can show that for X_t^ε given by (2.5) the process $H(X_t^\varepsilon)$ converges weakly in the space of continuous functions $\mathbf{C}_{[0,T]}$ to the diffusion process defined by the equation

$$\dot{Z}_t = \overline{\beta}(Z_t) + \tilde{\beta}(Z_t) + \tilde{\sigma}(Z_t)\dot{\tilde{W}}_t, \qquad Z_0 = H(X_0).$$

Here $\overline{\beta}(z)$ is the same as in (2.4), and

$$\tilde{\beta}(z) = \frac{1}{2T(z)} \oint_{C(z)} \sum_{i,j} a_{ij}(x) \frac{\partial^2 H}{\partial x^i \partial x^j} \frac{d\ell}{|\nabla H(x)|},$$

$$\sigma^2(z) = \frac{1}{T(z)} \int_{G(z)} \operatorname{div}\big(a(x)\,\nabla H(x) \cdot \nabla H(x)\big)\, dx.$$

R.Z.Khasminskii was the first who rigorously considered averaging for stochastic differential equations [16], [17].

One can consider perturbed systems of the form

$$\begin{aligned} \dot{X}_t^\varepsilon &= \frac{1}{\varepsilon}\overline{\nabla} H(X_t^\varepsilon) + \beta_1(X_t^\varepsilon, Y_t^\varepsilon), \\ \dot{Y}_t^\varepsilon &= \beta_2(X_t^\varepsilon, Y_t^\varepsilon), \end{aligned} \tag{2.7}$$

with the component X_t^ε two-dimensional, and Y_t^ε one-dimensional. In this case, the fast component is the same as before, but the phase space of the slow component is an open book being the direct product of the slow-component phase space of the system (2.1) and the y-axis, since now the coordinate y is a first integral of the corresponding non-perturbed system.

Another type of perturbations is systems with a small delay. Let us consider a perturbation of (1.1) of the following form:

$$\ddot{q}_t^h = f(q_t^h, q_{t-h}^h), \qquad t \geq h. \tag{2.8}$$

This equation should be solved with the initial condition q_t^h given for t between 0 and h.

Suppose $f(\ ,\ )$ is a smooth function; let us introduce the function $f(q)$ of one variable that is equal to the value of the function $f(\ ,\ )$ on the "diagonal": $f(q) = f(q,q)$. Then for small positive h we can write $\ddot{q}_t^h = f(q_t^h, q_{t-h}^h) = f(q_t^h) - h \cdot f_2(q_t^h, q_t^h)\,\dot{q}_t^h + o(h)$. So, to some extent, we can treat small delay as a small additive perturbation.

Let us give a sketch of the proof of (2.3). Consider an auxiliary function $u(x)$, $x \in \mathbb{R}^2$, that satisfies the equation

$$Lu(x) := \overline{\nabla} H(x) \cdot \nabla u(x) = \nabla H(x) \cdot \beta(x) - \overline{\beta}(z) \tag{2.9}$$

on each curve $C(z) = \{x\colon\ H(x) = z\}$, $z > \min_x H(x)$. We assume that $H(x)$ has just one well as in Figure 1, so that $C(z)$ has one connected component, and the non-perturbed system has just one normalized invariant measure concentrated on $C(z)$. For each z, the equation (2.9) has a solution, since the right-hand side of (2.9) is orthogonal to this measure. This solution is unique up to an additive constant $a = a(z)$; and the constants $a(z)$ can be chosen so that $u(x)$ has continuous derivatives in $\mathbb{R}^2$.

If X_t^ε is the solution of the equation (2.1), then

$$u(X_t^\varepsilon) - u(X_0^\varepsilon) = \frac{1}{\varepsilon}\int_0^t \nabla u(X_s^\varepsilon) \cdot \overline{\nabla} H(X_s^\varepsilon)\, ds + \int_0^t \nabla u(X_s^\varepsilon) \cdot \beta(X_s^\varepsilon)\, ds.$$

From this equality and (2.9) we conclude that for any $T > 0$ there exists a constant C_1 such that

$$\max_{0 \leq t \leq T}\left| \int_0^t \left[\beta(X_s^\varepsilon) \cdot \overline{\nabla} H(X_s^\varepsilon) - \overline{\beta}(H(X_s^\varepsilon))\right] ds \right| \leq C_1 \varepsilon. \tag{2.10}$$

It follows from (2.2), (2.4) and (2.10) that

$$|H(X_t^\varepsilon) - Z_t| \le K \int_0^t |H(X_s^\varepsilon) - Z_s|\, ds + C_1\varepsilon, \quad 0 \le t \le T,$$

where K is the Lipschitz constant for $\overline{\beta}(z)$. This inequality implies (2.3).

At the first glance, using such an auxiliary function $u(x)$ is just a trick, but actually, similar proofs of the averaging principle work in many situations.

In the example mentioned above, deterministic motion along the loops $C(z)$ played the part of the fast motion. Consider now perturbations of the stochastic equation (1.4) in $\mathbb{R}^{2n}$. After an appropriate time change, the perturbed equation takes the form

$$\dot{X}_t^\varepsilon = \frac{1}{\varepsilon}\overline{\nabla}H(X_t^\varepsilon) + \frac{1}{\sqrt{\varepsilon}}\sigma(X_t^\varepsilon) * \dot{W}_t + \beta(X_t^\varepsilon) \tag{2.11}$$

(for brevity's sake, we consider only deterministic perturbations). Let the assumptions made at the end of Section 1 be satisfied, and the Hamiltonian $H(x)$, $x \in \mathbb{R}^{2n}$, has just one well. On each $(2n-1)$-dimensional manifold $C(z) = \{x \in \mathbb{R}^{2n} : H(x) = z\}$ the process defined by (1.4) has a unique invariant density with respect to the $(2n-1)$-dimensional surface area dS

$$m_z(x) = \big(T(z)|\nabla H(x)|\big)^{-1}, \qquad T(z) = \oint_{C(z)} |\nabla H(y)|^{-1}\, dS.$$

Let $G(z)$ be the region in $\mathbb{R}^{2n}$ bounded by $C(z)$, and define the function $\overline{\beta}(z)$ by

$$\overline{\beta}(z) = \frac{1}{T(z)} \int_{G(z)} \operatorname{div}\beta(x)\, dx.$$

For each z, let us consider the problem (similar to (2.9)):

$$Lu(x) = \nabla H(x) \cdot \beta(x) - \overline{\beta}(z), \qquad x \in C(z). \tag{2.12}$$

Here $Lu(x) = \frac{1}{2}\operatorname{div}\big(a(x)\nabla u(x)\big) + \overline{\nabla}H(x) \cdot \nabla u(x)$ is the generator of the non-perturbed diffusion process on $C(z)$. Since the right-hand side of (2.12) is orthogonal to $m_z(x)$ (which is the solution of the adjoint equation $L^* m_z(x) = 0$, $x \in C(z)$, unique up to a constant factor), the problem (2.12) is solvable (by the Fredholm alternative). The solution of (2.12) is unique up to an additive constant $\alpha = \alpha(z)$. Using the uniqueness of the normalized invariant density, one can prove that the constants $\alpha = \alpha(z)$ can be chosen so that the function $u(x)$ has continuous second derivatives in $\mathbb{R}^{2n}$. Then, applying Itô's formula to $u(X_t^\varepsilon)$ one can obtain bounds allowing to prove convergence of $H(X_t^\varepsilon)$ in probability to the solution Z_t

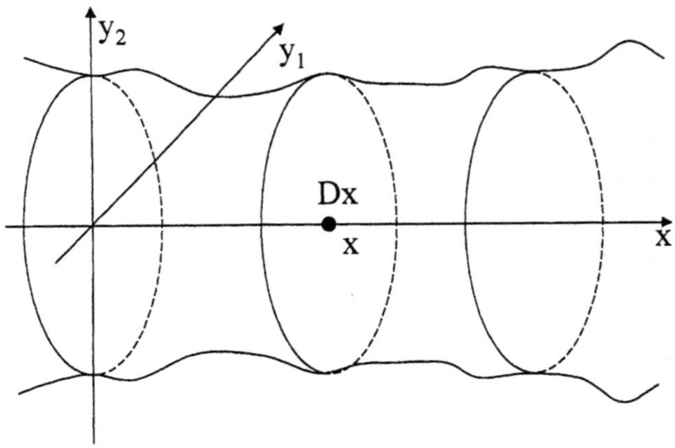

FIG. 4.

of the equation (2.4) with the initial condition $Z_0 = H(X\varepsilon_0)$, uniformly in every finite time interval (we assume that the initial condition for the process X_0^ε is taken the same for all $\varepsilon > 0$).

Let us mention one more averaging problem (see [4]). Let D be a set in $\mathbb{R}^3$ shown in Figure 4: D contains the x-axis; for each $x \in \mathbb{R}^1$, the cross-section $D_x = \{y = (y_1, y_2) \in \mathbb{R}^2 : (x, y_1, y_2) \in D$ is a bounded connected region; D has a smooth boundary ∂D, and the normal $n(x,y)$ to is not parallel to the x-axis for any $(x,y) \in \partial D$.
Let us make the region D narrower in the y-directions by the factor ε: $D^\varepsilon = \{(x,y) : x \in \mathbb{R}^1,\ y \in \mathbb{R}^2,\ (x, \varepsilon^{-1}y) \in D\}$. Consider the Wiener process $Z_t^\varepsilon = \big(X^\varepsilon(t), Y_1^\varepsilon(t), Y_2^\varepsilon(t)\big)$ in D^ε with normal reflection at the boundary ∂D^ε. Let $\gamma_1^\varepsilon(x,y)$, $(x,y) \in D^\varepsilon$, be the projection of the unit inward normal vector to ∂D^ε on the x-axis, and $\gamma_2^\varepsilon(x,y)$ its projection onto the (y_1, y_2)-plane. Then the process Z_t^ε can be described by the stochastic equations

$$\begin{aligned} \dot{X}_t^\varepsilon &= \dot{W}_t^1 + \gamma_1^\varepsilon(X_t^\varepsilon, Y_t^\varepsilon)\dot{l}_t^\varepsilon, \\ \dot{Y}_t^\varepsilon &= \dot{W}_t^2 + \gamma_2^\varepsilon(X_t^\varepsilon, Y_t^\varepsilon)\dot{l}_t^\varepsilon, \end{aligned} \tag{2.13}$$

where W_t^1, W_t^2 are, respectively, a one-dimensional and a two-dimensional independent Wiener processes, and l_t^ε is the local time of the process Z_t^ε on ∂D^ε. Here the fast component of Z_t^ε is a two-dimensional Wiener process in the cross-section with reflection at its boundary, while X_t^ε is the slow component. One can expect that the slow component converges as $\varepsilon \downarrow 0$ to a one-dimensional diffusion process X_t; the local-time term in the first of the equations (2.13) leads to a drift term in the process X_t. To calculate the drift term we can consider the following problem (that corresponds, in this case, to the problem (2.12)):

$$\Delta u(y) = A, \quad y \in D_x, \qquad \left.\frac{\partial u}{\partial \tilde{n}}\right|_{y \in D_x} = \frac{\gamma_1^1(x,y)}{|\gamma_2^1(x,y)|}, \tag{2.14}$$

where $x \in \mathbb{R}^1$ is a parameter, and $\tilde{n} = \tilde{n}(y)$ is the unit inward normal vector to ∂D_x. The constant A can be chosen in a unique way to make the problem (2.14) solvable:

$$A = -\frac{1}{S(x)} \oint_{\partial D_x} \frac{\gamma_1^1(x,y)}{|\gamma_2^1(x,y)|}\, d\ell,$$

where $S(x)$ is the area of the cross-section D_x. Using the Itô formula, one can obtain that the component X_t^ε converges in distribution as $\varepsilon \downarrow 0$ to the diffusion process X_t in $\mathbb{R}^1$ governed by the operator

$$Lv(x) = \frac{1}{2}\frac{d^2v}{dx^2} + \frac{S'(x)}{2S(x)}\frac{dv}{dx}.$$

3. Multiwell Hamiltonians. In the previous section we mentioned several examples in which using the averaging principle one can calculate the characteristics of the limiting slow motion. The fast motion in those examples had exactly one normalized invariant measure on each level set $C(z) = \{x \in \mathbb{R}^{2n} : H(x) = z\}$: we assumed that $H(x)$ had just one well (if the fast motion was deterministic, we assumed that $n = 1$, and for $n > 1$ we considered the case of the fast motion being a non-degenerate diffusion process on each manifold $C(z)$).

Now we are going to consider the case of non-unique invariant measure. Let, at first, the system have one degree of freedom, and suppose that the Hamiltonian has more than one well like in Fig. 2. Consider a perturbed system described by equation (2.1). In this case, the fast motion is again the motion along the non-perturbed trajectories. But now the fast motion, at least for some z, has several normalized invariant measures. For example, in the case shown in Fig. 2, at least two such measures exist for $H(O_3) \leq z < H(O_2)$: one concentrated on $C(z,1)$, another on $C(z,2)$.

The slow component Y_t^ε is the projection of X_t^ε onto the graph Γ associated with $H(x)$: $Y_t^\varepsilon = Y(X_t^\varepsilon) = \big(H(X_t^\varepsilon), k(X_t^\varepsilon)\big)$. One can show that $\lim_{\varepsilon \downarrow 0} Y_t^\varepsilon$, in general, may not exist (see, for instance, [3]). So in the multiwell case, in general, the averaging principle does not hold at least in the classical sense.

To establish a weaker version of the averaging principle in the multiwell case, one can consider a stochastic regularization of the problem. Let us denote by $X_t^{x,\varepsilon}$ the solution of (2.1) with initial condition $X_0^{x,\varepsilon} = x \in \mathbb{R}^2$. Suppose the initial point is stochastically perturbed: we consider the solution $X_t^{x+\delta\zeta,\varepsilon}$ of (2.1) with the initial condition $X_0^{x+\delta\zeta,\varepsilon} = x+\delta\zeta$, where ζ is uniformly distributed in a unit circle centered at the origin, and δ is a positive parameter (that we'll take eventually to 0). Then the solution is a stochastic process. We can consider, first, the limiting behavior of the slow component $Y(X_t^{x+\delta\zeta,\varepsilon})$ of $X_t^{x+\delta\zeta,\varepsilon}$ as $\varepsilon \downarrow 0$, and then the limit as $\delta \downarrow 0$. If such a double limit exists in some sense, it can be considered as the limiting slow component in our original problem. This approach is natural since the initial point is usually known with some error. This way of stochastic regularization of a deterministic problem was, actually, used by D.Anosov [1] for justification of the averaging principle for a class of multidimensional problems, in particular, for perturbations of completely integrable systems in a region in which the angle-action coordinates can be introduced. The strongest results for perturbations of systems in the angle-action coordinates were obtained by A.Neistadt [19].

It was shown in [3] that this approach can be used for justification of the averaging principle for a class of one-degree of freedom multiwell systems. In particular, it works if the Hamiltonian has just one saddle point. But in the case of more than one saddle point, the double limit does not exist for a broad class of perturbations.

Let us explain what the double limit passage yields in the case of a Hamiltonian with just one saddle point.

Suppose $H(x)$ has one saddle point, as shown in Fig. 2; Γ is the graph corresponding to H. To be specific, assume that $\operatorname{div}\beta(x) < 0$ for $x \in \mathbb{R}^2$. Then for every small positive δ and every $x \in \mathbb{R}^2$ with $H(x) > H(O_2)$ there exists a $t_0 = t_0(x)$ such that $Y_{t_0}^{\varepsilon,\delta} = Y(X_{t_0}^{x+\delta\zeta,\varepsilon})$ belongs alternately to the edges I_2 or I_3 as $\varepsilon \to 0$. Consider a stochastic process Y_t^δ, $t \geq 0$, on Γ that, while inside the edges I_i, $i \in \{1,2,3\}$, moves in accordance with the equation (2.4) with $T(z)$ and $\overline{\beta}(z)$ replaced with $T_i(z)$, $\overline{\beta}_i(z)$ calculated by integrating, respectively, over $C(z,i)$ and $G(z,i)$. When the trajectory Y_t^δ comes to the vertex O_2 (from the edge I_1), it goes without any delay to I_2 or I_3 with probabilities $p_i = \int_{G_i} \operatorname{div}\beta(x)\,dx \Big/ \int_{G_1\cup G_2} \operatorname{div}\beta(x)\,dx$, $i = 2, 3$, respectively (the regions G_2 and G_3 are shown in Fig. 2).

One can prove [3] that the slow motion $Y_t^{\varepsilon,\delta}$ converges weakly as $\varepsilon \downarrow 0$ in the space $\mathbb{C}_{[0,T]}$ of continuous functions on the interval $[0,T]$ with values in Γ to the process Y_t^δ in Γ. Now we can take the limit of Y_t^δ as $\delta \downarrow 0$: Y_t^δ converges weakly in
$mathbbC_{[0,T]}$ as $\delta \downarrow 0$ to the process Y_t that is governed by the same differential equations within the edges, and passes from I_1 to I_2 or I_3 with the same probabilities as Y_t^δ, only its initial condition is no longer $\big(H(x+\delta\zeta),k\big)$, but just $Y_0 = \big(H(x),k\big)$. It is this stochastic process Y_t, the double limit of $Y(X_t^{x+\delta\zeta,\varepsilon})$ as first ε and then δ are taken to zero, that should be considered as the limit of the (purely deterministic) slow component $Y(X_t^\varepsilon)$.

As was mentioned above, if the Hamiltonian has more than one saddle point, such a regularization works only for a special class of perturbations (see [3]). So to justify the averaging principle in the multiwell case we should consider some stochastic perturbations of a different kind. It turns out that for this end it's good to consider white-noise type stochastic perturbations (of course, such perturbations are of interest by themselves and not as a device for regularization of deterministic perturbations).

Let $X_t^{\varepsilon,\varkappa}$ be defined by the equation

$$\begin{aligned} \dot{X}_t^{\varepsilon,\varkappa} &= \frac{1}{\varepsilon}\,\overline{\nabla} H(X_t^{\varepsilon,\varkappa}) + \beta(X_t^{\varepsilon,\varkappa}) + \sqrt{\varkappa}\,\sigma(X_t^{\varepsilon,\varkappa}) * \dot{W}_t, \\ X_0^{\varepsilon,\varkappa} &= x \in \mathbb{R}^2, \quad 0 < \varepsilon,\ \varkappa \ll 1, \end{aligned} \tag{3.1}$$

so that the generator $L^{\varepsilon,\varkappa}$ of $X_t^{\varepsilon,\varkappa}$ is given by

$$\begin{aligned} L^{\varepsilon,\varkappa}u(x) &= \frac{1}{\varepsilon}\,\overline{\nabla} H(x)\cdot\nabla u(x) + \beta(x)\cdot\nabla u(x) + \frac{\varkappa}{2}\operatorname{div}\big(a(x)\nabla u(x)\big), \\ a(x) &= \sigma(x)\,\sigma^*(x). \end{aligned}$$

Let us consider the case of $\beta(x) \equiv 0$. It follows from [12], Ch. 8 that in this case the processes $Y_t^{\varepsilon,\varkappa} = Y(X_t^{\varepsilon,\varkappa})$ on Γ converge weakly in $\mathbf{C}_{[0,T]}$ as $\varepsilon \downarrow 0$ to a diffusion process $Y_t^{\varkappa}$ on the graph Γ. For each $\varkappa > 0$, the process $Y_t^{\varkappa}$ is determined by a set of operators $\tilde{L}_i^{\varkappa}$ defined by $\tilde{L}_i^{\varkappa}v_i(z) = \dfrac{\varkappa}{2T_i(z)}\dfrac{d}{dz}\big(\overline{a}_i(z)\dfrac{dv_i(z)}{dz}\big)$, one operator on each edge of Γ, and by the gluing conditions at the vertices; $T_i(z)$ are the same as before, and $\overline{a}_i(z) = \dfrac{1}{T_i(z)}\displaystyle\int_{G(z,i)} \operatorname{div}\big(a(x)\nabla H(x)\big)\,dx$. The operators $\tilde{L}_i^{\varkappa}$ determine the behavior of $Y_t^{\varkappa}$ inside the edges, while the gluing conditions determine it at the vertices.

The vertices of the graph Γ are classified into *interior* and *exterior* ones. For example, for the Hamiltonian shown in Figure 2, just one interior vertex exists, namely, O_2. Exterior vertices are inaccessible, and no gluing conditions are given at them; at every interior vertex a gluing condition is prescribed, in addition to the requirement of continuity of $v_i(z)$ and $\tilde{L}_i^{\varkappa}v_i(z)$ as functions of $y = (z,i)$. For the example of Fig. 2, the gluing condition has the form $\sum_{i=1}^3 \alpha_i \dfrac{dv_i}{dz}(O_2) = 0$, where $\alpha_i = \displaystyle\int_{G_i} \operatorname{div}\big(a(x)\nabla H(x)\big)\,dx$, $i = 2, 3$, and $\alpha_1 = -\alpha_2 - \alpha_3$.

If $\beta(x) \not\equiv 0$ in (3.1), the gluing conditions for $Y_t^{\varkappa}$ remain the same (assuming that $\det a(x) \neq 0$); this is proved using the Girsanov formula. The operators $L_i^{\varkappa}$ in this case have the form $L_i^{\varkappa} = \tilde{L}_i^{\varkappa} + \overline{\beta}_i(z)\dfrac{d}{dz}$.

Now we can look for the limit of $Y_t^{\varkappa}$, $0 \le t \le T$, as $\varkappa \downarrow 0$. It turns out that the weak limit of $Y_t^{\varkappa}$ exists, and it is the same process Y_t as was considered above. So if the regularization by random perturbation of the initial point is possible, its result coincides with that of regularization by means of white-noise type perturbations of the equation (this last is possible in the general situation). Moreover, from what was said it's clear that the limiting process Y_t does not depend on the diffusion matrix $a(x)$. This means that the stochasticity of the limiting slow motion is an intrinsic property of the unperturbed system (1.1): a result of its instability near saddle points; the characteristics of the noise that we added for regularization do not influence the result.

Some degenerate (but not *completely* degenerate) stochastic perturbations of Hamiltonian systems that are natural in certain problems are considered in [7], [3]. A system of the form (2.8) with a small delay and multiwell potential $F(q)$ is studied in [10].

Results concerning weak convergence of the slow component in the multiwell case are of special interest when motion of a large number of particles, or motion of a fluid is considered. For example, motion of particles in an incompressible flow that is close to a planar motion can be described by the equations (2.7). Stochasticity of such a motion and the characteristics of the corresponding stochastic process are studied in [5], [6]. In the case of equations (2.13) in a tube D^ε with branching, the limiting slow motion is also a diffusion process on the corresponding graph [11].

Finally, we'll mention in this section perturbations of a diffusion process with conservation law $H(x)$ (equation (2.11)). Let Γ be the graph homeomorphic to the set of connected components of level sets of $H(x)$. Under the non-degeneracy assumptions mentioned at the end of Section 1, a unique invariant measure is concentrated on each connected component $C(z, k)$. One can prove (see [8], [9]) that the slow component of the perturbed motion, which is again the projection $Y(X_t^\varepsilon)$ of X_t^ε onto the graph, converges weakly as $\varepsilon \downarrow 0$ to a diffusion process on Γ. It is worth mentioning that in the multidimensional case the branching of the graph occurs at vertices corresponding to critical points of the Hamiltonian (which is assumed to be generic) at which the quadratic main part of $H(x)$ has only one plus or only one minus; to all other vertices are attached only one or two edges.

4. Systems with many degrees of freedom. Consider a Hamiltonian system with $n > 1$ degrees of freedom. Suppose it has l smooth first integrals H_1, ..., $H_l(x)$. If the system is ergodic on each (almost each) connected component of the level sets $C(z) = \{x \in \mathbb{R}^{2n} : H_1(x) = z_1, \dots, H_l(x) = z_l\}$, $z = (z_1, \dots, z_l) \in \mathbb{R}^l$, one can expect that a version of the averaging principle holds for perturbations of this system, and the space (an open book) obtained by identifying all points of each connected component of $C(z)$ is the phase space of the limiting slow motion for the perturbed system (see [13]).

We have seen that in the case of one degree of freedom, in a region where the Hamiltonian has no critical points, averaging is relatively simple. At the same time, considering such parts of the phase space (parts corresponding to the interior corresponding to the interior of one page of the open book) is a necessary first step to understanding the general situation. In a region without critical points, under some additional assumptions, one can introduce the angle-action coordinates (see [2], Section 50). So we'll start with a completely integrable system with n degrees of freedom, which can be written in the action-angle coordinates. In the case of many degrees of freedom the situation with the averaging principle is not so simple even in the case of a region without critical points.

After an appropriate time rescaling, the perturbed system in this case has the form:

$$\begin{aligned} \dot{I}_t^\varepsilon &= \beta_1(I_t^\varepsilon, \varphi_t^\varepsilon), \\ \dot{\varphi}_t^\varepsilon &= \frac{1}{\varepsilon}\,\Omega(I_t^\varepsilon) + \beta_2(I_t^\varepsilon, \varphi_t^\varepsilon), \end{aligned} \tag{4.1}$$

where $I_t^\varepsilon = \big(I_1^\varepsilon(t), ..., I_n^\varepsilon(t)\big) \in \mathbb{R}^n$, and $\varphi_t^\varepsilon = \big(\varphi_1^\varepsilon(t), ..., \varphi_n^\varepsilon(t)\big)$ belongs to an n-dimensional torus $\mathbb{T}^n$ with coordinates $\varphi_i \in [0, 2\pi)$; $\Omega(I) = \big(\omega_1(I), ..., \omega_n(I)\big)$, and $\beta_1(I,\varphi)$, $\beta_2(I,\varphi)$ are vector functions on $\mathbb{R}^n \times \mathbb{T}^n$. The fast motion is a motion on $\mathbb{T}^n$ with velocity vector that is constant for every value of $I = (I_1, ..., I_n)$. The slow motion is characterized by the evolution of the action vector I_t^ε.

If, for a given $I \in \mathbb{R}^n$, a non-zero vector $k = (k_1, ..., k_n)$ with integer components k_j exists such that $\sum_{j=1}^n k_j\, \omega_j(I) = 0$, the frequencies $\omega_j(I)$ are called rationally dependent. It is well known that if the frequencies $\omega_j(I)$ are rationally independent, then the uniform distribution on $\mathbb{T}^n$ is the unique invariant measure for the non-perturbed flow on $\mathbb{T}^n$ for the value of I considered. If $\omega_j(I)$ are rationally dependent, there exists an infinite set of such measures. In the generic case, the set A ($\subset \mathbb{R}^n$) consisting of all I for which $\omega_j(I)$ are rationally dependent (this set is called the resonance set) is dense in $\mathbb{R}^n$. One can expect that if this set is small in some sense, the slow component converges to the motion $\overline{I}_t$ with velocity obtained by averaging with respect to the Lebesgue measure:

$$\dot{\overline{I}}_t = \overline{\beta}_1(I_t), \qquad \overline{\beta}_1(I) = \frac{1}{(2\pi)^n}\int_{\mathbb{T}^n} \beta_1(I,\varphi)\, d\varphi.$$

This seems natural since the normalized Lebesgue measure is the unique invariant measure for most of the tori. Results in this direction can be found in [1], [18], [19] that can be interpreted as concerning the regularization of the original problem using stochastic perturbations of the initial point. The convergence of $\max_{0\le t\le T} |I_t^\varepsilon - \overline{I}_t|$ to zero as $\varepsilon \downarrow 0$ (the averaged trajectory being taken with the same initial condition $\overline{I}_0$ as I_0^ε) is proved to take place in the sense of convergence in Lebesgue measure on the space of initial points (which is the same as convergence in probability).

As we have already mentioned in the previous section, regularization by a stochastic perturbation of the initial point does not work, in general, for systems with multiwell Hamiltonians, so we should study regularization by stochastic perturbations of the equation. As the first step in this direction, such a regularization should be considered in a region where the action-angle coordinates can be introduced. This problem (in a slightly more general setting) was studied in [14]:

Suppose that stochastic perturbations are added to equation (4.1):

$$\begin{aligned} \dot I_t^{\varepsilon,\varkappa} &= \beta_1(I_t^{\varepsilon,\varkappa},\varphi_t^{\varepsilon,\varkappa}) + \sqrt{\varkappa}\,\sigma_1(I_t^{\varepsilon,\varkappa},\varphi_t^{\varepsilon,\varkappa})\,\dot W_t, \\ \dot I_t^{\varepsilon,\varkappa} &= \frac{1}{\varepsilon}\,\Omega(I_t^{\varepsilon,\varkappa}) + \beta_2(I_t^{\varepsilon,\varkappa},\varphi_t^{\varepsilon,\varkappa}) + \sqrt{\varkappa}\,\sigma_2(I_t^{\varepsilon,\varkappa},\varphi_t^{\varepsilon,\varkappa})\,\dot W_t. \end{aligned} \tag{4.2}$$

Here $\varkappa \ge 0$ is a parameter, W_t a Wiener process in $\mathbb{R}^{2n}$, and $\sigma_1(I,\varphi)$, $\sigma_1(I,\varphi)$ are matrices of the appropriate dimensions with smooth entries that are 2π-periodic in φ_j. Take $a(I,\varphi) = \big(a_{ij}(I,\varphi)\big) = \sigma_1(I,\varphi)\,\sigma_1^*(I,\varphi)$, $\overline{a}_{ij}(I) = (2\pi)^{-n}\int_{\mathbb{T}^n} a_{ij}(I,\varphi)\,d\varphi$, $\overline{a}(I) = \big(\overline{a}_{ij}(I)\big)$, $\overline{\beta}_1(I) = (2\pi)^{-n}\int_{\mathbb{T}^n}\beta(I,\varphi)\,d\varphi$. Let $A \subset \mathbb{R}^n$ be the resonance set for the frequencies $\omega_j(I)$.

It was proved in [14] that if the matrix $a(I,\varphi)$ is non-degenerate and the resonance set A has zero Lebesgue measure, then, whatever the initial conditions $I_0^{\varepsilon,\varkappa} = I_0$, $\varphi_0^{\varepsilon,\varkappa} = \varphi_0$ for the equation (4.2) are, the slow component $I_t^{\varepsilon,\varkappa}$ converges weakly as $\varepsilon \downarrow 0$ to the diffusion process $\overline{I}_t^{\varkappa}$ in $\mathbb{R}^n$ described by the equation

$$\dot{\overline{I}}_t^{\varkappa} = \overline{\beta}_1(\overline{I}_t^{\varkappa}) + \sqrt{\varkappa}\,\overline{\sigma}(\overline{I}_t^{\varkappa})\,\dot W_t, \quad \overline{I}_0^{\varkappa} = I_0, \tag{4.3}$$

where the matrix function $\overline{\sigma}(I)$ satisfies the equality $\overline{\sigma}(I)\,\overline{\sigma}^*(I) = \overline{a}(I)$. This means that under the conditions mentioned above the averaging is performed in the natural way. It is easy to understand that this is not true, in general, if the set A contains a non-empty open set.

After this, we can consider the limiting behavior of the processes $\overline{I}_t^{\varkappa}$ defined by (4.3) as $\varkappa \downarrow 0$. It can be proved that they converge in probability uniformly on every finite time interval $[0,T]$ to the solution $\overline{\overline{I}}_t$ of the equation

$$\dot{\overline{\overline{I}}}_t = \beta_1(\overline{\overline{I}}_t), \qquad \overline{\overline{I}}_0 = I_0.$$

This means that the averaging principle holds for equation (4.1) in the sense of weak convergence after regularization by means of a non-degenerate white-noise-type perturbation of the equation, if the resonance set A has zero Lebesgue measure.

In Hamiltonian systems, it seems to be natural to consider white-noise-type perturbations added only to the "velocity" coordinates. This leads to a degenerate matrix $a(I, \varphi)$. Some results in this case can also be found in [14].

Now let us consider small deterministic perturbations of the simple completely integrable system (1.3), possibly with multiwell Hamiltonians $H_k(x_k)$(see [15]). After an appropriate time change, the perturbed system has the form

$$\dot{X}_k^\varepsilon(t) = \frac{1}{\varepsilon}\overline{\nabla} H_k(X_k^\varepsilon(t)) + \beta_k(X_1^\varepsilon(t), ..., X_n^\varepsilon(t)). \qquad (4.4)$$

In the perturbed system, the oscillators are no longer independent, therefore we called them weakly coupled oscillators. The fast component of $X_t^\varepsilon = (X_1^\varepsilon(t), ..., X_n^\varepsilon(t))$ is once again the motion along the non-perturbed trajectories (with the time change). The phase space of the slow component is the open book $\Pi = \Gamma_1 \times ... \times \Gamma_n$ (see the notations in Section 1), and the slow component is the projection $Y(X_t^\varepsilon)$ of X_t^ε onto Π. The averaging principle should describe the limiting behavior of $Y(X_t^\varepsilon)$ as $\varepsilon \downarrow 0$. In general, $\lim_{\varepsilon\downarrow 0} Y(X_t^\varepsilon)$ does not exist; and one has to consider regularization of (4.4) by a stochastic perturbation of the equations:

$$\dot{X}_k^{\varepsilon,\varkappa}(t) = \frac{1}{\varepsilon}\overline{\nabla} H_k(X_k^{\varepsilon,\varkappa}(t)) + \beta_k(X_1^{\varepsilon,\varkappa}(t), ..., X_n^{\varepsilon,\varkappa}(t)) + \sqrt{\varkappa}\sigma_k \dot{W}_t^k. \qquad (4.5)$$

We choose here as σ_k some non-degenerate (2×2)-matrices with constant entries; the two-dimensional Wiener processes W_t^k are assumed independent.

Inside each n-dimensional page of the open book Π (each such page is the product of n edges $I_{1\,i_1} \times I_{2\,i_2} \times ... \times I_{n\,i_n}$ of the graphs Γ_1, ..., Γ_n), one can introduce action-angle coordinates. Thus the limiting as $\varepsilon \downarrow 0$ (in the sense of weak convergence) slow motion $Y_t^\varkappa$ exists inside the pages, and is a diffusion process – at least if the corresponding resonance set $A \subset \mathbb{R}^n$ has Lebesgue measure 0. This diffusion process, in general, hits the binding of the open book in a finite time.

It was shown in [15] that if the process $Y_t^\varkappa$ starts at any interior point of a page, then with probability 1 it hits the binding at some point of its $(n-1)$-dimensional part: the parts of the binding of smaller dimensions, corresponding to the ends of the edges $I_{k\,i_k}$ being taken for more than one k, are inaccessible. Thus gluing conditions for $Y_t^\varkappa$ should be prescribed only at the $(n-1)$-dimensional part of the binding. These gluing conditions, roughly speaking, determine the behavior of the process after it hits the binding. The differential operators determining the process $Y_t^\varkappa$ inside the pages, as well as the gluing conditions, are to be found in [15]. It turns out that, for any $\varkappa > 0$, the gluing conditions are independent from the drift terms $\beta_k(x_1, ..., x_n)$, and the diffusion coefficients of $Y_t^\varkappa$ in each page are proportional to $\varkappa$.

Now we take $\varkappa$ to zero, and find the limiting slow motion Y_t on Π. Since the diffusion coefficients of $Y_t^\varkappa$ go to 0 as $\varkappa \downarrow 0$, the process Y_t will be deterministic inside the pages. The stochasticity of Y_t is concentrated on the binding only and can be described by the probabilities of going from one page to this or that other page that come together at a point belonging to the binding; these probabilities may depend on the point of the binding considered. It's worth mentioning that the probabilities of passing from one page to another at the binding are determined by the deterministic part of the perturbations and are independent of the matrices σ_k responsible for the white-noise part of the perturbations. In this sense, we can say that the stochasticity of the limiting slow motion for system (4.4) is an intrinsic property of this deterministic system (caused, actually, by the instabilities in the original non-perturbed system). The stochastic terms in (4.5) serve just for the purpose of regularization of the problem.

REFERENCES

[1] D. Anosov, *Averaging in systems of differential equations with rapidly oscillating solutions*, Izv. Akad. Nauk SSSR Math., **24** (1960), 721–742.

[2] V. Arnold, *Mathematical Methods of Classical Mechanics*, Springer, 1978.

[3] M. Brin and M. Freidlin, *On stochastic behavior of perturbed Hamiltonian systems*, Ergodic Th. Dynam. Sys., **20** (2000), 55–76.

[4] M. Freidlin, *Markov Processes and Differential Equations: Asymptotic Problems*, Birkhauser, 1996.

[5] M. Freidlin, *Deterministic 3D-perturbations of planar incompressible flow lead to stochasticity*, Journ. Statistical Phys., **111** (2003), 1209–1218.

[6] M. Freidlin, *Reaction-convection in incompressible 3D-fluid: a homogenization problem*, [incomplete].

[7] M. Freidlin and M. Weber, *Random perturbations of nonlinear oscillators*, Ann. of Probability, **26**(3) (1998), 925–967.

[8] M. Freidlin and M. Weber, *On random perturbations of Hamiltonian systems with many degrees of freedom*, Stoch. Processes and Their Applications, **94** (2001), 199–239.

[9] M. Freidlin and M. Weber, *Random perturbations of dynamical systems with conservation laws*, Probab. Theory and Related Fields, **128** (2004), 441–466.

[10] M. Freidlin and M. Weber, *On stochasticity of solutions of differential equations with a small delay*, Stochastics and Dynamics, **5**(3) (2005), 475–486.

[11] M. Freidlin and A. Wentzell, *Diffusion processes on graphs and the averaging principle*, Ann. of Probability, **21**(4) (1993), 2215–2245.

[12] M. Freidlin and A. Wentzell, *Random Perturbations of Dynamical Systems*, Springer, Second edition, 1998.

[13] M. Freidlin and A. Wentzell, *Diffusion processes on an open book and the averaging principle*, Stoch. Processes and Their Applications, **113** (2004), 101–126.

[14] M. Freidlin and A. Wentzell, *Averaging principle for stochastic perturbations of multifrequency systems*, Stochastics and Dynamics, **3**(3) (2003), 393–408.

[15] M. Freidlin and A. Wentzell, *Long-time behavior of weakly coupled oscillators*, Journal of Statistical Physics, **123**(6) (2006), 1311–1337.

[16] R. Khasminskii, *Ergodic properties of recurrent diffusion processes and stabilization of solutions of the Cauchy problem for parabolic equations* (Russian), Teor. veroyatnost. i primen., **5** (1960), 179–196.

[17] R. Khasminskii, *Averaging principle for stochastic differential equations* (Russian), Kybernetica, Academia, Praha, **4** (1968), 260–279.

[18] P. Lochak and S. Meunier, *Multiphase Averaging for Classical Systems*, Springer, 1988.

[19] A. Neistadt, *Averaging in multi-frequency systems*, Dokl. Akad. Nauk SSSR, **223** (1975), 314–317; Averaging in multi-frequency systems II, ibid. **226** (1976), 1295–1298.

CRAMER'S THEOREM FOR NONNEGATIVE MULTIVARIATE POINT PROCESSES WITH INDEPENDENT INCREMENTS

FIMA KLEBANER* AND ROBERT LIPTSER†

Abstract. We consider a continuous time version of Cramer's theorem with nonnegative summands $S_t = \frac{1}{t}\sum_{i:\tau_i \le t} \xi_i$, $t \to \infty$, where $(\tau_i, \xi_i)_{i \ge 1}$ is a sequence of random variables such that tS_t is a random process with independent increments.

Key words. Nonnegative summands, boundary effect.

AMS(MOS) subject classifications. Primary 60F10, 60J27.

1. Introduction and main result. The following version of the Cramer theorem, [1], can be extracted from Dembo and Zeitouni, [2].

THEOREM 1.1. *Let $(\xi_i)_{i \ge 1}$ be a sequence of nonnegative identically distributed and independent random variables with ξ_1 admitting the Laplace transform:*

$$\mathscr{L}(\lambda) = \mathsf{E}e^{\lambda\xi_1}, \ \lambda \in (-\infty, \Lambda), \quad \exists\ \Lambda = \inf\{\lambda > 0 : \mathscr{L}(\lambda) = \infty\}.$$

Then, the family

$$S_n = \frac{1}{n}\sum_{i=1}^{n} \xi_i, \quad n \to \infty,$$

obeys the Large Deviation Principle (LDP) in the metric space $(\mathbb{R}_+, \varrho)$ (ϱ is Euclidean's metric) with the rate $\frac{1}{n}$ and the rate function

$$I(u) = \begin{cases} \sup\limits_{\lambda \in (-\infty,\Lambda)} \big[\lambda u - g(\lambda)\big], & u > 0 \\ -\log \mathsf{P}(\xi_1 = 0), & u = 0, \end{cases}$$

where $g(\lambda)$ is the log moment generation function,

$$g(\lambda) = \log \mathsf{E}e^{\lambda\xi_1}, \ \lambda < \Lambda.$$

In this paper, we study a "continuous time version" of this theorem. For $t \ge 0$, set

$$S_t = \frac{1}{t}\sum_{i:\tau_i \le t} \xi_i,$$

*School of Mathematical Sciences, Building 28M, Monash University, Clayton Campus, Victoria 3800, Australia (fima.klebaner@sci.monash.edu.au).

†Department of Electrical Engineering Systems, Tel Aviv University, 69978 Tel Aviv, Israel (liptser@eng.tau.ac.il).

where $(\xi_i, \tau_i)_{i\geq 1}$ is a sequence of random pairs, where ξ_i's and τ_i's are random variables:

$$\xi_i \geq 0 \quad \text{and} \quad \tau_0 = 0 < \tau_1 < \tau_2 < \ldots \tau_i < \ldots$$

defined on a probability space $(\Omega, \mathcal{F}, \mathsf{P})$. Let $(\mathscr{G}_n)_{n\geq 0}$ be the filtration with $\mathscr{G}_0 = (\varnothing, \Omega)$ and $\mathscr{G}_n := \sigma\{(\tau_i, \xi_i)_{i\leq n}\}$. Random variables ξ_i are assumed to be identically distributed and independent of $\mathscr{G}_{i-1}$ with the distribution function

$$G(x) = \mathsf{P}(\xi_1 \leq x), \ x \geq 0.$$

The conditional distribution of τ_i given $\mathscr{G}_{i-1}$ is exponential:

$$\mathsf{P}\big(\tau_i \leq t|\mathscr{G}_{i-1}\big) = \big(1 - e^{-r(t-\tau_{i-1})}\big), \ t \geq \tau_{i-1},$$

where r is a positive number. Moreover, we assume that

$$\mathsf{P}\big(\xi_i \leq x, \tau_i \leq t|\mathscr{G}_{i-1}\big) = G(x)\big(1 - e^{-r(t-\tau_{i-1})}\big), \ i \geq 1.$$

The following theorem is an analogue of Theorem 1.1.

THEOREM 1.2. *The family*

$$S_t = \frac{1}{t}\sum_{i:\tau_i\leq t}\xi_i, \quad t \to \infty,$$

obeys the LDP in the metric space $(\mathbb{R}_+, \varrho)$ *with the rate* $\frac{1}{t}$ *and the rate function*

$$I(u) = \begin{cases} \sup\limits_{\lambda\in(-\infty,\Lambda)} \big[\lambda u - r\int_0^\infty (e^{\lambda z} - 1)dG(z)\big], & u > 0 \\ r[1 - G(0+)], & u = 0. \end{cases} \tag{1.1}$$

We give two examples illustrating compatibility with Theorems 1.1 and 1.2. For both discrete and continuous time cases, let

$$\mathsf{P}(\xi_1 \leq x) = 1 - e^{-x}, \ x \geq 0,$$

so that, ξ_1 has the Laplace transform with $\Lambda = 1$. The log moment generating function is

$$g(\lambda) = -\log(1 - \lambda), \ \lambda < 1.$$

For $G(x) = 1 - e^{-x}$, $x \geq 0$.

$$\int_0^\infty r(e^{\lambda z} - 1)dG(z) = \frac{r\lambda}{1-\lambda}, \ \lambda < 1.$$

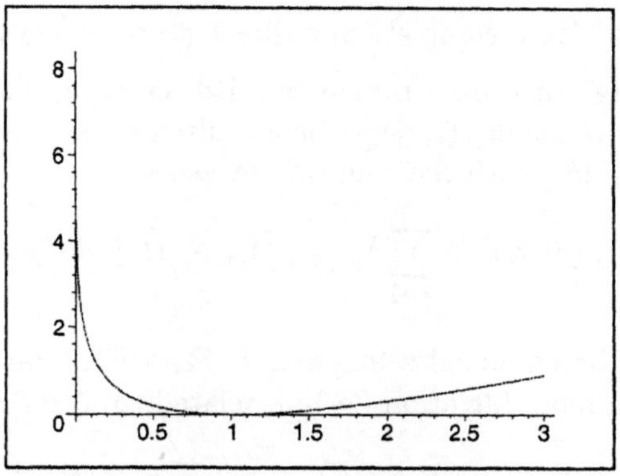

FIG. 1. *The rate function* $I^d(u)$

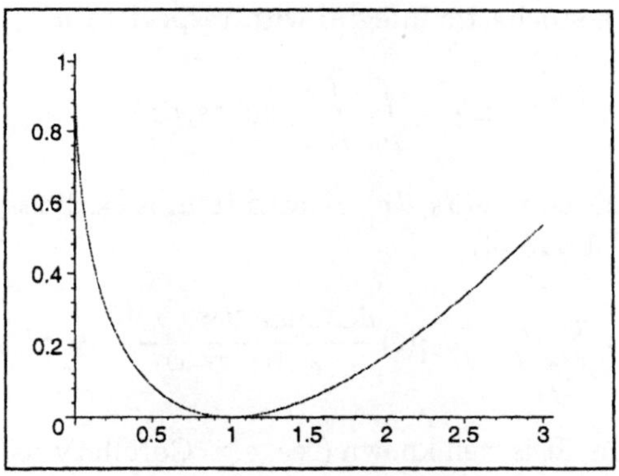

FIG. 2. *The rate function* $I^c(u)$ *for* $r = 1$

In both cases, rate functions are explicitly computable (see also Figures 1 and 2),

$$I^d(u) = \begin{cases} u - 1 - \log(u), & u > 0 \\ \infty, & u = 0 \end{cases} \quad \text{(discrete time case)}$$

$$I^c(u) = \begin{cases} \left(\sqrt{r} - \sqrt{u}\right)^2, & u > 0 \\ r, & u = 0. \end{cases} \quad \text{(continuous time case)}$$

REMARK 1.1. Related topics to Theorem 1.2 can be found e.g. in Georgii and Zessin, [3], serving a class of marked point random fields. Probably, the proof of Theorem 1.2 can be adapted with arguments from proofs in [3] provided that many details not related to our setting have to be omitted and other ones concerning to the boundary effect have to be added.

We prefer to give a complete and direct proof of Theorem 1.2.

2. Counting random measure, its compensator. Laplace transform. We consider $(\xi_i, \tau_i)_{i\geq 1}$ as a multivariate (marked) point process (see, e.g. [4], [5]) with the counting measure

$$\mu(dt, dy) = \sum_{i\geq 1} \mathbf{I}_{\{\tau_i<\infty\}}\delta_{\{\tau_i,\xi_i\}}(t, y)dtdy,$$

where $\delta_{\{\tau_i,\xi_i\}}$ is the Dirac delta-function on $\mathbb{R}_+ \times \mathbb{R}_+$. Parallel to $(\mathscr{G}_n)_{n\geq 0}$, we introduce one more filtration $(\mathscr{G}_t)_{t\geq 0}$ related to $(\xi_i, \tau_i)_{i\geq 1}$:

$$\mathscr{G}_t := \sigma(\mu([0, t'] \times \Gamma) : t' \leq t, \Gamma \in \mathscr{B}(\mathbb{R}_+)),$$

where $\mathscr{B}(\mathbb{R}_+)$ is the Borel σ-algebra on $\mathbb{R}_+$, and assume that $\mathscr{G}_0$ is augmented by P-zero sets from $\mathcal{F}$ (notice that, then, $(\mathscr{G}_t)_{t\geq 0}$ satisfies the general conditions). With the help of the counting measure μ, one can present tS_t in a form of a stochastic integral with respect to μ:

$$tS_t = \int_0^t \int_{x>0} x\mu(ds, dx). \tag{2.1}$$

Then, the Levy measure $\nu(ds, dx)$, related to μ, is explicitly computed (see, e.g. Theorem III.1.33, [5])

$$\nu(ds, dx) = \sum_{i\geq 1} I_{]\tau_{i-1},\tau_i]}(t)\frac{dG(x)de^{-r(s-\tau_{i-1})}}{e^{-r(s-\tau_{i-1})}} = rdsdG(x) \tag{2.2}$$

and is not random. It is well known (see, e.g. Corollary to Theorem 1 in §4, Ch.4, [6], that under the deterministic Levy measure the random process tS_t has independent increments from non-overlapping time intervals. We recall a useful property: for any nonnegative and $(\mathscr{G}_t)$-predictable function $f(\omega, x, t)$,

$$\mathsf{E}\int_0^t \int_{x>0} f(\omega, x, s)\mu(ds, dx) = \mathsf{E}\int_0^t \int_{x>0} f(\omega, x, s)\nu(ds, dx)$$

with "$\infty = \infty$".

LEMMA 2.1. [Laplace transform] *For any $\lambda < \Lambda$ and $t > 0$,*

$$\mathsf{E}e^{\lambda tS_t} = e^{rt\int_{x>0}[e^{\lambda z}-1]dG(z)}.$$

Proof. Though the direct computation of Laplace's transform is permissible, we prefer to apply the stochastic calculus. The process $U_t = e^{\lambda tS_t}$ has right-continuous piece-wise constant paths with jumps

$$\triangle U_s = (U_s - U_{s-}) = U_{s-}\int_{x>0} [e^{\lambda x} - 1]\mu(\{s\}, dx),$$

so that, for any $t > 0$,

$$U_t = 1 + \int_0^t \int_{x>0} U_{s-}[e^{\lambda x} - 1]\mu(ds, dx).$$

Since the function $f(\omega, x, s) := U_{s-}[e^{\lambda x} + 1]$ is nonnegative and predictable, the following equality with $\lambda < \Lambda$ holds true:

$$E\int_0^t \int_{x>0} U_{s-}[e^{\lambda x} + 1]\mu(ds, dx) = E\int_0^t \int_{x>0} U_{s-}[e^{\lambda x} + 1]\nu(ds, dx) \ (< \infty).$$

Then, we also have

$$E\int_0^t \int_{x>0} U_{s-}[e^{\lambda x} - 1]\mu(ds, dx) = E\int_0^t \int_{x>0} U_{s-}[e^{\lambda x} - 1]\nu(ds, dx) \ (\in \mathbb{R}).$$

By $\nu(ds, dx) = rdsdG(x)$ (see (2.2)), the later provides the integral equation $(\mathsf{E}U_t) = 1 + \int_0^t \int_{x>0}(EU_s)[e^{\lambda x} - 1]dG(x)rds$ which is equivalent to the differential equation $\frac{d(\mathsf{E}U_t)}{dt} = (\mathsf{E}U_t)\int_{x>0}[e^{\lambda x} - 1]dG(x)r$ subject to $(\mathsf{E}U_0) = 1$.

Thus, the desired result holds. □

3. The proof of Theorem 1.2. We verify the necessary and sufficient conditions for the LDP to hold (for more details, see Puhalskii, [7]):

1) exponential tightness,

$$\lim_{j\to\infty} \overline{\lim_{t\to\infty}} \frac{1}{t} \log \mathsf{P}\big(S_t \in \mathbb{R}_+ \setminus \mathcal{K}_j\big) = -\infty,$$

where $\mathcal{K}_j$'s are compacts increasing to $\mathbb{R}_+$;

2) local LDP, defining the rate function $I(u)$, $u \in \mathbb{R}_+$

$$\lim_{\delta\to 0} \lim_{t\to\infty} \frac{1}{t} \log \mathsf{P}\big(|S_t - u| \le \delta\big). = -I(u).$$

3.1. The exponential tightness. By choosing

$$K_j = \{x \in \mathbb{R}_+ : x \in [0, j]\}$$

and applying Chernoff's inequality with parameter 0.5Λ, we find that

$$\mathsf{P}(S_t > j) \le e^{-0.5\Lambda j + \log \mathsf{E}e^{0.5\Lambda t S_t}}$$

By Lemma 2.1, we have $\mathsf{E}e^{0.5\Lambda t S_t} = e^{rt\int_{x>0}[e^{0.5\Lambda x} - 1]dG(x)}$ and, therefore,

$$\frac{1}{t} \log \mathsf{P}\big(S_t > j\big) \le -0.5\Lambda j + r\int_{x>0} [e^{0.5\Lambda x} - 1]dG(x) \xrightarrow[j\to\infty]{} -\infty.$$

□

3.2. The local LDP. We begin with computation of $I(0)$ and prove

$$\begin{aligned} &\varliminf_{\delta\to 0}\varliminf_{t\to\infty}\frac{1}{t}\log \mathsf{P}(S_t\le\delta)\ge -r[1-G(0+)] \\ &\varlimsup_{\delta\to 0}\varlimsup_{t\to\infty}\frac{1}{t}\log \mathsf{P}(S_t\le\delta)\le -r[1-G(0+)]. \end{aligned} \tag{3.1}$$

By (2.2), $\{tS_t=0\}=\{\mu((0,t]\times\{x>0\})=0$. Consequently for any $t>0$,

$$\mathsf{P}(S_t\le\delta)\ge \mathsf{P}(S_t=0)=\mathsf{P}(tS_t=0)=\mathsf{P}\big(\mu((0,t],\{x>0\})=0\big).$$

The counting process $\pi_t:=\mu\big((0,t],\{x>0\}\big)$ has independent increments and the rate $\mathsf{E}\mu\big((0,t],\{x>0\}\big)=\nu\big((0,t],\{x>0\}\big)=r[1-G(0+)]t$.

So, π_t is the counting process with the compensator

$$\nu\big((0,t],\{x>0\}\big)=r[1-G(0+)]t.$$

Therefore, by the Watanabe theorem, [8], π_t is Poisson process with parameter $r[1-G(0+)]$. Hence, due to well known property of the Poisson process $\mathsf{P}(\pi_t=0)=e^{-tr[1-G(0+)]}$, we find that

$$\frac{1}{t}\log \mathsf{P}(S_t\le\delta)\ge\frac{1}{t}\log \mathsf{P}(\pi_t=0)=-r[1-G(+)]$$

and the lower bond from (3.1).

The upper bound from (3.1) is derived with the help of Laplace's transform with $0<\lambda<\Lambda$. To this end, we use identity

$$1=\mathsf{E}\exp\Big(\lambda tS_t-tr\int_{x>0}[e^{\lambda x}-1]dG(x)\Big)$$

implying the inequality $1\ge \mathsf{E}I_{\{S_t\le\delta\}}\exp\big(t\big[\lambda\delta-r\int_{x>0}[e^{\lambda x}-1]dG(x)\big]\big)$ being equivalent to $\frac{1}{t}\log\mathsf{P}(S_t\le\delta)\le-\lambda\delta+r\int_{x>0}[e^{-\lambda x}-1]dG(x)$.

Now, passing $t\to\infty$, we obtain the following upper bound depending on δ and λ: $\varlimsup_{t\to\infty}\frac{1}{t}\log\mathsf{P}(S_t\le\delta)\le-\lambda\delta+r\int_{x>0}[e^{\lambda x}-1]dG(x)$ and, then, passing $\delta\to 0$ and λ to $-\infty$ we find that

$$\varlimsup_{\delta\to 0}\varlimsup_{t\to\infty}\frac{1}{t}\log\mathsf{P}(S_t\le\delta)\le -r\int_{x>0}dG(x)=-r[1-G(0+)].$$

We continue the proof by checking the validity of the $I(u)$, $u>0$, from (1.1). In other words, we show that

$$\begin{aligned} &\varlimsup_{\delta\to 0}\varlimsup_{t\to\infty}\frac{1}{t}\log\mathsf{P}(|S_t-u|\le\delta)\le -I(u) \\ &\varliminf_{\delta\to 0}\varliminf_{t\to\infty}\frac{1}{t}\log\mathsf{P}(|S_t-u|\le\delta)\ge -I(u), \end{aligned}$$

with $I(u) = \sup_{\lambda\in(-\infty,\Lambda)} \big[\lambda u - r\int_0^\infty (e^{\lambda x}-1)dG(x)\big]$.

The Laplace transform

$$1 = \mathsf{E}\exp\Big(\lambda t S_t - tr\int_{x>0}[e^{\lambda x}-1]dG(x)\Big),\ \lambda<\Lambda,$$

implies the inequality

$$1 \ge \mathsf{E}I_{\{|S_t-u|\le\delta\}}\exp\Big(-t\delta u + \lambda t u - tr\int_{x>0}[e^{\lambda x}-1]dG(x)\Big)$$

and, in turn, the upper bound, depending on λ:

$$\overline{\lim_{\delta\to0}}\ \overline{\lim_{t\to\infty}}\ \frac{1}{t}\log\mathsf{P}(|S_t-u|\le\delta) \le -\Big(\lambda u - r\int_{x>0}[e^{\lambda x}-1]dG(x)\Big).$$

The desired upper bound is obtained by further maximization in λ over $(-\infty,\Lambda)$ of $\big(\lambda u - r\int_{x>0}[e^{\lambda x}-1]dG(x)\big)$.

The lower bound proof uses a standard approach of changing "probability measure". Denote by $\lambda^* = \operatorname{argmax}_{\lambda<\Lambda}\big(\lambda u - r\int_{x>0}[e^{\lambda x}-1]dG(x)\big)$. Since λ^* solves the equation (with $u>0$)

$$u = r\int_{x>0} x e^{\lambda x}dG(x), \tag{3.2}$$

λ^* is a proper positive number strictly less than Λ. Set

$$\mathfrak{L}_t(\lambda^*) = \exp\left(\lambda^* t S_t - \int_0^t\int_{x>0} r[e^{\lambda^* x}-1]dG(x)ds\right). \tag{3.3}$$

First of all we notice that $\mathsf{E}\mathfrak{L}_t(\lambda^*)=1$. Moreover, taking into account (2.1) and applying the Itô formula to $\mathfrak{L}_t(\lambda^*)$ one can see that $(\mathfrak{L}_t(\lambda^*),\mathscr{G}_t)_{t\ge0}$ is a positive local martingale with paths from the Skorokhod space $\mathbb{D}_{[0,\infty)}$. Then, a measure $\widetilde{\mathsf{P}}_t$, defined by $d\widetilde{\mathsf{P}}_t = \mathfrak{L}_t(\lambda^*)d\mathsf{P}_t$, where P_t is a restriction of P on $\mathscr{G}_t$, is the probability measure. We introduce the probability space $(\Omega,\mathcal{F},\widetilde{\mathsf{P}}_t)$. Since $\mathfrak{L}_t(\lambda^*)>0$, P-a.s., not only $\widetilde{\mathsf{P}}_t \ll \mathsf{P}_t$ but also $\widetilde{\mathsf{P}}_t \ll \mathsf{P}_t$ with $d\mathsf{P}_t = \mathfrak{L}_t^{-1}(\lambda^*)\widetilde{\mathsf{P}}_t$. This property and (3.3) provide a lower bound

$$\begin{aligned}\mathsf{P}\big(|S_t-u|\le\delta\big) = \mathsf{P}_t\big(|S_t-u|\le\delta\big) &= \int_{\{|S_t-u|\le\delta\}}\mathfrak{L}_t^{-1}(\lambda^*)d\widetilde{\mathsf{P}}_t\\ &\ge e^{-\lambda^* t\delta - tI(u)}\widetilde{\mathsf{P}}_t\big(|S_t-u|\le\delta\big).\end{aligned}$$

Therefore, we find that

$$\underline{\lim}_{t\to\infty}\frac{1}{t}\log\mathsf{P}\big(|S_t-u|\le\delta\big) \ge -I(u) - \lambda^*\delta + \underline{\lim}_{t\to\infty}\frac{1}{t}\log\widetilde{\mathsf{P}}_t\big(|S_t-u|\le\delta\big).$$

It is clear, the desired lower bound to obtain it is left to prove that

$$\varliminf_{t\to\infty} \widetilde{\mathsf{P}}_t\big(|S_t - u| \le \delta\big) = 1$$

or, equivalently,

$$\lim_{t\to\infty} \widetilde{\mathsf{P}}_t\big(|S_t - u| > \delta\big) = 0. \tag{3.4}$$

Thus the last step of proof deals with (3.4). To this end, we show that ($\widetilde{\mathsf{E}}_t$ denotes the expectation relative to $\widetilde{\mathsf{P}}_t$)

$$\lim_{t\to\infty} \widetilde{\mathsf{E}}_t|S_t - u|^2 = 0.$$

Since $\mathsf{E}\mathfrak{L}_t(\lambda) = 1$, it holds

$$\begin{aligned}
0 &= \frac{\partial^2 \mathsf{E}\mathfrak{L}_t(\lambda)}{\partial\lambda^2}\Big|_{\lambda=\lambda^*} \\
&= t^2\mathsf{E}\Big(S_t - r\int_{\{x>0\}} x e^{\lambda^* x} dG(x)\Big)^2 \mathfrak{L}_t(\lambda^*) - \underbrace{t\, r\int_{\{x>0\}} x^2 e^{\lambda^* x} dG(x)}_{=u \text{ (see (3.2))}} \\
&= t^2\widetilde{\mathsf{E}}\Big(S_t - r\int_{\{x>0\}} x e^{\lambda^* x} dG(x)\Big)^2 - tu.
\end{aligned}$$

Hence,

$$\widetilde{\mathsf{E}}\big(S_t - u\big)^2 = \frac{1}{t} r\mathsf{E}\int_{\{x>0\}} x^2 e^{\lambda^* x} dG(x) \xrightarrow[t\to\infty]{} 0.$$

□

REFERENCES

[1] Cramer H. (1938), Sur un nouveau theoreme-limite de la theoriedes probabilites. *Actualites Santifiques et Industrielles. Colloque cosacrea la throrie des probabilitrs*, **3**(736): 2–23. Hermann, Paris.

[2] Dembo A. and Zeitouni O. (1993), *Large Deviations Techniques and Applications*, Jones and Bartlet, 1993.

[3] Georgii Hans-Otto and Zessin Hans (1993), Large deviations and the maximum entropy principle of marked point random fields. *Probability Theory and Related Fields*, **96**: 177–204.

[4] Jacod J., Multivariate point processes: predictable projection, Radon–Nikodym derivatives, representation martingales. Z. Wahrsch. Verw. Gebiete (1975), **31**: 235–53.

[5] Jacod J. and Shiryaev A.N., *Limit Theorems for Stochastic Processes*. Springer-Verlag, New York, Heidelberg, Berlin (1987).

[6] Liptser R.S. and Shiryaev A.N. (1989), Theory of Martingales. Kluwer, Dordrecht (Russian edition 1986).

[7] Puhalskii A. (2001), Large Deviations and Idempotent Probability, Chapman & Hall/CRC Press.

[8] Watanabe S. (1964), On discontinuous additive functionals and Lévy measures of Markov process. *Japan J. Math.*, **34**: 53–70.

ON BOUNDED SOLUTIONS OF THE BALANCED GENERALIZED PANTOGRAPH EQUATION

LEONID BOGACHEV*, GREGORY DERFEL†, STANISLAV MOLCHANOV‡, AND JOHN OCKENDON§

Abstract. The question about the existence and characterization of bounded solutions to linear functional-differential equations with both advanced and delayed arguments was posed in the early 1970s by T. Kato in connection with the analysis of the pantograph equation, $y'(x) = ay(qx) + by(x)$. In the present paper, we answer this question for the *balanced* generalized pantograph equation of the form $-a_2 y''(x) + a_1 y'(x) + y(x) = \int_0^\infty y(\alpha x)\,\mu(\mathrm{d}\alpha)$, where $a_1 \geq 0$, $a_2 \geq 0$, $a_1^2 + a_2^2 > 0$, and μ is a probability measure. By setting $K := \int_0^\infty \ln\alpha\,\mu(\mathrm{d}\alpha)$, we prove that if $K \leq 0$ then the equation does not have nontrivial (i.e., nonconstant) bounded solutions, while if $K > 0$ then such a solution exists. The result in the critical case, $K = 0$, settles a long-standing problem. The proof exploits the link with the theory of Markov processes, in that any solution of the balanced pantograph equation is an $\mathcal{L}$-harmonic function relative to the generator $\mathcal{L}$ of a certain diffusion process with "multiplication" jumps. The paper also includes three "elementary" proofs for the simple prototype equation $y'(x)+y(x) = \frac{1}{2}\,y(qx) + \frac{1}{2}\,y(x/q)$, based on perturbation, analytical, and probabilistic techniques, respectively, which may appear useful in other situations as efficient exploratory tools.

Key words. Pantograph equation, functional-differential equations, integro-differential equations, balance condition, bounded solutions, WKB expansion, q-difference equations, ruin problem, Markov processes, jump diffusions, $\mathcal{L}$-harmonic functions, martingales.

AMS(MOS) subject classifications. Primary 34K06, 45J05, 60Jxx; secondary 34K12.

1. Introduction.

The classical *pantograph equation* is the linear first-order functional-differential equation (with rescaled argument) of the form

$$y'(x) = ay(qx) + by(x), \tag{1.1}$$

where a, b are constant coefficients (real or complex) and $q > 0$ is a rescaling parameter. Historically,[1] the term "pantograph" dates back to the seminal paper of 1971 by Ockendon and Tayler [24], where such equations[2] emerged

*Department of Statistics, University of Leeds, Woodhouse Lane, Leeds LS2 9JT, UK (bogachev@maths.leeds.ac.uk).

†Department of Mathematics, Ben Gurion University of the Negev, Beer Sheva, Israel (derfel@math.bgu.ac.il).

‡Department of Mathematics, University of North Carolina at Charlotte, Charlotte NC 28223, USA (smolchan@uncc.edu).

§University of Oxford, Centre for Industrial and Applied Mathematics, Mathematical Institute, 24-29 St Giles, Oxford OX1 3LB, UK (ock@maths.ox.ac.uk).

[1] The name "pantograph equation" was not in wide use until it was coined by Iserles [14] for a more general class of functional-differential equations.

[2] To be more precise, a certain vector analog of Eq. (1.1).

in a mathematical model for the dynamics of an overhead current collection system on an electric locomotive (with the physically relevant value $q < 1$). At about the same time, a systematic analysis of solutions to the pantograph equation was started by Fox *et al.* [11], where various analytical, perturbation, and numerical techniques were discussed at length (for both $q < 1$ and $q > 1$).

It is interesting to note that an equation of the form (1.1) (with $q > 1$) was derived more than 25 years earlier by Ambartsumian [2] to describe the absorption of light by the interstellar matter. Some particular cases of Eq. (1.1) are also found in early work by Mahler [19] on a certain partition problem in number theory (where Eq. (1.1), with $a = 1$, $b = 0$, $q < 1$, appears as a limit of a similar functional-difference equation) and by Gaver [12] on a special ruin problem (with $a = 1$, $b = -1$, $q > 1$). Subsequently, the pantograph equation has appeared in numerous applications ranging from the problem of coherent states in quantum theory [26] to cell-growth modeling in biology [28] (see further references in Refs. [7, 8, 14, 20]). These and other examples suggest that, typically, the pantograph equation and similar functional-differential equations with rescaling are relevant as long as the systems in question possess some kind of self-similarity.

Since its introduction into the mathematical literature in the early 1970s, the theory of the pantograph equation (and some of its natural generalizations) has been the subject of persistent attention and research effort, yielding over years a number of significant developments. In particular, the classification of Eq. (1.1) with regard to various domains of the parameters,[3] including existence and uniqueness theorems, and an extensive asymptotic analysis of the corresponding solutions have been given by Kato and McLeod [17] and Kato [16]. The investigation of such equations in the complex domain was initiated by Morris *et al.* [22] and Oberg [23] and continued by Derfel and Iserles [5] and Marshall *et al.* [20]. A systematic treatment of the generalized first-order pantograph equation (with matrix coefficients and also allowing for a term with rescaled derivative) is contained in the influential paper by Iserles [14], where in particular a fine geometric structure of almost-periodic solutions has been described. Asymptotics for equations with variable coefficients have been studied by Derfel and Vogl [6].

Higher-order generalizations of the pantograph equation (1.1) lead to the class of linear functional-differential equations with rescaling,

$$\sum_{j=1}^{\ell}\sum_{k=0}^{m} a_{jk}\, y^{(k)}(\alpha_j x + \beta_j) = 0 \tag{1.2}$$

[3]Depending on whether $q < 1$ or $q > 1$ and also on the cases $\Re b < 0$, $\Re b > 0$, and $\Re b = 0$.

(see Ref. [4] and references therein).[4] Kato [16] posed a problem of asymptotic analysis of Eq. (1.2), including the question of existence and characterization of *bounded* solutions. Some partial answers to the latter question have been given by Derfel [7, 4] and Derfel and Molchanov [8].

In particular, Derfel [7] considered the "balanced" generalized first-order pantograph equation of the form

$$y'(x) + y(x) = \sum_{j=1}^{\ell} p_j\, y(\alpha_j x), \tag{1.3}$$

subject to the condition

$$\sum_{j=1}^{\ell} p_j = 1, \qquad p_j > 0 \quad (j = 1, \dots, \ell), \tag{1.4}$$

so that the weights p_j of the rescaled y-terms on the right-hand side of Eq. (1.3) match the unit coefficient of the $y(x)$ on the left. Note that, owing to the balance condition (1.4), Eq. (1.3) always has a trivial solution $y =$ const. The question of existence of *nontrivial* (i.e., nonconstant) *bounded solutions* is most interesting (and most difficult) in the case where the right-hand side of Eq. (1.3) involves both "advanced" ($\alpha_j > 1$) and "delayed" ($0 < \alpha_j < 1$) arguments. It turns out that the answer depends crucially on the quantity

$$K := \sum_{j=1}^{\ell} p_j \ln \alpha_j\, . \tag{1.5}$$

Derfel [7] has proved that if $K < 0$ then Eq. (1.3) has no nontrivial bounded solutions, whereas if $K > 0$ then such a solution always exists. In the "critical" case $K = 0$, this question has remained open as yet.

In the present paper, we consider a more general integro-differential equation[5] of the pantograph type, namely,

$$-a_2 y''(x) + a_1 y'(x) + y(x) = \int_0^\infty y(\alpha x)\, \mu(\mathrm{d}\alpha), \tag{1.6}$$

where $a_1 \ge 0$, $a_2 \ge 0$, $a_1^2 + a_2^2 > 0$ (so that a_1, a_2 do not vanish simultaneously), and μ is a probability measure on $(0, \infty)$,

$$\mu(0, \infty) = \int_0^\infty \mu(\mathrm{d}\alpha) = 1. \tag{1.7}$$

[4]Note that the theory of such equations is closely related to the theory of q-difference equations developed by Birkhoff [3] and Adams [1] (see also Section 3 below).

[5]In fact, the results of the paper [7] mentioned above include first-order equations of the form (1.6), i.e., with $a_2 = 0$. Let us also remark that more general first-order integro-differential equations (but with delayed arguments only, i.e., $\alpha \in (0, 1)$) were considered by Iserles and Liu [15].

The parameter α in Eq. (1.6) can be viewed as a random variable, with values in $(0,\infty)$ and the probability distribution given by the measure μ, i.e., $\mathsf{P}\{\alpha \in A\} = \mu(A)$, $A \subset (0,\infty)$. Note that Eq. (1.6) is balanced in the same sense as Eq. (1.3), since the mean contribution of the distributed rescaled term $y(\alpha x)$ is matched by that of $y(x)$. Moreover, Eq. (1.6) reduces to Eq. (1.3) when $a_1 = 1$, $a_2 = 0$, and the measure μ is discrete, with atoms $p_j = \mu(\alpha_j)$, $j = 1, \dots, \ell$ (i.e., α is a discrete random variable, with the distribution $\mathsf{P}\{\alpha = \alpha_j\} = p_j$, $j = 1, \dots, \ell$).

As already mentioned, due to the balance condition (1.7) any constant satisfies Eq. (1.6), and by linearity of the equation one can assume, without loss of generality, that $y(0) = 0$. Moreover, if $x > 0$ ($x < 0$) then the right-hand side of Eq. (1.6) is determined solely by the values of the function $y(u)$ with $u > 0$ (respectively, $u < 0$). Therefore, the two-sided equation (1.6) is decoupled at $x = 0$ into two one-sided boundary value problems,

$$\begin{aligned} &-a_2 y''(x) + a_1 y'(x) + y(x) = \int_0^\infty y(\alpha x)\,\mu(\mathrm{d}\alpha), \qquad x \gtrless 0, \\ &y(0) = 0. \end{aligned} \tag{1.8}$$

For Eq. (1.6), the analog of Eq. (1.5) is given by

$$K := \int_0^\infty \ln\alpha\,\mu(\mathrm{d}\alpha) = \mathsf{E}[\ln\alpha]. \tag{1.9}$$

Our main result is the following theorem, which resolves the problem of nontrivial bounded solutions in the critical case, $K = 0$ (and also recovers and extends the result of Ref. [7] for the case $K < 0$).

THEOREM 1.1. *Assume that* $0 \neq \mathsf{E}|\ln\alpha| < \infty$, *so that* K *in Eq. (1.9) is well defined and the measure* μ *is not concentrated at the point* $\alpha = 1$, *i.e., the random variable* α *does not degenerate to the constant* 1. *Under these hypotheses, the condition* $K \le 0$ *implies that any bounded solution of equation (1.6) is trivial, i.e.,* $y(x) \equiv \mathrm{const}$, $x \in \mathbb{R}$.

The apparent probabilistic structure of Eq. (1.6) is crucial for our proof of this result. The main idea is to construct a certain diffusion process X_t, with negative drift and "multiplication" jumps (i.e., of the form $x \mapsto \alpha x$), such that Eq. (1.6) can be rewritten as $\mathcal{L}y = 0$, where $\mathcal{L}$ is the infinitesimal generator of the Markov process X_t. That is to say, the class of bounded solutions of Eq. (1.6) coincides with the set of bounded $\mathcal{L}$-harmonic functions. This link brings in the powerful tool kit of Markov processes; particularly instrumental is the well-known fact (see, e.g., Ref. [9]) that for any $\mathcal{L}$-harmonic function $f(x)$, the random process $f(X_t)$ is a martingale, and hence, for any $t \ge 0$,

$$f(x) = \mathsf{E}[f(X_t) | X_0 = x], \qquad x \in \mathbb{R}. \tag{1.10}$$

On the other hand, due to the multiplication structure of independent consecutive jumps of the process X_t, its position after n jumps is expressed in terms of a background random walk $S_k = \xi_1 + \cdots + \xi_k$ $(0 \le k \le n)$, where ξ_i's are independent random variables with the same distribution as $\ln \alpha$. The hypothesis $K \le 0$ of Theorem 1.1 implies that, almost surely (a.s.), the random walk S_n travels arbitrarily far to the left. Using an optional stopping theorem (whereby the boundedness of $f(x)$ is important), we can apply the martingale identity (1.10) at the suitably chosen stopping (first-passage) times, which eventually leads to the conclusion that $f'(x) \equiv 0$ and hence $f(x) \equiv \text{const}$.

This approach also allows us to give an example of a nontrivial bounded solution to equation (1.6) in the case $K > 0$ (thus extending the result by Derfel [7] to the second-order pantograph equation).

THEOREM 1.2. *Suppose that $K > 0$, and set*

$$f_\infty(x) := \mathsf{P}\Big\{\liminf_{t\to\infty} X_t = +\infty \,|\, X_0 = x\Big\}, \qquad x \in \mathbb{R}, \tag{1.11}$$

where X_t is the random process constructed in the proof of Theorem 1.1. Then the function $f_\infty(x)$ is $\mathcal{L}$-harmonic and such that $f_\infty(x) \to 0$ as $x \to -\infty$ and $f_\infty(x) \to 1$ as $x \to +\infty$.

When $a_2 = 0$, Eq. (1.6) becomes

$$a_1 y'(x) + y(x) = \int_0^\infty y(\alpha x)\, \mu(\mathrm{d}\alpha), \qquad x \in \mathbb{R}. \tag{1.12}$$

In this case, the diffusion component of the random process X_t is switched off, and it follows, due to the negative drift and multiplication jumps (see details in Section 4), that if $X_0 = x \le 0$, then $X_t \le 0$ for all $t \ge 0$. That is to say, the negative semi-axis $(-\infty, 0]$ is an absorbing set for the process X_t, and hence the function $f_\infty(x)$, defined by Eq. (1.11) as the probability to escape to $+\infty$ starting from x, vanishes for all $x \le 0$. This leads to the following interesting specification of the example in Theorem 1.2.

COROLLARY 1.1. *If $a_2 = 0$ in Eq. (1.6) then $f_\infty(x) \equiv 0$ for all $x \le 0$. Moreover, Eq. (1.6) implies that all derivatives of the function $f_\infty(x)$ vanish at zero, $f_\infty^{(k)}(0) = 0$ $(k = 1, 2, \dots)$.*

Before elaborating on the ideas outlined above, we would like to make a short digression in order to consider the simple prototype example of Eq. (1.3), namely,

$$y'(x) + y(x) = \frac{1}{2}\, y(qx) + \frac{1}{2}\, y(q^{-1}x) \qquad (q \ne 1), \tag{1.13}$$

and to give several different "sketch" proofs of Theorem 1.1 in this case. Note that, according to Eq. (1.5), we have

$$K = \frac{1}{2} \ln q + \frac{1}{2} \ln \frac{1}{q} = 0,$$

so Eq. (1.13) falls in the (most interesting) critical case. In fact, this example was the starting point of our work and a kind of mathematical test-tube to try various approaches and ideas. Although not strictly necessary for the exposition, after some deliberation we have cautiously decided to include our early proofs (based on perturbation, analytical, and probabilistic arguments,[6] respectively), partly because this will hopefully equip the reader with some insight into validity of the result, and also because these methods may appear useful as exploratory tools in other situations.

The rest of the paper is laid out as follows. In Sections 2, 3, and 4, we discuss the three approaches to equation (1.13) as just mentioned. In Section 5 we start a more systematic treatment by describing the construction of a suitable diffusion process with multiplication jumps. In Section 6, we discuss the corresponding $\mathcal{L}$-harmonic functions and obtain an *a priori* bound for the derivative of a solution. Finally, in Section 7 we prove our main Theorems 1.1 and 1.2.

2. Perturbative proof. Following the ideas used by Ockendon and Tayler [24] and Fox *et al.* [11] in the case of the original pantograph equation (1.1), we start by observing that if $q = 1$ then Eq. (1.13) is reduced to the equation $y' = 0$, which has constant solutions only. Therefore, when the parameter q is close to 1, it is reasonable to seek solutions of Eq. (1.13) in a form that involves "superposition" of (exponentially) small oscillations (*fast variation*) on top of an almost constant (polynomial) function (*slow variation*). This leads to a WKB-type asymptotic expansion of the solution in terms of perturbation parameter $\varepsilon \approx 0$ (see Ref. [25]), which in the first-order approximation yields two first-order differential equations: a nonlinear equation (called the *eikonal equation*) for the fast variation and a linear equation (called the *transport equation*) for the slow variation. For simplicity of presentation, we will restrict ourselves to the first-order approximation, but in principle one can go on to the analysis of higher-order terms, which are described by linear equations and therefore can be determined without much trouble.

To implement this approach, set $q = 1 \pm \varepsilon$ $(\varepsilon > 0)$, $x = \varepsilon^{-1}u$, and $y(x) = y(\varepsilon^{-1}u) =: f(u)$. Then

$$y'(x) = \frac{\mathrm{d}f}{\mathrm{d}u} \cdot \frac{\mathrm{d}u}{\mathrm{d}x} = \varepsilon f'(u),$$

and Eq. (1.13) takes the form

$$\varepsilon f'(u) + f(u) = \frac{1}{2} f\big((1 \pm \varepsilon)u\big) + \frac{1}{2} f\big((1 \pm \varepsilon)^{-1}u\big). \tag{2.1}$$

As explained in the Introduction, without loss of generality we may assume that $f(0) = 0$.

[6]It is amusing that these three methods represent nicely the traditional organization of British mathematics into Applied Mathematics, Pure Mathematics, and Statistics, which is reflected in the names of mathematical departments in most universities in the UK.

Now, suppose that, for small ε, the function $f(u)$ admits a WKB-type expansion,

$$f(u) \sim \bigl(A_0(u) + \varepsilon A_1(u) + \cdots\bigr) \exp\bigl(\varepsilon^{-1} V(u)\bigr), \tag{2.2}$$

which in principle should be valid uniformly for all u, including the limiting values $u \to 0$ and $u \to \infty$. Differentiation of Eq. (2.2) yields

$$\varepsilon f'(u) \sim \bigl(A_0(u)V'(u) + \varepsilon\bigl(A_0'(u) + A_1(u)V'(u)\bigr) + \cdots\bigr) \exp\bigl(\varepsilon^{-1}V(u)\bigr). \tag{2.3}$$

From Eq. (2.2) we also obtain

$$\begin{aligned} f\bigl((1 \pm \varepsilon)u\bigr) &\sim \bigl(A_0(u) + \varepsilon\bigl(A_1(u) \pm uA_0'(u)\bigr) + \cdots\bigr) \\ &\qquad \times \exp\left(\frac{V(u)}{\varepsilon} \pm uV'(u) + \frac{\varepsilon u^2 V''(u)}{2} + \cdots\right), \end{aligned} \tag{2.4}$$

and

$$\begin{aligned} f\bigl((1 \pm \varepsilon)^{-1}u\bigr) &\sim \bigl(A_0(u) + \varepsilon\bigl(A_1(u) \mp uA_0'(u)\bigr) + \cdots\bigr) \\ &\qquad \times \exp\left(\frac{V(u)}{\varepsilon} \mp uV'(u) + \varepsilon\left(uV'(u) + \frac{u^2\, V''(u)}{2}\right) + \cdots\right). \end{aligned} \tag{2.5}$$

Substituting the expansions (2.3), (2.4) and (2.5) into Eq. (2.1), canceling out the common factor $\exp\bigl(V(u)/\varepsilon\bigr)$, and collecting the terms that remain after setting $\varepsilon = 0$, we get

$$A_0(u)V'(u) + A_0(u) = \frac{1}{2}\, A_0(u) \exp\bigl(\pm uV'(u)\bigr) + \frac{1}{2}\, A_0(u) \exp\bigl(\mp uV'(u)\bigr).$$

Assuming that $A_0(u) \neq 0$, this gives the equation

$$1 + V'(u) = \cosh\bigl(uV'(u)\bigr), \tag{2.6}$$

or equivalently

$$u = \frac{uV'(u)}{\cosh\bigl(uV'(u)\bigr) - 1}. \tag{2.7}$$

Similarly, equating the terms of order of ε and noting that $A_1(u)$ cancels out owing to Eq. (2.6), we obtain

$$\begin{aligned} A_0'(u) = A_0'(u)\, u \sinh\bigl(uV'(u)\bigr) &+ \frac{1}{2}\, A_0(u)\, u^2\, V''(u) \cosh\bigl(uV'(u)\bigr) \\ &+ \frac{1}{2}\, A_0(u)\, uV'(u) \exp\bigl(\mp uV'(u)\bigr) \end{aligned} \tag{2.8}$$

We can now check that the formal expansion (2.2) is compatible with the zero initial condition, $f(0) = 0$. Equation (2.7) implies that if $u \to 0$ then $uV'(u) \to \infty$, and moreover

$$uV'(u) \sim -\ln u + \ln\ln\frac{1}{u} + \cdots, \tag{2.9}$$

whence

$$V(u) \sim -\frac{\ln^2 u}{2} + \ln u \cdot \ln \ln \frac{1}{u} + \cdots \tag{2.10}$$

Furthermore, differentiation of Eq. (2.9) gives

$$u^2 V''(u) \sim \ln u - \ln \ln \frac{1}{u} + \cdots \tag{2.11}$$

Inserting formulas (2.9) and (2.11) into Eq. (2.8), we obtain for $A_0(u)$ the asymptotic differential equation

$$\frac{A_0'(u)}{A_0(u)} \sim \mp \frac{1}{2u} \left(\ln u - \ln \ln \frac{1}{u} + \cdots \right),$$

which solves to

$$\ln A_0(u) \sim \mp \frac{\ln u}{4} \left(\ln u - 2 \ln \ln \frac{1}{u} + \cdots \right). \tag{2.12}$$

Finally, substituting the expansions (2.10) and (2.12) into Eq. (2.2) we obtain that $f(u) \to 0$ as $u \to 0$, as required.

Let us now explore the behavior of the solution as $u \to \infty$. In this limit, Eq. (2.7) gives $uV'(u) \to 0$, and moreover

$$u \sim \frac{2}{uV'(u)} - \frac{uV'(u)}{6} + \cdots,$$

whence

$$uV'(u) \sim \frac{2}{u} - \frac{2}{3u^3} + \cdots, \tag{2.13}$$

$$u^2 V''(u) \sim -\frac{4}{u} + \frac{8}{3u^3} + \cdots \tag{2.14}$$

From (2.13), we also find

$$V(u) \sim C - \frac{2}{u} + \frac{2}{9u^3} + \cdots, \tag{2.15}$$

where $C = \text{const}$. Inserting the expansions (2.13) and (2.14) into Eq. (2.8), we obtain

$$\frac{A_0'(u)}{A_0(u)} \sim \frac{1}{u} \pm \frac{2}{u^2} + \cdots,$$

and hence

$$A_0(u) \sim C_0 \, u \exp\left(\mp \frac{2}{u} + \cdots \right). \tag{2.16}$$

The case $C_0 \neq 0$ is unsuitable, since Eq. (2.16) would imply that $A_0(u) \to \infty$ as $u \to \infty$ and, in view of formulas (2.15) and (2.2), the solution $f(u)$ appears to be unbounded, which contradicts our assumption. Therefore, $C_0 = 0$ and hence $A_0(u) = 0$, thus reducing the expansion (2.2) to

$$f(u) \sim \varepsilon\big(A_1(u) + \varepsilon A_2(u) + \cdots\big) \exp\big(\varepsilon^{-1} V(u)\big).$$

Arguing as above, we successively obtain $A_1(u) = 0$, $A_2(u) = 0$, etc. This indicates that $f(u) = 0$, which was our aim.

3. Analytical proof. In this section, we demonstrate how the theory of q-difference equations (see Refs. [1, 3]) can be used to show that Eq. (1.13) has no nontrivial bounded solutions. In what follows, we assume that $q \neq 1$.

As explained in the Introduction (see Eq. (1.8)), Eq. (1.13) splits into two (similar) one-sided equations, so it suffices to consider the boundary-value problem

$$\begin{aligned} &y'(x) + y(x) = \frac{1}{2}\, y(qx) + \frac{1}{2}\, y(q^{-1}x), \qquad x \geq 0,\\ &y(0) = 0. \end{aligned} \tag{3.1}$$

Assume that $y(x)$ is a bounded solution of Eq. (3.1), $|y(x)| \leq B$ $(x \geq 0)$, and let $\hat{y}(s)$ be the Laplace transform of $y(x)$,

$$\hat{y}(s) := \int_0^\infty \mathrm{e}^{-sx} y(x)\, \mathrm{d}x,$$

then $\hat{y}(s)$ is analytic in the right half-plane, $\Re s > 0$, and

$$|\hat{y}(s)| \leq \frac{B}{\Re s}, \qquad \Re s > 0. \tag{3.2}$$

On account of the boundary condition $y(0) = 0$, Eq. (3.1) transforms into

$$(1 + s)\hat{y}(s) = \frac{1}{2q}\, \hat{y}(q^{-1}s) + \frac{q}{2}\, \hat{y}(qs), \tag{3.3}$$

or, after the substitution $\varphi(s) := s\,\hat{y}(s)$,

$$(1 + s)\varphi(s) = \frac{1}{2}\varphi(q^{-1}s) + \frac{1}{2}\varphi(qs). \tag{3.4}$$

Note that the estimate (3.2) implies

$$|\varphi(s)| \leq \frac{B|s|}{\Re s}, \qquad \Re s > 0, \tag{3.5}$$

and in particular $\varphi(s)$ is bounded in the vicinity of the origin.

Let us rewrite Eq. (3.4) in the form

$$\varphi(q^2 s) - 2(1+qs)\varphi(qs) + \varphi(s) = 0. \tag{3.6}$$

Equation (3.6) is a linear q-difference equation of order 2. According to the general theory of such equations (see Ref. [1]), the characteristic equation for Eq. (3.6) (in the vicinity of $s = 0$) reads

$$q^{2\rho} - 2q^{\rho} + 1 = 0,$$

and $\rho_{1,2} = 0$ is its multiple root. The corresponding fundamental set of solutions to Eq. (3.6) is given by

$$\varphi_1(s) = P_1(s),$$
$$\varphi_2(s) = P_2(s) + \frac{\ln s}{\ln q}\, P_3(s),$$

where $P_1(s)$, $P_2(s)$, and $P_3(s)$ are generic power series convergent in some neighborhood of zero, and $\ln s$ denotes the principal branch of the logarithm. Note that the solution $\varphi_2(s)$ is unsuitable because it is unbounded near $s = 0$ (see Eq. (3.5)). On the other hand, the function $\varphi_1(s)$ is analytic in the vicinity of zero and, moreover, it can be analytically continued, step by step, into the whole complex plane $\mathbb{C}$ by means of Eq. (3.4). For instance, if $q > 1$ then the analytic continuation from a disk $|s| \le a$ to the bigger disk $|s| \le qa$ is furnished by the formula $\varphi(qs) = 2(1+s)\varphi(s) - \varphi(q^{-1}s)$ (see Eq. (3.4)), and so on.

That is to say, $\varphi_1(s)$ can be extended to an entire function $\varphi(s)$, which by construction satisfies Eq. (3.6) for all $s \in \mathbb{C}$. But then, according to one result by Mason [21], the entire function $\varphi(s)$ must be of *zero order*, and consequently (see Ref. [27, §8.7.3]) it is *unbounded* on any ray (in particular, for $s \in (0, \infty)$), unless it is a constant. However, the unboundedness for real $s > 0$ contradicts the estimate (3.5). Hence, $\varphi(s) = \text{const}$, so that $\hat{y}(s) = \text{const}\cdot s^{-1}$, and by the uniqueness theorem for the Laplace transform this implies that $y(x) \equiv \text{const}$, i.e, $y(x) \equiv y(0) = 0$, as claimed.

4. Probabilistic proof. In this section, we give a probabilistic interpretation of Eq. (1.13) via a certain ruin problem, and prove that the corresponding solution is constant using elementary probabilistic considerations. Although our argument does not cover the whole class of bounded solutions, it contains some ideas that we will use in the second half of the paper to give a complete proof of our general result.

Let us consider the following "double-or-half" gambling model (in continuous time). Suppose that a player spends his initial capital, x, at rate v per unit time, so that after time t he is left with capital $x - vt$. However, at a random time τ (with exponential distribution), he gambles by putting the remaining capital at stake, whereby he can either *double* his money or lose *half* of it, both with probability 1/2. After that, the process continues in a similar fashion, independently of the past history. If the capital reaches zero and then moves down to become negative, this is interpreted as borrowing, so the process proceeds in the same way without termination. In that case, gambling will either double or halve the debt, and in particular the capital will remain negative forever.

More generally, if X_t denotes the player's capital at time $t \geq 0$, starting with the initial amount $X_0 = x$, then the random process X_t moves with constant negative drift $(-v)$, interrupted at random time instants σ_i by random multiplication jumps from its location $x_i = X_{\sigma_i - 0}$ (i.e., immediately before the jump) to either qx_i or $q^{-1}x_i$ $(q \neq 1)$, both with probability $1/2$. We assume that the jumps occur at the arrival times $\sigma_1, \sigma_2, \ldots$ of an auxiliary Poisson process with parameter $\lambda > 0$, so that the waiting times until the next jump, $\tau_i = \sigma_i - \sigma_{i-1}$ $(\sigma_0 := 0)$, are independent identically distributed (i.i.d.) random variables, each with the exponential distribution

$$\mathsf{P}\{\tau > t\} = \mathrm{e}^{-\lambda t}, \qquad t > 0.$$

According to this description, (X_t) is a Markov process, in that the probability law of its future development is completely determined by its current state, but not by the past history ("lack of memory") (see, e.g., Refs. [9, 10]).

We are concerned with the *ruin problem* for this model.[7] Namely, consider the probability $f_0(x)$ of becoming bankrupt starting with the initial capital x,

$$f_0(x) := \mathsf{P}\Big\{\liminf_{t\to\infty} X_t \leq 0 \,|\, X_0 = x\Big\} \equiv \mathsf{P}_x\{T_0 < \infty\}, \qquad x \in \mathbb{R},$$

where $T_0 := \min\{t \geq 0 : X_t \leq 0\}$ is the random time to bankruptcy and P_x denotes the probability measure conditioned on the initial state $X_0 = x$.

From the definition of the process X_t, it is clear that if $x \leq 0$ then $T_0 = 0$ and so $f_0(x) = 1$. For $x > 0$, we note that if the first jump does not occur prior to time x/v, then the process will simply drift down to 0, in which case $T_0 = x/v < \infty$. Otherwise (i.e., if a jump does happen before time x/v), the ruin problem may be reformulated by treating the landing point after the jump as a new starting point (thanks to the Markov property). More precisely, by conditioning on the first jump instant σ_1 $(= \tau_1)$ and using the (strong) Markov property, we obtain

[7] A similar ruin problem for the process with deterministic multiplication jumps of the form $x_i \mapsto qx_i$ $(q > 1)$, was first considered by Gaver [12], leading to the equation $y'(x) + y(x) = y(qx)$ (i.e., with advanced argument, cf. Eq. (1.3)). The systematic theory of general processes with multiplication jumps was developed by Lev [18].

$$
\begin{aligned}
f_0(x) &= \mathsf{P}_x\{\sigma_1 > x/v\} + \int_0^{x/v} \lambda \mathrm{e}^{-\lambda s}\, \mathsf{P}_x(T_0 < \infty \,|\, \sigma_1 = s)\, \mathrm{d}s \\
&= \mathrm{e}^{-\lambda x/v} + \lambda \int_0^{x/v} \mathrm{e}^{-\lambda s}\, \frac{1}{2} \Bigl(f_0\bigl(q(x-vs)\bigr) + f_0\bigl(q^{-1}(x-vs)\bigr) \Bigr)\, \mathrm{d}s \qquad (4.1)\\
&= \mathrm{e}^{-\lambda x/v} \left(1 + \frac{\lambda}{2v} \int_0^{x} \mathrm{e}^{\lambda u/v} \bigl(f_0(qu) + f_0(q^{-1}u) \bigr)\, \mathrm{d}u \right),
\end{aligned}
$$

where in the last line we have made the substitution $u = x - vs$. The representation (4.1) implies that the function $f_0(x)$ is continuous and, moreover, (infinitely) differentiable, and by differentiation of Eq. (4.1) with respect to x, it follows that the function $y = f_0(x)$ satisfies the generalized pantograph equation (cf. Eq. (1.13))

$$
\frac{v}{\lambda}\, y'(x) + y(x) = \frac{1}{2}\, y(qx) + \frac{1}{2}\, y(q^{-1}x). \qquad (4.2)
$$

It is easy to see that, in fact, this equation is satisfied on the whole axis, $x \in \mathbb{R}$. As we have mentioned, $f_0(x) = 1$ for all $x \le 0$, and it is now our aim to show that the same is true for all $x > 0$, which would mean that the solution $y = f_0(x)$ to equation (1.13) is a constant, $f_0(x) \equiv 1$, $x \in \mathbb{R}$.

To this end, note that the position of the process after n jumps is given by

$$
X_{\sigma_n} = (X_{\sigma_{n-1}} - v\tau_n)\, q^{\xi_n}, \qquad n = 1, 2, \dots,
$$

where ξ_n's are i.i.d. random variables taking the values ± 1 with probabilities $1/2$. By iterations (using that $X_0 = x$), we obtain[8]

$$
\begin{aligned}
X_{\sigma_n} &= \Bigl(\bigl((x - v\tau_1)\, q^{\xi_1} - v\tau_2 \bigr)\, q^{\xi_2} - \dots - v\tau_n \Bigr)\, q^{\xi_n} \\
&= (x - v\tau_1)\, q^{\xi_1 + \xi_2 + \dots + \xi_n} - v\tau_2\, q^{\xi_2 + \dots + \xi_n} - \dots - v\tau_n\, q^{\xi_n} \qquad (4.3)\\
&= q^{S_n} \left(x - v \sum_{i=1}^{n} \tau_i\, q^{-S_{i-1}} \right),
\end{aligned}
$$

where $S_n := \xi_1 + \xi_2 + \dots + \xi_n$, $S_0 := 0$. Note that S_n can be interpreted as a (simple) random walk, which in our case is symmetric (i.e., $\mathsf{P}\{\xi_i = 1\} = \mathsf{P}\{\xi_i = -1\} = 1/2$) and therefore *recurrent* (see, e.g., Ref. [10]). In particular, the events $A_n := \{S_{n-1} = 0\}$ ($n = 1, 2, \dots$) occur infinitely often, with probability 1. Furthermore, setting $B_n := \{\tau_n > 1\}$, we note that the events $A_n \cap B_n$ ($n = 1, 2, \dots$) are conditionally independent, given the realization of the random walk $\{S_k,\, k \ge 1\}$. Since the random variables τ_n (and therefore the events B_n) are independent of $\{S_k\}$, we have, with probability 1,

[8] Similar random sums as in Eq. (4.3) arise in products of certain random matrices in relation to random walks on the group of affine transformations of the line (see Ref. [13]).

$$\sum_{n=1}^{\infty} \mathsf{P}(A_n \cap B_n | \{S_k\}) = \sum_{n=1}^{\infty} \mathsf{P}(B_n)\, \mathbf{1}_{A_n}$$
$$= \mathrm{e}^{-\lambda} \#\{n : A_n \text{ occurs}\} = \infty,$$

where $\mathbf{1}_{\{\dots\}}$ denotes the indicator of an event. Hence, Borel-Cancelli's lemma (see, e.g., Ref. [10]) implies

$$\mathsf{P}(A_n \cap B_n \text{ occur infinitely often} | \{S_k\}) = 1 \qquad \text{a.s.,} \tag{4.4}$$

and by taking the expectation in Eq. (4.4) (with respect to the distribution of the sequence $\{S_k\}$), the same is true in the unconditional form,

$$\mathsf{P}(A_n \cap B_n \text{ occur infinitely often}) = 1.$$

As a consequence, the terms in the random series

$$\sum_{n=1}^{\infty} \tau_n q^{-S_{n-1}}$$

will infinitely often exceed the value 1, all other terms being nonnegative. Therefore, the series diverges to $+\infty$ a.s., and from (4.3) it follows that

$$\liminf_{n\to\infty} X_{\sigma_n} \le 0 \qquad \text{a.s.}$$

In turn, this implies that $T_0 < \infty$ a.s., and so $f_0(x) = 1$ for all $x > 0$, as claimed.

5. Jump diffusions. We now pursue a more general (and more systematic) approach. Equations of the form (1.2) are linked in a natural way with certain continuous-time Markov processes (more specifically, diffusions with multiplication jumps). To describe this class of processes, let us consider a Brownian motion $B_t^{\kappa,v}$, starting at the origin, with diffusion coefficient $\kappa \ge 0$ and nonpositive (constant) drift $-v \le 0$,

$$\mathrm{d}B_t^{\kappa,v} = \kappa \mathrm{d}B_t - v\mathrm{d}t, \qquad B_0^{\kappa,v} = 0,$$

or equivalently

$$B_t^{\kappa,v} = \kappa B_t - vt, \qquad t \ge 0,$$

where $B_t = B_t^{1,0}$ is a standard Brownian motion (with continuous sample paths). We assume that $\kappa^2 + v^2 > 0$, so that $B_t^{\kappa,v}$ does not degenerate to a (zero) constant.

The random process $B_t^{\kappa,v}$ determines the underlying diffusion dynamics for a process with jumps, $(X_t,\, t \ge 0)$, which is defined as follows. Suppose that the jump instants are given by the arrival times $\sigma_1, \sigma_2, \dots$ of an auxiliary Poisson process with parameter $\lambda > 0$, so that $\tau_i = \sigma_i - \sigma_{i-1}$ $(i = 1, 2, \dots)$ are i.i.d. random variables with exponential distribution,

$$\mathsf{P}\{\tau > t\} = \mathrm{e}^{-\lambda t}, \qquad t > 0$$

(we set formally $\sigma_0 := 0$). Furthermore, suppose that the successive jumps are determined by the rescaling coefficients α_i of the form $\alpha_i = \mathrm{e}^{\xi_i}$, where ξ_i's are i.i.d. random variables. Then, the (right-continuous) sample paths of the process X_t are defined inductively by

$$X_t = \begin{cases} x + B_t^{\kappa,v}, & 0 = \sigma_0 \le t < \sigma_1, \\ \mathrm{e}^{\xi_i} X_{\sigma_i - 0} + B_t^{\kappa,v} - B_{\sigma_i}^{\kappa,v}, & \sigma_i \le t < \sigma_{i+1}, \qquad i = 1, 2, \dots \end{cases} \tag{5.1}$$

That is to say, the process $(X_t,\, t \ge 0)$ starts at point x and moves as $X_t = x + B_t^{\kappa,v}$ until a random time σ_1, when it jumps to a random point

$$X_{\sigma_1} = \mathrm{e}^{\xi_1} X_{\sigma_1 - 0} = \mathrm{e}^{\xi_1}(x + B_{\sigma_1}^{\kappa,v}) = \mathrm{e}^{\xi_1} x + \mathrm{e}^{\xi_1}\zeta_1,$$

where $\zeta_1 := B_{\sigma_1}^{\kappa,v} = B_{\sigma_1}^{\kappa,v} - B_{\sigma_0}^{\kappa,v}$. Thereafter, the process proceeds in a diffusive way as $X_t = X_{\sigma_1} + B_t^{\kappa,v} - B_{\sigma_1}^{\kappa,v}$ until a random time σ_2, when it makes the next jump to

$$\begin{aligned} X_{\sigma_2} &= \mathrm{e}^{\xi_2} X_{\sigma_2 - 0} = \mathrm{e}^{\xi_2}(X_{\sigma_1} + B_{\sigma_2}^{\kappa,v} - B_{\sigma_1}^{\kappa,v}) \\ &= \mathrm{e}^{\xi_2+\xi_1} x + \mathrm{e}^{\xi_2+\xi_1}\zeta_1 + \mathrm{e}^{\xi_2}\zeta_2, \end{aligned}$$

where $\zeta_2 := B_{\sigma_2}^{\kappa,v} - B_{\sigma_1}^{\kappa,v}$, and so on. Iterating, we obtain that the n-th jump, occurring at time σ_n, lands at the point

$$\begin{aligned} X_{\sigma_n} &= \mathrm{e}^{\xi_n} X_{\sigma_n - 0} = \mathrm{e}^{\xi_n}(X_{\sigma_{n-1}} + B_{\sigma_n}^{\kappa,v} - B_{\sigma_{n-1}}^{\kappa,v}) \\ &= \mathrm{e}^{S_n}(x + \zeta_1 + \zeta_2\,\mathrm{e}^{-S_1} + \dots + \zeta_n\,\mathrm{e}^{-S_{n-1}}), \end{aligned} \tag{5.2}$$

where

$$\begin{aligned} \zeta_n &:= B_{\sigma_n}^{\kappa,v} - B_{\sigma_{n-1}}^{\kappa,v} \\ &= \kappa(B_{\sigma_n} - B_{\sigma_{n-1}}) - v(\sigma_n - \sigma_{n-1}) \qquad (n = 1, 2, \dots) \end{aligned} \tag{5.3}$$

is a sequence of i.i.d. random variables.

6. $\mathcal{L}$-harmonic functions. Let us study the properties of the process X_t in greater detail.

DEFINITION 6.1. *The* infinitesimal operator *(*generator*)* $\mathcal{L}$ *of a Markov random process* $(X_t,\, t \ge 0)$ *is defined by*

$$(\mathcal{L}f)(x) := \lim_{h \to 0+} \frac{\mathsf{E}_x[f(X_h)] - f(x)}{h}, \tag{6.1}$$

with the domain $\mathcal{D}(\mathcal{L})$ *consisting of functions* f *for which the limit in Eq. (6.1) exists.*

In a standard way, by considering possible scenarios for the process (X_t) up to an infinitesimal time h (see Ref. [9]), one obtains the following.

PROPOSITION 6.1. *For the random diffusion with jumps (X_t) defined above, its generator $\mathcal{L}$ acts on bounded C^2-smooth functions as*

$$(\mathcal{L}f)(x) = \frac{\kappa^2}{2} f''(x) - vf'(x) + \lambda\bigl(\mathsf{E}[f(\mathrm{e}^{\xi}x)] - f(x)\bigr), \qquad (6.2)$$

where ξ is a random variable with the same distribution as any one of the i.i.d. random variables $\xi_1, \xi_2, \ldots$.

DEFINITION 6.2. *A function f is called $\mathcal{L}$*-harmonic *if $\mathcal{L}f = 0$.*

Note that the expectation in Eq. (6.2) can be written as a Stieltjes integral,

$$\mathsf{E}[f(\mathrm{e}^{\xi}x)] = \int_0^\infty f(zx)\,\mathrm{d}F(z),$$

where $F(z)$ is the cumulative distribution function of the random variable $Z = \mathrm{e}^{\xi}$, i.e., $F(z) := \mathsf{P}\{\mathrm{e}^{\xi} \le z\} = \mathsf{P}\{\xi \le \ln z\}$ $(0 < z < \infty)$. Hence, the equation $\mathcal{L}f = 0$, with $\mathcal{L}$ given by Eq. (6.2), is equivalent to

$$-\frac{\kappa^2}{2} f''(x) + vf'(x) + \lambda f(x) = \lambda \int_0^\infty f(zx)\,\mathrm{d}F(z),$$

which is a balanced generalized pantograph equation of the form (1.6).

Our aim is to study the class of bounded $\mathcal{L}$-harmonic functions. Let us denote by $\|\cdot\|$ the standard sup-norm on $\mathbb{R}$:

$$\|f\| := \sup_{x\in\mathbb{R}} |f(x)|.$$

As a first step, we estimate the derivative of an $\mathcal{L}$-harmonic function.

PROPOSITION 6.2. *Suppose that $v > 0$. If $\|f\| < \infty$ and $\mathcal{L}f = 0$ then $\|f'\| < \infty$.*

Proof. If $\kappa = 0$ then, according to Eq. (6.2), the equation $\mathcal{L}f = 0$ takes the form

$$vf'(x) = \lambda\bigl(\mathsf{E}[f(\mathrm{e}^{\xi}x)] - f(x)\bigr),$$

which gives

$$\|f'\| \le \frac{2\lambda}{|v|}\,\|f\| < \infty.$$

For $\kappa > 0$, the condition $\mathcal{L}f = 0$ is equivalent to

$$f''(x) - \gamma f'(x) = -g(x), \qquad (6.3)$$

where

$$\gamma := \frac{2v}{\kappa^2} > 0, \qquad g(x) := \frac{2\lambda}{\kappa^2}\bigl(\mathsf{E}[f(\mathrm{e}^{\xi}x)] - f(x)\bigr).$$

Solving equation (6.3), we obtain

$$\begin{aligned} f'(x) &= \int_x^\infty g(u)\,\mathrm{e}^{-\gamma(u-x)}\,\mathrm{d}u + C\mathrm{e}^{\gamma x} \\ &= \int_0^\infty g(u+x)\,\mathrm{e}^{-\gamma u}\,\mathrm{d}u + C\mathrm{e}^{\gamma x}, \end{aligned} \tag{6.4}$$

where $C = \text{const}$. Since $\|g\| \le 4\lambda\kappa^{-2}\|f\| < \infty$, Eq. (6.4) implies that if $C \neq 0$ then

$$f'(x) = O(1) + C\mathrm{e}^{\gamma x} \to \infty \qquad (x \to +\infty),$$

so that $\lim_{x\to+\infty} f(x) = \infty$, which contradicts the assumption $\|f\| < \infty$. Therefore, $C = 0$ and

$$\|f'\| \le \|g\| \int_0^\infty \mathrm{e}^{-\gamma u}\,\mathrm{d}u = \frac{\|g\|}{\gamma} < \infty.$$

The proof is completed. □

7. Proof of the main results. Let $(\mathcal{F}_t,\, t \ge 0)$ be the natural filtration generated by the process (X_t), i.e., $\mathcal{F}_t = \sigma\{X_s,\, s \le t\}$ is the minimal σ-algebra containing all "level" events $\{X_s \le c\}$ $(c \in \mathbb{R},\, s \le t)$. Intuitively, $\mathcal{F}_t$ is interpreted as the collection of all the information that can be obtained by observation of the random process (X_s) up to time t. As is well known (see, e.g., Ref. [9, Ch. 4]), if a function f is $\mathcal{L}$-harmonic then the random process $f(X_t)$ is a *martingale* relative to $(\mathcal{F}_t)$, i.e., for any $0 \le s \le t$, with probability 1,

$$\mathsf{E}_x[f(X_t)\,|\,\mathcal{F}_s] = f(X_s). \tag{7.1}$$

In words, this means that if we are trying to predict the mean value of the martingale $f(X_t)$ at some future time t using its past history up to time $s \le t$, then the best estimate is given by $f(X_s)$, i.e., the value of the process at the latest available time instant s.

In particular, taking expectation of both sides of Eq. (7.1) at $s = 0$ gives

$$\mathsf{E}_x[f(X_t)] = f(x), \qquad t \ge 0. \tag{7.2}$$

In addition, if f is bounded then, by Doob's optional stopping theorem (see, e.g., Ref. [29, §8]), Eq. (7.2) extends to

$$\mathsf{E}_x[f(X_T)] = f(x), \tag{7.3}$$

where T is a random *stopping time*, i.e., such that $\{T \le t\} \in \mathcal{F}_t$ for each $t \ge 0$. In words, one should be able to decide whether the random time T has occurred by observing the process up to a given time. The martingale property (7.3) is crucial in the proof of our main theorem below, where we will apply it to the special sequence of stopping times.

THEOREM 7.1 (cf. Theorem 1.1). *Assume that* $\mathsf{E}[\xi] \le 0$ *and* $\mathsf{P}\{\xi \ne 0\} > 0$. *If* $\mathcal{L}f = 0$ *and* $\|f\| < \infty$, *then* $f(x) \equiv \mathrm{const}$, $x \in \mathbb{R}$.

Proof. For $r > 0$, set $N_r := \min\{n : S_n \le -r\}$, where $S_n = \xi_1 + \cdots + \xi_n$, and define

$$T_r := \sigma_{N_r} = \sum_{n=1}^{\infty} \sigma_n \mathbf{1}_{\{N_r = n\}}, \tag{7.4}$$

where (σ_i) are the time instants of successive jumps (see Section 3). The assumption $\mathsf{E}[\xi] \le 0$ implies (see, e.g., Ref. [10]) that $\liminf_{n\to\infty} S_n = -\infty$ a.s., so that $N_r < \infty$ a.s. and therefore the random variable T_r is well defined.

Note that T_r is a stopping time for the random process (X_t), i.e., $\{T_r \le t\} \in \mathcal{F}_t$ for each $t \ge 0$. Indeed, suppose that we are given a sample path of the process X_s up to time t, and in particular we know the time instants σ_i and the magnitudes of all the jumps prior to t. Then, using Eq. (5.2), we can reconstruct the corresponding values[9] $\xi_i = \ln(X_{\sigma_i}/X_{\sigma_i - 0})$. In turn, this allows us to determine if the threshold $(-r)$ has been reached by the associated random walk S_n and, therefore, whether or not the condition $T_r \le t$ holds, as required.

Now, applying the optional stopping theorem in the form (7.3), we obtain

$$f(x) = \mathsf{E}_x[f(X_{T_r})] = \mathsf{E}\big[f(\mathrm{e}^{S_{N_r}} x + \text{terms independent of } x)\big]. \tag{7.5}$$

First, suppose that $v > 0$. Differentiation of Eq. (7.5) with respect to x gives

$$f'(x) = \mathsf{E}\big[\mathrm{e}^{S_{N_r}} f'(\mathrm{e}^{S_{N_r}} x + \cdots)\big], \tag{7.6}$$

whence, using Proposition 6.2, we get

$$\|f'\| \le \mathrm{e}^{-r} \|f'\| < \infty.$$

Letting here $r \to +\infty$, we conclude that $\|f'\| = 0$ and so $f = \mathrm{const}$.

If $v = 0$ then, differentiating Eq. (7.6) we get

$$f''(x) = \mathsf{E}\big[\mathrm{e}^{2S_{N_r}} f''(\mathrm{e}^{S_{N_r}} x + \cdots)\big],$$

so that

$$\|f''\| \le \mathrm{e}^{-2r} \|f''\| < \infty. \tag{7.7}$$

[9] There is a slight problem if $X_{\sigma_i - 0} = X_{\sigma_i} = 0$, but this only happens with zero probability, so may be ignored.

Equation (6.3) (with $\gamma = 0$) implies $\|f''\| \le \|g\| < \infty$, so taking $r \to +\infty$ in Eq. (7.7) yields $f''(x) \equiv 0$. Therefore, $f(x) = c_1 x + c_0$, but since $\|f\| < \infty$, we must have $c_1 = 0$, so that $f(x) \equiv c_0 = \text{const}$. □

PROPOSITION 7.1. *Denote* $A_\infty := \{\liminf_{t\to\infty} X_t = +\infty\}$, *and consider the probability of the event* A_∞ *as a function of the initial point of the process* X_t,

$$f_\infty(x) := \mathsf{P}_x(A_\infty) = \mathsf{P}(A_\infty | X_0 = x), \qquad x \in \mathbb{R}.$$

Then the function $f_\infty(x)$ *is* $\mathcal{L}$*-harmonic, i.e.,*

$$(\mathcal{L} f_\infty)(x) = 0, \qquad x \in \mathbb{R},$$

where $\mathcal{L}$ *is the generator of the random process* (X_t) *given by Eq. (6.2).*

Proof. Conditioning on X_h $(< +\infty)$, by the Markov property we obtain

$$\begin{aligned} f_\infty(x) &= \mathsf{E}_x[\mathsf{P}_x(A_\infty | X_h)] \\ &= \mathsf{E}_x[\mathsf{P}_{X_h}(A_\infty)] \\ &= \mathsf{E}_x[f_\infty(X_h)], \end{aligned}$$

whence, by the definition (6.1), it readily follows that $\mathcal{L} f_\infty = 0$. □

THEOREM 7.2 (cf. Theorem 1.2). *Suppose that* $\mathsf{E}[\xi] > 0$. *Then the function* $f_\infty(x)$ *is a nontrivial bounded* $\mathcal{L}$*-harmonic function; in particular,* $f_\infty(x) \to 0$ *as* $x \to -\infty$ *and* $f_\infty(x) \to 1$ *as* $x \to +\infty$.

Proof. The function $f_\infty(x)$ is $\mathcal{L}$-harmonic by Proposition 7.1. In order to obtain the limits of $f_\infty(x)$ as $x \to \pm\infty$, note that

$$\begin{aligned} 1 - f_\infty(x) &= \mathsf{P}_x\Big\{\liminf_{t\to\infty} X_t < +\infty\Big\} \\ &= \mathsf{P}_x\Big\{\liminf_{n\to\infty} X_{\sigma_n} < +\infty\Big\} \\ &= \lim_{M\to\infty} \mathsf{P}_x\{X_{\sigma_n} \le M \text{ infinitely often}\}. \end{aligned} \tag{7.8}$$

According to Eq. (5.2), the condition $X_{\sigma_n} \le M$ can be rewritten as

$$x + \zeta_1 + \zeta_2\,\mathrm{e}^{-S_1} + \dots + \zeta_n\,\mathrm{e}^{-S_{n-1}} \le M\,\mathrm{e}^{-S_n}. \tag{7.9}$$

Since $\mathsf{E}[\xi] > 0$, the strong Law of Large Numbers implies that, with probability 1,

$$S_n \sim n\mathsf{E}[\xi] \to +\infty \qquad (n \to \infty). \tag{7.10}$$

It follows that if the inequality (7.9) holds for infinitely many n, then

$$x + \liminf_{n\to\infty} \big(\zeta_1 + \zeta_2\,\mathrm{e}^{-S_1} + \dots + \zeta_n\,\mathrm{e}^{-S_{n-1}}\big) \le \lim_{n\to\infty} M\,\mathrm{e}^{-S_n} = 0. \tag{7.11}$$

Moreover, using the estimate (7.10) and recalling that ζ_i are i.i.d. random variables (see Eq. (5.3)), it is easy to show (e.g., using Kolmogorov's "three series" theorem, see Ref. [10]) that the random series

$$\eta := \sum_{n=1}^{\infty} \zeta_n \,\mathrm{e}^{-S_{n-1}} \tag{7.12}$$

converges with probability 1. Therefore, from Eqs. (7.9), (7.11) and (7.12) it follows that for any $M > 0$,

$$\mathsf{P}_x\{X_{\sigma_n} \le M \text{ infinitely often}\} \le \mathsf{P}\{\eta \le -x\}.$$

Returning to Eq. (7.8), we deduce that

$$\begin{aligned} 1 \ge f_\infty(x) &\ge 1 - \mathsf{P}\{\eta \le -x\} \\ &= \mathsf{P}\{\eta > -x\} \to 1 \qquad (x \to +\infty). \end{aligned}$$

On the other hand, writing the left-hand side of Eq. (7.9) as $x+\eta+\delta_n$, where $\delta_n \to 0$ a.s., we have, for any $\varepsilon > 0$, $M > 0$,

$$\begin{aligned} \mathsf{P}\{x+\eta \le -\varepsilon\} &\le \mathsf{P}\{x+\eta+\delta_n \le 0 \text{ for all } n \text{ large enough}\} \\ &\le \mathsf{P}\{x+\eta+\delta_n < M\mathrm{e}^{-S_n} \text{ infinitely often}\}, \end{aligned}$$

which in view of Eq. (7.8) implies

$$\begin{aligned} 0 \le f_\infty(x) &\le 1 - \mathsf{P}\{\eta \le -x-\varepsilon\} \\ &= \mathsf{P}\{\eta > -x-\varepsilon\} \to 0 \qquad (x \to -\infty). \end{aligned}$$

Thus, the proof is completed. □

Acknowledgments. Part of this research was done when the second author (G.D.) was visiting the University of Cambridge in May–June 2005, and his thanks are due to Arieh Iserles for stimulating discussions and useful remarks. The third author (S.M.) gratefully acknowledges the support from the Center of Advanced Studies in Mathematics of the Ben Gurion University during his visit in May–June 2006, and he would like to thank Michael Lin for kind hospitality.

REFERENCES

[1] C.R. Adams, *Linear q-difference equations*, Bull. Amer. Math. Soc. **37** (1931), 361–400.

[2] V.A. Ambartsumian, *On the theory of brightness fluctuations in the Milky Way*, (Russian) Doklady Akad. Nauk SSSR **44** (1944), 244–247; (English translation) Compt. Rend. (Doklady) Acad. Sci. URSS **44** (1944), 223–226.

[3] G.D. Birkhoff, *The generalized Riemann problem for linear differential equations and the allied problems for linear difference and q-difference equations*, Proc. Amer. Acad. Arts Sci. **49** (1913), 521–568.

[4] G. DERFEL, *Functional-differential and functional equations with rescaling*, in Operator Theory and Boundary Eigenvalue Problems (International Workshop, Vienna, July 27-30, 1993), Operator Theory: Advances and Applications, Vol. 80, I. Gohberg and H. Langer (eds.), Birkhäuser, Basel, 1995, pp. 100–111.

[5] ——— AND A. ISERLES, *The pantograph equation in the complex plane*, J. Math. Anal. Appl. **213** (1997), 117–132.

[6] ——— AND F. VOGL, *On the asymptotics of solutions of a class of linear functional-differential equations*, European J. Appl. Math. **7** (1996), 511–518.

[7] G.A. DERFEL, *Probabilistic method for a class of functional-differential equations*, (Russian) Ukrain. Mat. Zh. **41** (1989), 1322–1327; (English translation) Ukrainian Math. J. **41** (1990), 1137–1141.

[8] ——— AND S.A. MOLCHANOV, *Spectral methods in the theory of functional-differential equations*, (Russian) Mat. Zametki **47**(3) (1990), 42–51; (English translation) Math. Notes **47** (1990), 254–260.

[9] S.N. ETHIER AND T.G. KURTZ, *Markov Processes: Characterization and Convergence*, Wiley Series in Probability and Mathematical Statistics, John Wiley & Sons, New York, 1986.

[10] W. FELLER, *An Introduction to Probability Theory and Its Applications*, Vol. II, Wiley Series in Probability and Mathematical Statistics, 2nd ed., John Wiley & Sons, New York, 1971.

[11] L. FOX, D.F. MAYERS, J.R. OCKENDON, AND A.B. TAYLER, *On a functional differential equation*, J. Inst. Math. Appl. **8** (1971), 271–307.

[12] D.P. GAVER, JR., *An absorption probability problem*, J. Math. Anal. Appl. **9** (1964), 384–393.

[13] A.K. GRINTSEVICHYUS, *On the continuity of the distribution of a sum of dependent variables connected with independent walks on lines*, (Russian) Teor. Veroyatn. i Primenen. **19** (1974), 163–168; (English translation) Theory Probab. Appl. **19** (1974), 163–168.

[14] A. ISERLES, *On the generalized pantograph functional-differential equation*, European J. Appl. Math. **4** (1993), 1–38.

[15] ——— AND Y.K. LIU, *On pantograph integro-differential equations*, J. Integral Equations Appl. **6** (1994), 213–237.

[16] T. KATO, *Asymptotic behavior of solutions of the functional differential equation* $y'(x) = ay(\lambda x) + by(x)$, in Delay and Functional Differential Equations and Their Applications (Proc. Conf., Park City, Utah, March 6–11, 1972), K. Schmitt (ed.), Academic Press, New York, 1972, pp. 197–217.

[17] ——— AND J.B. MCLEOD, *The functional-differential equation* $y'(x) = ay(\lambda x) + by(x)$, Bull. Amer. Math. Soc. **77** (1971), 891–937.

[18] G.SH. LEV, *Semi-Markov processes of multiplication with drift*, (Russian) Teor. Veroyatn. i Primenen. **17** (1972), 160–166; (English translation) Theory Probab. Appl. **17** (1972), 159–164.

[19] K. MAHLER, *On a special functional equation*, J. London Math. Soc. **15** (1940), 115–123.

[20] J.C. MARSHALL, B. VAN-BRUNT, AND G.C. WAKE, *A natural boundary for solutions to the second order pantograph equation*, J. Math. Anal. Appl. **299** (2004), 314–321.

[21] T.E. MASON, *On properties of the solutions of linear q-difference equations with entire function coefficients*, Amer. J. Math. **37** (1915), 439–444.

[22] G.R. MORRIS, A. FELDSTEIN, AND E.W. BOWEN, *The Phragmén-Lindelöf principle and a class of functional differential equations*, in Ordinary Differential Equations (Proc. Conf., Math. Res. Center, Naval Res. Lab., Washington, D.C., 1971), L. Weiss (ed.), Academic Press, New York, 1972, pp. 513–540.

[23] R.J. OBERG, *Local theory of complex functional differential equations*, Trans. Amer. Math. Soc. **161** (1971), 302–327.

[24] J.R. Ockendon and A.B. Tayler, *The dynamics of a current collection system for an electric locomotive*, Proc. Royal Soc. London A **322** (1971), 447–468.

[25] B.K. Shivamoggi, *Perturbation Methods for Differential Equations*, Birkhäuser, Boston, 2003.

[26] V. Spiridonov, *Universal superpositions of coherent states and self-similar potentials*, Phys. Rev. A **52** (1995), 1909–1935.

[27] E.C. Titchmarsh, *The Theory of Functions*, 2nd ed., Oxford University Press, Oxford, 1939.

[28] G.C. Wake, S. Cooper, H.K. Kim, and B. van-Brunt, *Functional differential equations for cell-growth models with dispersion*, Commun. Appl. Anal. **4** (2000), 561–573.

[29] J. Yeh, *Martingales and Stochastic Analysis*, Series on Multivariate Analysis, Vol. 1, World Scientific, Singapore, 1995.

NUMERICAL METHODS FOR NON-ZERO-SUM STOCHASTIC DIFFERENTIAL GAMES: CONVERGENCE OF THE MARKOV CHAIN APPROXIMATION METHOD*

HAROLD J. KUSHNER†

Abstract. The Markov chain approximation method is an efficient and popular collection of methods for the numerical solution of stochastic control problems in continuous time, for reflected-jump-diffusion-type models and the convergence proofs have been extended to zero-sum stochastic differential games. We apply it to a class of non-zero-sum stochastic differential games with a diffusion system model where the controls for the two players are separated, It is shown that equilibrium values for the approximating chain converge to equilibrium values for the original process and that any equilibrium value for the original process can be approximated by an ϵ-equilibrium for the chain for arbitrarily small $\epsilon > 0$. The actual numerical algorithm is that for a stochastic game for a finite-state Markov chain.

1. Introduction. The paper is concerned with the convergence proofs of the Markov chain approximation method to non-zero-sum stochastic differential games. The method is widely used for the numerical solution of stochastic control and optimal control problems in continuous time, for controlled reflected-jump-diffusion type models. The method was extended to zero-sum stochastic differential games in [15, 16, 17]. The Markov chain approximation method has been used to solve non-zero-sum differential games [11, 12], but there have not been results concerning convergence of the numerical equilibrium values to the equilibrium values for the original diffusion model. Papers such as [4] deal with approximations to non-zero-sum games in normal form, and do not apply to the diffusion-type system models or to the type of approximations that we use.

We will work with a discounted cost problem for a diffusion model in a compact set G, with absorption on the boundary. This particular state space and cost function are chosen to simplify the development, allowing us to concentrate on the issues that are particular to the non-zero-sum case. One can replace the boundary absorption by boundary reflection, with the boundary and reflection directions satisfying the conditions in [18, 15]. We will work with two-player games for notational simplicity, but the methods allow the use of any number. The non-zero-sum game is difficult because, as opposed to the zero-sum case, the players are not strictly competitive and have their own value functions. Weak convergence methods as in [18] are still employed, but many of the key techniques that were used previously cannot be carried over and new methods must be applied. The proof for the two-person zero-sum game in [15] has the advantage that the controls are

*This work was partially supported by NSF grant DMS-0506928 and ARO contract W911NF-05-10928.

†Applied Mathematics Department, Brown University, Providence, RI 02912 (hjk@dam.brown.edu).

determined by a minmax operation and that there is a single cost function, so that one player's gain is another's loss, properties that the non-zero-sum game does not have. This difference creates difficulties for the non-zero-sum case that require considerable modification of the proofs. The methods that are employed require the use of strong-sense, rather than with weak-sense solutions as in [18]. Unlike the single player problem. one must work with strategies and not simply controls.

The Markov chain approximation technique first approximates the controlled diffusion by a Markov chain on a finite state space with a discretization parameter h. The cost function is then approximated to apply to the chain. The numerical procedure is to solve the game problem for the chain model. Then one proves convergence, namely, that the equilibrium or ϵ-equilibrium values for the two players for the approximating chain model converge to equilibrium or ϵ-equilibrium values for the diffusion model as the approximating parameter h goes to zero. The methods of proof are purely probabilistic, no PDE techniques are required. The key condition that the approximating chain must satisfy is a simple "local consistency" condition. Getting such approximations is straightforward and many methods are in [18]. The approximating chains are similarly obtained for the game problem. The numerical approximations are processes which are close to the original, which allows great flexibility.

In Section 2, the model and the cost functions for the players are defined, the boundary conditions discussed and a review of some background material is given. The convergence proofs use the fact that the player's strategies for the original diffusion process can be simplified (uniformly in the controls), with various approximations to the controls. The first of such results is given in Section 3, which shows that the diffusion model can be uniformly approximated by a discrete-time model, and the relation between randomized and relaxed controls is developed. Section 4 contains the main results concerning approximation of the policies of the players. It is shown the general relaxed controls be discretized in time, approximated by finite-valued ordinary controls, and delayed slightly, without changing the values very much. The definition of equilibrium is in Section 5, where it is also shown that strategies for the players can be well-approximated by a "smooth" conditional probability, depending only on selected samples of the driving Wiener process (and not on the entire Wiener process). This representation will be of great help in the proofs. The Markov chain approximation is introduced in Section 6

The proofs depend on certain representations of approximations to the approximating chains that allow us to show that the costs for the chain itself change little if the controls are approximated. Some such results are in Section 7. These results are new and should be useful in dealing with numerical approximations. The approximations in Section 7 are applied in Section 8 to show that the 'approximate" equilibrium (values or strategies) for the diffusion are approximate equilibrium (values or strategies) for the chain for small h. If the ϵ-equilibrium values for the chain are unique for small $\epsilon > 0$, then the convergence proof is complete since an "approximate" equilibrium value for the chain is also one for the diffusion. If the value is not unique then the proof of this last fact is more difficult, and we restrict attention to the case where the diffusion coefficient is constant. The proof uses a representation of the chain that is close to a discrete-time system, and the proof of this representation exploits a strong approximation theorem. This is also new and is developed in Section 9. The final part of the convergence proof is in Section 10.

Part of the proof uses a weak convergence analysis as in [18], and the paper is organized to take advantage of those results to the extent possible so that we can avoid duplicating details. For S a topological space, let $D[S;0,\infty)$ denote the S-valued functions on $[0,\infty)$ that are right continuous and have left-hand limits, and with the Skorokhod topology [6, 18] used. If $S = I\!R^v$, then we write $D[S;0,\infty) = D^v[0,\infty)$.

2. The model. The system model is

$$x(t) = x(0) + \int_0^t \sum_{i=1}^{2} b^i(x(s), u_i(s))ds + \int_0^t \sigma(x(s))dw(s), \tag{2.1}$$

where $x(t) \in I\!R^v$, Euclidean v-space, Player i, $i = 1$ uses control $u_i(\cdot)$, and $w(\cdot)$ is a standard vector-valued Wiener process. Let $\beta > 0$ and define E_x^u as the expectation conditioned on the use of control $u(\cdot) = (u_1(\cdot), u_2(\cdot))$ and initial condition x. The cost function for Player i is

$$W_i(u) = E_x^u \int_0^\tau e^{-\beta t} k_i(x(s), u_i(s))ds + E_x^u e^{-\beta\tau} g_i(x(\tau)), \tag{2.2}$$

where τ is the first time that the boundary ∂G of a compact set G is hit ($\tau = \infty$ if the boundary is never reached). The set G satisfies (A2.1) below. Define $b(\cdot) = b^1(\cdot) + b^2(\cdot)$ and $k(\cdot) = k_1(\cdot) + k_2(\cdot)$. Let $\mathcal{U}_i$ denote the set of admissible controls for Player i: $u_i(\cdot) \in \mathcal{U}_i$ if it is measurable, non-anticipative with respect to $w(\cdot)$, and U_i-valued.

The following condition on the regularity of the functions is assumed.

A2.1. *$b^i(\cdot)$, and $\sigma(\cdot)$ are bounded and continuous and Lipschitz continuous in x, uniformly in u. The controls $u_i(\cdot)$ for Player i take values in U_i, a compact set in some Euclidean space, and $k_i(\cdot)$ and $g_i(\cdot)$ are bounded and continuous.*

The first boundary hitting time τ. The proof of convergence, whether here or for the problems in [13, 18], generates a sequence of process approximations (continuous-time interpolations of the approximating chain) and the exit or boundary hitting times of this sequence has to converge to the exit time of the original diffusion process (2.1). For the convergence of any numerical procedure, one must assure that something analogous takes place. If the costs for the approximating problem are to converge to the costs for (2.1), (2.2), then we need to assure (at least with probability one) that the paths of the process $x(\cdot)$ are not "tangent" to ∂G at the moment τ of first hitting the boundary. For $\phi(\cdot)$ in $D^v[0,\infty)$ (with the Skorokhod topology used), define the function $\hat{\tau}(\phi)$ with values in $\overline{I\!R}^+ = [0,\infty]$ by: $\hat{\tau}(\phi) = \infty$, if $\phi(t) \in G^0$, the interior of G, for all $t < \infty$, and otherwise use

$$\hat{\tau}(\phi) = \inf\{t : \phi(t) \notin G^0\}.$$

A2.2. *For a continuous real-valued function $\Phi(\cdot)$ on $I\!R^v$, define $G = \{x : \Phi(x) \leq 0\}$, and suppose that it is the closure of its interior $\{x : \Phi(x) < 0\}$. For each initial condition and control, the function $\hat{\tau}(\cdot)$ is continuous (as a map from $D^v[0,\infty)$ to the compactified interval $[0,\infty]$) with probability one relative to the measure induced by the solution to* (2.1).

The tangency problem would be a concern with any numerical method, since they all depend on some sort of approximation. For example, the convergence theorems for the classical finite difference methods for elliptic and parabolic equations generally use a nondegeneracy condition on $a(x)$ in order to (implicitly) guarantee (A2.2). This issue is discussed in [13, Section 4.4] and in [18, Section 10.2]. In [18, Section 10.2] a randomized stopping criterion is described, which assures that (A2.2) holds and has minimal effect on the costs. See also [18, p 280, sec ed.] where it is shown that the Girsanov transformation method can play a useful role in the verification of (A2.2). Note that typical methods for the solution of parabolic or elliptic equations with Dirichlet boundary conditions require nondegeneracy and smooth boundaries, which assures (A2.2).

Relaxed controls: Review. When proving the convergence of sequences or approximations in control theory, there is a great advantage in working with "relaxed" controls in lieu of ordinary controls. They are used only for the purposes of the convergence proofs and have no role in practice. Let $\{\mathcal{F}_t, t < \infty\}$ be the filtration on the probability space and let $w(\cdot)$ be a standard vector-valued $\mathcal{F}_t$-Wiener process. Let $r_i(\cdot), i = 1, 2$, be measures on the Borel sets of $U_i \times [0,\infty)$ such that $r_i(U_i \times [0,t]) = t$ and the process $r_i(A \times [0,\cdot])$ is measurable and non-anticipative for each Borel set $A \subset U_i$. Then $r_i(\cdot)$ is an *admissible relaxed control* for Player i [7, 18]. We will use $\mathcal{U}_i$ for the set of admissible relaxed controls as well as for the admissible ordinary controls. For Borel sets $A \subset U_i$, write $r_i(A \times [t_0, t_1]) = r_i(A, [t_0, t_1])$, and use $r_i(A, t_1)$ if $t_0 = 0$. Define $U = U_1 \times U_2$ and $\mathcal{U} = \mathcal{U}_1 \times \mathcal{U}_2$.

For almost all (ω, t) and each Borel $A \subset U_i$, one can define the left derivative (w.p.1)

$$r_i'(A,t) = \lim_{\delta \to 0} \frac{r_i(A,t) - r_i(A,t-\delta)}{\delta}.$$

We can suppose that the limit exists for all (ω, t), with no loss of generality.

Then for all (ω, t), $r_i'(\cdot, t)$ is a probability measure on the Borel sets of U_i and for any Borel set B in $U_i \times [0, \infty)$,

$$r_i(B) = \int_0^\infty \int_{U_i} I_{\{(\alpha_i,t)\in B\}} r_i'(d\alpha_i, t)dt.$$

If $u_i(\cdot)$ is an ordinary control, then it has the relaxed control representation $r_i(\cdot)$ define by $r_i'(A, t) = I_A(u_i(t))$, where I_A is the indicator function of the set A. The weak topology [18] will be used on the space of admissible relaxed controls. In this topology, any sequence of relaxed controls is compact; hence it has a weakly convergent subsequence. The use of relaxed controls does not change the range of values of the cost functions.

The system and cost functions with relaxed controls. Define the "product" relaxed control $r(\cdot)$ by its derivative $r'(\cdot, t) = r_1'(\cdot, t)r_2'(\cdot, t)$. Thus $r(\cdot)$ is a product measure, with marginals $r_i(\cdot), i = 1, 2$. We will usually write $r(\cdot) = (r_1(\cdot), r_2(\cdot))$ without ambiguity. The pair $(w(\cdot), r(\cdot))$ is called an *admissible pair* if each of the $r_i(\cdot)$ is admissible with respect to $w(\cdot)$. In relaxed control terminology, (2.1) and (2.2) are written as

$$x(t) = x(0) + \int_0^t \int_{U_i} b(x(s), \alpha_i) r'(d\alpha_i, s)ds + \int_0^t \sigma(x(s))dw(s). \quad (2.1r)$$

$$W_i(x, r) = E_x^r \int_0^\tau e^{-\beta t} \int_{U_i} k_i(x(s), \alpha_i) r_i'(d\alpha_i, s)ds + E_x^r e^{-\beta\tau} g_i(x(\tau)). \quad (2.2r)$$

3. A discrete time approximation and randomized controls. A relaxed control can be approximated by a randomized ordinary control. This is best illustrated for a discrete time system, for which we will also have need. Let $\Delta > 0$ and suppose that $r(t, \cdot)$ is adapted to $\mathcal{F}_{n\Delta-}$ for $t \in [n\Delta, n\Delta + \Delta)$. Then define $x^\Delta(n\Delta)$ recursively by $x^\Delta(0) = x(0)$ and for $n \geq 0$,

$$\begin{aligned} x^\Delta(n\Delta + \Delta) = x^\Delta(n\Delta) &+ \int_{n\Delta}^{n\Delta+\Delta} \int_U b(x^\Delta(n\Delta), \alpha) r'(d\alpha, s)ds \\ &+ \sigma(x^\Delta(n\Delta))[w(n\Delta + \Delta) - w(\Delta)]. \end{aligned} \quad (3.1)$$

For $t \in [n\Delta, n\Delta + \Delta)$, the continuous time interpolation can be defined either by $x^\Delta(t) = x^\Delta(n\Delta)$ or by

$$\begin{aligned} x^\Delta(t) = \; & x^\Delta(n\Delta) + \int_{n\Delta}^t \int_U b(x^\Delta(n\Delta), \alpha) r'(d\alpha, s)ds \\ & + \int_{n\Delta}^t \sigma(x^\Delta(n\Delta))dw(t), \end{aligned} \quad (3.2)$$

The cost functions $W_i^\Delta(x, r)$ for the interpolated processes are given by (2.2r), with $x^\Delta(\cdot)$ replacing $x(\cdot)$ and the relaxed control in (3.1) or (3.2) used.

A note on convergence. Note the following fact concerning the approximation. Let $\{r^\Delta(\cdot)\}$ and $r(\cdot)$ be admissible relaxed controls with $r^\Delta(\cdot) \to r(\cdot)$ w.p.1 (in the weak topology) as $\Delta \to 0$, and $r^\Delta(\cdot, n\Delta)$ adapted to $\mathcal{F}_{n\Delta-}$, as in (3.1). Then it is easy to see that, as $\Delta \to 0$, the sequence of solutions $\{x^\Delta(\cdot)\}$ converges w.p.1, uniformly on any bounded time interval and the limit $(x(\cdot), r(\cdot), w(\cdot))$ solves (2.1r). By the boundary continuity (A2.2), concerning the w.p.1 continuity of the hitting times, the sequence of first hitting times of the boundary for $x^\Delta(\cdot)$ converges w.p.1. to that of the limit. These facts imply that the costs converge to those for the limit processes.

Randomized controls. We now show that a relaxed control can be viewed as a limit of randomized controls, a fact that will be needed in the convergence proofs. Recall that $r_i(\cdot, [n\Delta, n\Delta + \Delta])/\Delta$ is a probability distribution on the Borel sets of U_i. Define the randomized control version of this distribution, as follows. Let $r(\cdot)$ be an admissible relaxed control and let $E_{n\Delta}$ denote the conditional expectation given $\mathcal{F}_{n\Delta-}$. Let $\tilde{u}^\Delta_{i,n}$ be a random variable with the conditional distribution $r^\Delta_{i,n}(\cdot) = E_{n\Delta} r_i(\cdot, [n\Delta, n\Delta + \Delta])/\Delta$. Define $\tilde{u}^\Delta_n = (\tilde{u}^\Delta_{1,n}, \tilde{u}^\Delta_{2,n})$, and let $\tilde{u}^\Delta(\cdot)$ denote the continuous time interpolation, with interpolation intervals Δ. Finally, define $\tilde{x}^\Delta(\cdot)$ by $\tilde{x}^\Delta(0) = x^\Delta(0) = x(0)$ and

$$\begin{aligned}\tilde{x}^\Delta(n\Delta + \Delta) &= \tilde{x}^\Delta(n\Delta) + \Delta b(\tilde{x}^\Delta(n\Delta), \tilde{u}^\Delta_n) \\ &\quad + \sigma(\tilde{x}^\Delta(n\Delta))[w(n\Delta + \Delta) - w(n\Delta)].\end{aligned} \tag{3.3}$$

with continuous time interpolation $\tilde{x}^\Delta(t)$. Define $r^\Delta_n(\cdot) = r^\Delta_{1,n}(\cdot) r^\Delta_{2,n}(\cdot)$, and let $r^\Delta(\cdot)$ be the relaxed control with derivative $r^\Delta_n(\cdot)$ on $[n\Delta, n\Delta + \Delta)$. Then we have the following result, where $r^\Delta(\cdot)$ is used for $x^\Delta(\cdot)$ in (3.2). The theorem shows that relaxed controls are approximations of randomized controls.

THEOREM 3.1. *Under* (A2.1) *and using* $r^\Delta_n(\cdot)$ *for* $r'(\cdot, s), s \in [n\Delta, n\Delta + \Delta)$, *in* (3.1) *and* (3.2), *for any* $T < \infty$,

$$\lim_{\Delta \to 0} \sup_{x(0) \in G} \sup_{r \in \mathcal{U}} E \sup_{t \le T} \left| x^\Delta(t) - x(t) \right|^2 = 0, \tag{3.4a}$$

$$\lim_{\Delta \to 0} \sup_{x(0) \in G} \sup_{r \in \mathcal{U}} E \sup_{t \le T} \left| x^\Delta(t) - \tilde{x}^\Delta(t) \right|^2 = 0. \tag{3.4b}$$

If we add the condition (A2.2) *then the costs for* (3.1) *and* (3.3) *converge (uniformly in* $x(0), r(\cdot)$*) to those for* (2.1r).

Proof. Define $\delta x^\Delta_n = x^\Delta(n\Delta) - \tilde{x}^\Delta(n\Delta)$. Then

$$\begin{aligned}\delta x^\Delta_{n+1} &= \delta x^\Delta_n + \Delta \int_U \left[b(x^\Delta(n\Delta), \alpha) - b(\tilde{x}^\Delta(n\Delta), \alpha) \right] r^\Delta_n(d\alpha) \\ &\quad + \left[\sigma(x^\Delta(n\Delta)) - \sigma(\tilde{x}^\Delta(n\Delta)) \right] \left[w(n\Delta + \Delta) - w(n\Delta) \right] + M^\Delta_n,\end{aligned} \tag{3.5}$$

where

$$M_n^\Delta = \Delta \left[\int_U b(\tilde{x}^\Delta(n\Delta), \alpha) r_n^\Delta(d\alpha) - b(\tilde{x}^\Delta(n\Delta), \tilde{u}_n^\Delta) \right].$$

By the definition of $\tilde{u}_n^\Delta(\cdot)$ via a conditioning on $\mathcal{F}_{n\Delta-}$, the random variable M_n^Δ is an $\mathcal{F}_{n\Delta-}$- martingale difference and its variance is $O(\Delta^2)$, uniformly in the controls. With these facts in hand, the Lipschitz condition can be used in the usual way to show (3.4b) and the fact that

$$\lim_{\Delta\to 0} \sup_{x(0)\in G} \sup_{r\in\mathcal{U}} \sup_{t\le T} E\left| \int_0^t e^{-\beta t} k(\tilde{x}^\Delta(s), \tilde{u}^\Delta(s)) ds - \int_0^t \int_U e^{-\beta t} k(x^\Delta(s), \alpha)) r^{\Delta,\prime}(d\alpha, s) ds \right| = 0.$$

The proof of (3.4a) is similar and is omitted. Now (3.4) and (A2.2) imply that the sequence of first hitting times also converges, uniformly in the controls Then the last expression and (3.4) imply that the costs converge, also uniformly in the controls. □

4. Approximating the controls. In the convergence proofs we exploit the fact that, without loss of generality in the proofs, the controls can be restricted to be finite-valued, piecewise constant and slightly delayed. The (uniform) approximation facts are developed in this section. Theorem 4.1 shows that we can represent the strategies in a particular "conditional probability" form, which will be convenient in the proofs.

We start with a canonical form for the control approximation. For each admissible relaxed control $r(\cdot)$ and each $\epsilon > 0$, let $r_i^\epsilon(\cdot)$ be an admissible relaxed control such that, for each $T < \infty$ and bounded and continuous real-valued function $\phi_i(\cdot)$,

$$\lim_{\epsilon\to 0} \sup_{r_i\in\mathcal{U}_i} E \sup_{t\le T} \left| \int_0^t \int_{U_i} \phi_i(\alpha_i) \left[r_i'(d\alpha_i, s) - r_i^{\epsilon,\prime}(d\alpha_i, s) \right] ds \right| = 0, \quad i = 1, 2, \tag{4.1}$$

Let $x(\cdot)$ and $x^\epsilon(\cdot)$ denote the solutions to (2.1r) corresponding to $r(\cdot)$ and $r^\epsilon(\cdot)$, respectively. The Wiener processes are the same, but the initial conditions might be different. In particular, define $x^\epsilon(\cdot)$ by[1]

$$x^\epsilon(t) = x^\epsilon(0) + \int_0^t \int_U b(x^\epsilon(s), \alpha) r^{\epsilon,\prime}(d\alpha, s) ds + \int_0^t \sigma(x^\epsilon(s)) dw(s). \tag{4.2}$$

The next theorem shows that the solution $x(\cdot)$ process and costs are (uniformly) continuous in the controls in the sense of (4.3) and (4.4).

[1] The processes $x(\cdot)$ and $x^\epsilon(\cdot)$ depend on $r(\cdot)$ and $r^\epsilon(\cdot)$, but this dependence is suppressed in the notation.

THEOREM 4.1. *Assume* (A2.1) *and let* $(r(\cdot), r^\epsilon(\cdot))$ *satisfy* (4.1). *Then for each* t

$$\lim_{\epsilon\to 0} \sup_{x(0),x^\epsilon(0):|x^\epsilon(0)-x(0)|\to 0} \sup_{r\in\mathcal{U}} E \sup_{s\le t} |x^\epsilon(t) - x(t)|^2 = 0. \tag{4.3}$$

If we add the condition (A2.2), *then*

$$\lim_{\epsilon\to 0} \sup_{x(0),x^\epsilon(0):|x^\epsilon(0)-x(0)|\to 0} \sup_{r\in\mathcal{U}} |W_i(x,r) - W_i(x,r^\epsilon)| = 0, \quad i = 1,2. \tag{4.4}$$

Proof. We comment only on the use of (4.1), since the proof is close to that of Theorem 3.1. Define $\delta x^\epsilon(t) = x^\epsilon(t) - x(t)$. Then

$$\begin{aligned}\delta x^\epsilon(t) = \delta x^\epsilon(0) &+ \int_0^t \int_U [b(x^\epsilon(s),\alpha) - b(x(s),\alpha)]\, r'(d\alpha,s)ds \\ &+ \int_0^t [\sigma(x^\epsilon(s)) - \sigma(x(s))]\, dw(s) \\ &+ \int_0^t \int_U b(x^\epsilon(s),\alpha)\, [r^{\epsilon,\prime}(d\alpha,s) - r'(d\alpha,s)]\, ds.\end{aligned} \tag{4.5}$$

We need to show that the last term is small, uniformly in the control for small ϵ. By (4.1), this would be the case if $b(\cdot)$ was a continuous function that depended only on time. The fact that $x^\epsilon(\cdot)$ is equicontinuous, with a high probability will imply the result; in particular, we will use the fact that

$$\lim_{\lambda\to 0} \sup_r \sup_\epsilon \sup_{l\lambda\le t} E \sup_{s\le\lambda} |x^\epsilon(l\lambda+s) - x^\epsilon(l\lambda)|^2 = 0. \tag{4.6}$$

Write the last term of (4.5) as[2]

$$\begin{aligned}&\sum_{l=0}^{[t/\lambda]-1} \int_{l\lambda}^{l\lambda+\lambda} \int_U b(x^\epsilon(l\lambda),\alpha)\, [r^{\epsilon,\prime}(d\alpha,s) - r'(d\alpha,s)]\, ds + O(\lambda) \\ &+ \sum_{l=0}^{[t/\lambda]-1} \int_{l\lambda}^{l\lambda+\lambda} [b(x^\epsilon(s),\alpha) - b(x^\epsilon(l\lambda),\alpha)]\, [r^{\epsilon,\prime}(d\alpha,s) - r'(d\alpha,s)]\, ds.\end{aligned} \tag{4.7}$$

(4.6) implies that, as $\lambda \to 0$, the expectation of the square of the sup of the last term over any finite time interval goes to zero, uniformly in $r(\cdot), r^\epsilon(\cdot), x(0), x^\epsilon(0)$, as $\epsilon \to 0$. Now assumption (4.1) implies that first term of (4.7) goes to zero as $\epsilon \to 0$, uniformly in λ and the controls and initial conditions. (4.3) is a consequence of these facts and the Lipschitz condition. The convergence of the costs follows from the convergence of the paths and controls, which implies the convergence of the exit times (via (A2.2)). □

[2] $[t/\lambda]$ denotes the integer part of t/λ.

Finite-valued and piecewise constant approximations. Next, we choose the approximations in (4.1) that will be of interest. First we discretize the U_i. Suppose that $U_i \in I\!R^{c_i}$, Euclidean c_i-space.[3] For each $\mu > 0$, partition $I\!R^{c_i}$ into disjoint hypercubes $\{R_i^{\mu,l}\}$ with diameters μ. The boundaries can be assigned to the adjoining sets in any way at all. Define $U_i^{\mu,l} = U_i \cap R_i^{\mu,l}$ and let $U_i^{\mu,l}, i = 1, \ldots, p_i^{\mu}$, denote the finite number of non-empty intersections. Fix a point $\alpha_i^{\mu,l} \in U_i^{\mu,l}$. Now, for relaxed controls $(r_1(\cdot), r_2(\cdot))$, define approximating relaxed controls $r_i^{\mu}(\cdot)$, on the control value space $U_i^{\mu} = \{\alpha_i^{\mu,l}, l \le p_i^{\mu}\}$, via their derivatives, $r_i^{\mu,\prime}(\alpha_i^{\mu,l}, t) = r_i'(U_i^{\mu,l}, t)$. The set of such finite-valued and piecewise constant controls is denoted by $\mathcal{U}_i(\mu)$. The next theorem follows from Theorem 4.1.

THEOREM 4.2. *Assume* (A2.1)–(A2.2), *and the above approximation of either one or both of the* $r_i(\cdot)$ *by* $r_i^{\mu}(\cdot) \in \mathcal{U}_i(\mu), i = 1, 2$. *Then* (4.1)*and Theorem* 4.1 *hold for* μ *replacing* ϵ, *and the approximation is uniform in the choice of* $\{\alpha_i^{\mu,l}\}$.

Delayed controls. Theorem 4.2 holds if the actions of the discretized controls are delayed slightly and approximated by piecewise constant ordinary controls To do this, let $\Delta > 0$ and $r_i^{\mu}(\cdot) \in \mathcal{U}_i(\mu)$. Define $\Delta_{i,k}^{\mu,l} = r_i^{\mu}(\alpha_i^{\mu,l}, k\Delta) - r_i^{\mu}(\alpha_i^{\mu,l}, k\Delta - \Delta)$, $l \le p_i^{\mu}$, $k = 1, \ldots$. The $\Delta_{i,k}^{\mu,l}$ is the total time that the measure defined by the derivative of the relaxed control is concentrated on α_i^{μ} on the time interval $[k\Delta - \Delta, k\Delta)$. Now define the piecewise constant ordinary controls $u_i^{\mu,\Delta}(\cdot)$ as follows. On the interval $[k\Delta, k\Delta + \Delta)$, they have values

$$u_i^{\mu,\Delta}(t) = \alpha_i^{\mu,l} \quad \text{for } t \in \left[k\Delta + \sum_{\nu=1}^{l-1} \Delta_{i,k}^{\mu,\nu},\ k\Delta + \sum_{\nu=1}^{l} \Delta_{i,k}^{\mu,\nu}\right). \tag{4.8}$$

On $[k\Delta, k\Delta + \Delta)$, $u_i^{\mu,\Delta}(\cdot)$ takes the value $\alpha_i^{\mu,l}$ on a time interval of length $\Delta_{i,k}^{\mu,l}$. Note also that the $u_i^{\mu,\Delta}(\cdot)$ are "delayed," in that the values of $r_i(\cdot)$ on $[k\Delta - \Delta, k\Delta)$ determine the values of $u_i^{\mu,\Delta}(\cdot)$ on $[k\Delta, k\Delta + \Delta)$. Thus $u_i^{\mu,\Delta}(t), t \in [k\Delta, k\Delta + \Delta)$, is $\mathcal{F}_{k\Delta-}$-measurable. Let $r_i^{\mu,\Delta}(\cdot)$ denote the relaxed control representation of $u_i^{\mu,\Delta}(\cdot)$, with time derivative $r_i^{\mu,\Delta,\prime}(\cdot)$. Let $\mathcal{U}_i(\mu, \delta)$ denote the subset of $\mathcal{U}_i(\mu)$ that are ordinary controls and constant on $[l\delta, l\delta + \delta), l = 0, 1, \ldots$.

It will be required that the approximating controls be constant on intervals that are multiples of some small $\delta > 0$, where Δ/δ is an integer. To modify the above construction to accomplish this, first divide $[k\Delta, k\Delta + \Delta)$ into Δ/δ subintervals of length δ. To each value $\alpha_i^{\mu,l}$, assign (in order $l = 1, 2 \ldots,$) $[\Delta_{i,k}^{\mu,l}/\delta]$ (the integer part) successive subintervals of length

[3]The U_i could be any compact set in a complete and separable metric space.

δ. On any time interval, the total unassigned time goes to zero as $\delta \to 0$, and whatever assignments are made to them is asymptotically (as $\delta \to 0$) unimportant. To eliminate any ambiguity, assign the remaining subintervals as follows. On the interval, $[k\Delta, k\Delta + \Delta)$, the unassigned length of time for value $\alpha_i^{\mu,l}$ is $L_{i,k}^{\mu,\delta,l} = \Delta_{i,k}^{\mu,l} - [\Delta_{i,k}^{\mu,l}/\delta]\delta$. The sum $S_{i,k}^{\mu,\delta} = \sum_l L_{i,k}^{\mu,\delta,l}$ is an integral (zero or positive) multiple of δ. Then assign each unassigned subinterval at random, with value $\alpha_{i,k}^{\mu,l}$ chosen with probability $L_{i,k}^{\mu,\delta,l}/S_{i,k}^{\mu,\delta}$. Let $\mathcal{U}_i(\mu,\delta,\Delta)$ denote the set of such controls.[4] Let $r_i^{\mu,\Delta,\delta,\prime}(\cdot)$ denote the time derivative of $r_i^{\mu,\Delta,\delta}(\cdot)$.

The next theorem is a consequence of Theorem 4.1. It states that for fixed μ and small δ the constructed controls $u_i^{\mu,\delta,\Delta}(\cdot)$ yield good approximations, uniformly in $r_i(\cdot)$ and $\{\alpha_i^{\mu,l}\}$. Equation (4.1) holds in that for each $\mu > 0$, $\Delta > 0$, and bounded and continuous $\phi_i(\cdot)$,

$$\lim_{\delta\to 0} \sup_{r_i\in\mathcal{U}_i} E \sup_{t\le T} \left| \int_0^t \int_{U_i} \phi_i(\alpha_i) \left[r_i^{\mu,\delta,\Delta,\prime}(d\alpha_i,s) - r_i^{\mu,\Delta,\prime}(d\alpha_i,s) \right] ds \right| = 0. \quad (4.9)$$

THEOREM 4.3. *Assume* (A2.1)–(A2.2). *and let* $r_i(\cdot) \in \mathcal{U}_i, i = 1,2$. *For* $(\mu,\delta,\Delta) > 0$, *construct* $r_i^{\mu,\delta,\Delta}(\cdot) \in \mathcal{U}_i(\mu,\delta,\Delta)$ *as above. Then* (4.1) *holds for* $r_i^{\mu,\delta,\Delta}(\cdot)$ *and* (μ,δ,Δ) *replacing* $r_i^\epsilon(\cdot)$ *and* ϵ, *respectively. Also,* (4.9) *holds: I.e., for any* $\epsilon > 0$, *there are* $\mu_\epsilon > 0, \delta_\epsilon > 0, \Delta_\epsilon > 0$ *and* $\kappa_\epsilon > 0$, *such that for* $\mu \le \mu_\epsilon, \delta \le \delta_\epsilon, \Delta \le \Delta_\epsilon$ *and* $\delta/\Delta \le \kappa_\epsilon$,

$$\sup_x \sup_{r_1} \sup_{r_2} \left| W_i(x,r_1,r_2) - W_i(x,r_1,u_2^{\mu,\delta,\Delta}) \right| \le \epsilon. \quad (4.10)$$

(4.10) *holds with the indices* 1 *and* 2 *interchanged or if both controls are approximated. Now consider the discrete-time system* (3.1). *Then the* $\mu_\epsilon > 0, \delta_\epsilon > 0, \Delta_\epsilon > 0$ *and* $\kappa_\epsilon > 0$ *can be defined so that*

$$\sup_x \sup_{r_1} \sup_{r_2} \left| W_i(x,r_1,r_2) - W_i^\Delta(x,r_1,u_2^{\mu,\delta,\Delta}) \right| \le \epsilon. \quad (4.11)$$

(4.11) *holds with the indices* 1 *and* 2 *interchanged or if both controls are approximated or delayed by* Δ.

5. Approximations to ϵ-equilibria. Studies in differential games for models such as (2.1) are usually based on the Elliott and Kalton definition of strategy [5, 8], which are generalizations of feedback controls. An Elliott-Kalton strategy $c_1(\cdot)$ for Player 1 is a mapping from $\mathcal{U}_2$ to $\mathcal{U}_1$ with the following property. If two admissible controls for Player 2 are identical up to time t, then the response of $c_1(\cdot)$ is the same to both until at least time t. It will not otherwise depend on what Player 2 might do in the

[4]The $u_i^{\mu,\delta,\Delta}(\cdot)$ are functions of $r_i(\cdot)$, but this dependence will be omitted in the notation.

future.[5] The definition for Player 2 is analogous. Let $\mathcal{C}_i$ denote the set of such strategies or mappings for Player i.

The above definition of strategy does not account for the possibility of randomized controls, where the sample response to a given control process of the other player might depend on the choice of the randomization. We extend the definition to allow randomized strategies that have the form of the second line in (5.2) for either one or both of the players. Theorem 4.1 shows the connection between relaxed and randomized controls.[6] The pair $\bar{c}_i(\cdot) \in \mathcal{C}_i, i = 1, 2$, is an ϵ-equilibrium strategy pair if

$$\begin{aligned} W_1(x, \bar{c}_1, \bar{c}_2) &\geq W_1(x, r_1, \bar{c}_2) - \epsilon, \\ W_2(x, \bar{c}_1, \bar{c}_2) &\geq W_2(x, \bar{c}_1, r_2) - \epsilon \end{aligned} \tag{5.1}$$

for all admissible controls $r_i(\cdot), i = 1, 2$. The notation $W_2(x, c_1, r_2)$ implies that Player 1 uses its strategy $c_1(\cdot)$ and Player 2 uses the relaxed control $r_2(\cdot)$.

We will require the following assumption.

A5.1. For each small $\epsilon > 0$ there is an ϵ-equilibrium Elliott-Kalton strategy $(\bar{c}_1^\epsilon(\cdot), \bar{c}_2^\epsilon(\cdot))$ under which the solution to (2.1) or (2.1r) is well defined.

That the processes are well defined is unrestrictive, since Theorem 4.3 shows that, for the purposes of constructing ϵ-optimal strategies, it is sufficient to restrict attention to strategies whose control functions are piecewise constant, finite-valued and need depend only on slightly delayed values of the other players control realizations. The control processes that a strategy might yield can be either ordinary or relaxed.

The following theorem gives a representation of an ϵ-equilibrium strategy as a conditional distribution, which plays an important role in the convergence proofs. It is not a practical control and, like all the other approximations to the controls, is used only for the proofs.

THEOREM 5.1. *Assume* (A2.1) *and* (A2.2). *Given* $\epsilon_1 > 0$, *there are* $(\mu, \delta, \Delta) > 0$, *where* Δ/δ *is an integer, such that the values for any strategy pair* $(c_1(\cdot), c_2(\cdot)), i = 1, 2$, *under which the solution to* (2.1r) *is well defined can be approximated within* ϵ_1 *by the following simpler strategy pairs* $(c_1^{\mu,\delta,\Delta}(\cdot), c_2^{\mu,\delta,\Delta}(\cdot))$ *for which the realizations of* $c_i^{\mu,\delta,\Delta}(\cdot)$ *are ordinary controls in* $\mathcal{U}_i(\mu, \delta, \Delta)$, *and denoted by* $u_i^{\mu,\delta,\Delta}(\cdot)$. *They are defined as follows.*

[5] The definition in [5] requires that the controls $r_i(A, \cdot)$ be progressively measurable, and not simply measurable and adapted, for each Borel set A. But due to the approximation results of Theorems 4.1–4.3, this added requirement is unnecessary in our case.

[6] As seen in the previous section, we need only consider strategies that give control processes that are piecewise constant, finite-valued and slightly delayed, (A5.1) is the weakest possible assumption concerning equilibria. If it does not hold, then there is no numerical problem, since there is no solution. Criteria for the existence of equilibria are in [1, 3, 10]. See also [9] and the survey and references in [19].

Let $k\delta \in [n\Delta, n\Delta + \Delta)$ for integers n, k. Then for $\alpha_i \in U_i^\mu$,

$$\begin{aligned} &P\left\{u_i^{\mu,\delta,\Delta}(k\delta) = \alpha_i \middle| w(s), s \le k\delta; u_j^{\mu,\delta,\Delta}(l\delta), j = 1,2, l < k\right\} \\ &= P\left\{u_i^{\mu,\delta,\Delta}(k\delta) = \alpha_i \middle| w(l\Delta), l \le n; u_j^{\mu,\delta,\Delta}(l\delta), j = 1,2, l\delta < n\Delta\right\} \\ &\equiv p_{i,k}\left(\alpha_i;\ w(l\Delta), l \le n; u_j^{\mu,\delta,\Delta}(l\delta), j = 1,2, l\delta < n\Delta\right). \end{aligned} \tag{5.2}$$

The functions $p_{i,k}(\cdot)$ can be assumed to be continuous in the $\{w(l\delta)\}$-arguments, for each value of the other arguments.

Proof. Theorem 4.3 says that we need only work with strategies $c_i^{\mu,\delta,\Delta}(\cdot)$ whose control process realizations are in $\mathcal{U}_i(\mu, \delta, \Delta)$, and that in the time interval $[n\Delta, n\Delta + \Delta)$ depend only on the control process values of the other player up to time $n\Delta - \Delta$. We will also need to define the response of such a strategy if the other player uses controls that are not discretized. We interpret (5.2), as applied to a non-discretized strategy of the other player, by applying it to a discretization of the realizations of the other players actions. For small μ, δ, Δ, Theorem 3.3 implies that this interpretation would have negligible effect on the values.

Let $u_i^{\mu,\delta,\Delta}(\cdot), i = 1,2$, denote the control realizations under $c_i^{\mu,\delta,\Delta}(\cdot)$, $i = 1,2$. The probability law of $\left(u_1^{\mu,\delta,\Delta}(\cdot), u_2^{\mu,\delta,\Delta}(\cdot); w(\cdot)\right)$ determines the probability law of $\left(x(\cdot), u_1^{\mu,\delta,\Delta}(\cdot), u_2^{\mu,\delta,\Delta}(\cdot); w(\cdot)\right)$ where $x(\cdot)$ is the solution to (2.1) that corresponds to these controls. Next, write the probability law of the controls as a conditional probability

$$P\left\{u_i^{\mu,\delta,\Delta}(k\delta) = \alpha_i \middle| w(s), s \le n\Delta; u_j^{\mu,\delta,\Delta}(l\delta), j = 1,2, l\delta < n\Delta\right\}, \tag{5.3}$$

where $k\delta \in [n\Delta, n\Delta + \Delta)$. The rule (5.3), for $i = 1,2$, derived from the rules $c_i^{\mu,\delta,\Delta}(\cdot)$, can be considered to be "randomized" Elliott-Kalton strategy pair.

The next step is to apply (5.3) to the piecewise-constant continuous-time interpolation of the discrete-time system $x^\Delta(\cdot)$ defined by (3.1). Since only discrete-time samples of the Wiener process are used in (3.1), the probability law of the solution to (3.1) on $[0, t]$ is determined by the law of

$$\left(u_1^{\mu,\delta,\Delta}(l\delta), u_2^{\mu,\delta,\Delta}(l\delta), l\delta < t; w(n\Delta), n\Delta \le t\right).$$

This implies that the probability law of the controls and paths for $x^\Delta(\cdot)$ can be written as

$$P\left\{u_i^{\mu,\delta,\Delta}(k\delta) = \alpha_i \middle| w(l\Delta), l \le n; u_j^{\mu,\delta,\Delta}(l\delta), j = 1,2, l\delta < n\Delta\right\}, \tag{5.4}$$

for $k\delta \in [n\Delta, n\Delta + \Delta)$. By Theorem 3.1, for small enough δ, Δ the paths $x^\Delta(\cdot)$ and $x(\cdot)$ and the associated costs are arbitrarily close, uniformly in the controls. This argument implies that, with minimal change in the costs, we can replace the rule (5.2) by the rule (5.4), in which the conditioning on $w(\cdot)$ is restricted to the samples $w(l\Delta)$.

Given the form (5.4), the assertion concerning continuity is proved by a "smoothing" procedure.[7] Let $\rho > 0$ and define the smoothed functions

$$\begin{aligned} &p^{\rho}_{i,k}\left(\alpha_i;\, w(l\Delta), l \le n; u_j^{\mu,\delta,\Delta}(l\delta), j = 1, 2, l\delta < n\Delta\right) \\ &= N(\rho) \int e^{-|z-w|^2/2\rho} p_{i,k}\left(\alpha_i; z; u_j^{\mu,\delta,\Delta}(l\delta), j = 1, 2, l\delta < n\Delta\right) dz \end{aligned} \tag{5.5}$$

where $N(\rho)$ is a normalizing constant and $w = \{w(l\Delta), l \le n\}$. Owing to the smoothing, the functions are continuous in the w-variables, uniformly in the values of the control variables (since there are only a finite number of values for the control variables). As $\rho \to 0$, (5.5) converges to

$$p_{i,k}\left(\alpha_i;\, w(l\Delta), l \le n; u_j^{\mu,\delta,\Delta}(l\delta), j = 1, 2, l\delta < n\Delta\right)$$

for almost all $\{w(l\Delta), l \le n\}$ values. Hence, outside of a set of arbitrarily small measure, it converges uniformly in $\{w(l\Delta), l \le n\}$. This argument implies that the smoothed conditional probability rule (5.5) and the original conditional probability rule (5.4) will choose the same control values with a probability that goes to unity as $\rho \to 0$. Hence we can suppose that the $p_{i,k}(\cdot)$ are smooth in the w-variables. □

A representation of the rule (5.2). The following method of realizing the random choices given by the rule (5.2) will be useful. Let $\{\theta_k\}$ be random variables that are i.i.d. and uniformly distributed on $[0, 1]$ with $\{\theta_k, k \ge l\}$ being independent of all data on the system before time $l\delta$. For each Player i, and each n and k in (5.2), divide the unit interval $[0, 1]$ into subintervals whose lengths are proportional to the conditional probability of the $\alpha_i^{\mu,l}, l \le p_i^{\mu}$, given by (5.2). Then choose the value $u_i^{\mu,\delta,\Delta}(k\delta) = \alpha_i^{\mu,l}$ for Player i if the random selection of θ_k on $[0, 1]$ falls into that subinterval. The same random variables $\{\theta_k\}$ are used for both players.

6. The Markov chain approximation: Brief review and approximations. First we will briefly describe the Markov chain approximation method [13, 14, 18] for a cost minimization problem, mainly to establish notation. Let $h > 0$ be an approximation parameter. First one determines a finite-state controlled Markov chain ξ_n^h that has a continuous-time interpolation that is an "approximation" of the process $x(\cdot)$. The numerical procedure is the solution of the optimization problem for the chain

[7]See also [18, Theorem 10.3.1] for a similar smoothing. The procedure for showing that it is sufficient for the control rule to depend only on the samples of the Wiener process is different here than in the reference. In the reference it was sufficient to approximate only an ϵ-optimal control. But here we have two players, and we need an approximation for each player that is valid no matter what the control of the other player.

and an approximating cost function. Under a simple (and minimal) local consistency condition, the minimal cost function for the chain converges to the minimal cost function for the original problem. The approximating chain and the local consistency conditions are the same for the game problem, except that the game problem for the approximating chain must be solved. The book [18] contains a complete discussion of many simple methods for getting the transition probabilities of the chain. The approximations have the basic properties of the original physical model and this connection can be exploited to improve the algorithms.

Since we are concerned mainly with the issues that are new to the game problem, we will use the simplest state space for the chain, one based on the regular h-grid S_h in $I\!R^v$. Define $G_h = S_h \cap G$ and $G_h^0 = S_h \cap G^0$. The numerical boundary points, called ∂G_h, are those points in $S_h - G_h^0$ that can be reached in one step from G_h^0 under some control. The process stops on first reaching them. Only $G_h^0 \cup \partial G_h$ is of interest. On G_h^0 the chain "approximates" the diffusion part of (2.1) or (2.1r).

Next we define local consistency. Let $u_n^h = (u_{1,n}^h, u_{2,n}^h)$ denote the controls at step n, define $\Delta\xi_n^h = \xi_{n+1}^h - \xi_n^h$, and let $E_{x,n}^{h,\alpha}$ denote the expectation given the data to step n[8] with $\xi_n^h = x$ and control value $\alpha = u_n^h$ to be used. For the game problem of this paper, $\alpha = (\alpha_1, \alpha_2)$ with $\alpha_i \in U_i$. Define the covariance matrix $a(x) = \sigma(x)\sigma'(x)$. The chain is locally consistent if (this defines the functions $b^h(\cdot)$ and $a^h(\cdot)$)

$$\begin{aligned} &E_{x,n}^{h,\alpha}\Delta\xi_n^h \equiv b^h(x,\alpha)\Delta t^h(x,\alpha) = b(x,\alpha)\Delta t^h(x,\alpha) + o(\Delta t^h(x,\alpha)), \\ &\text{cov}_{x,n}^{h,\alpha}[\Delta\xi_n^h - E_{x,n}^{h,\alpha}\Delta\xi_n^h] \\ &\qquad \equiv a^h(x,\alpha)\Delta t^h(x,\alpha) = a(x)\Delta t^h(x,\alpha) + o(\Delta t^h(x,\alpha)), \\ &\lim_{h\to 0} \sup_{x\in G,\alpha\in U} \Delta t^h(x,\alpha) = 0. \end{aligned} \tag{6.1}$$

The function $\Delta t^h(\cdot)$, called an interpolation interval, is obtained automatically when the transition probabilities are calculated; see [18]. Equation (6.1) shows that the approximating chain has the "local properties" (conditional mean change and conditional covariance) of the diffusion process. Let $p^h(x,y|\alpha)$ denote the probability that the next state is y given that the current state is x and control pair $\alpha = (a_1, \alpha_2)$ is used.

Under our condition that the controls for the two layers are separated in that $b(x,\alpha) = b^1(x,\alpha_1) + b^2(x,\alpha_2)$, one can construct the (still locally consistent) chain so that the controls are "separated" in that the one-step transition probability has the form

$$p^h(x,y|\alpha) = p_1^h(x,y|\alpha_1) + p_2^h(x,y|\alpha_2), \tag{6.2}$$

This might simplify the coding, but will not be necessary here.

[8]This is $\{\xi_l^h, l \le n, u_l^h, l < n\}$.

A one-dimensional example. Let h be small enough so that $h|b(\alpha, x)| \leq \sigma^2(x)$. Then

$$p^h(x, x \pm h|\alpha) = \frac{\sigma^2(x) \pm hb(x, \alpha)}{2\sigma^2(x)}, \quad \Delta t^h(x, \alpha) = \frac{h^2}{\sigma^2(x)} \tag{6.3}$$

is locally consistent [18, Chapter 5]

Admissible controls. Let $\mathcal{F}_n^h$ denote the minimal σ-algebra that measures the control and state data to step n (this does not include u_n^h, which is determined at step n), with E_n^h denoting the associated conditional expectation. An admissible control for Player i at step n is a U_i-valued and $\mathcal{F}_n^h$-measurable random variable. Let $\mathcal{U}_i^h$ denote the set of the admissible control processes for Player i. Next, define a relaxed control for the chain. Let $r_{i,n}^h(\cdot)$ be a measure-valued random variable that is a probability distribution on the Borel sets of U_i such that $r_{i,n}^h(A)$ is $\mathcal{F}_n^h$-measurable for each Borel set $A \in U_i$. Then the $r_{i,n}^h(\cdot)$ are called relaxed controls for Player i at step n. As done for the diffusion model (2.1r), an ordinary control can be represented by the relaxed control defined by $r_{i,n}^h(A) = I_{\{u_{i,n}^h \in A\}}$ for Borel sets $A \subset U_i$. Define the product relaxed control $r_n^h(\cdot)$ by $r_n^h(A_1 \times A_2) = r_{1,n}^h(A_1) r_{2,n}^h(A_2)$. The transition probability associated with a relaxed control is $\int_U p^h(x, y|a) r_n^h(d\alpha)$. Let C_i^h denote the set of control strategies for the chain ξ_n^h. Given an approximating cost function, the numerical problem is to solve the game problem for the approximating chain.

The approximating cost function for the chain. The cost functions are the analogs of (2.2) or (2.2r). The cost rate for Player i is $k_i(x, \alpha_i)\Delta t^h(x, \alpha)$. The stopping costs are $g_i(\cdot)$, and τ^h denotes the first time that the set G_h^0 is exited. Let $W_i^h(x, u_1^h, u_2^h)$ denote the expected cost for Player i under the control sequences $u_i^h = \{u_{i,n}^h, n \geq 0\}, i = 1, 2$.

A representation of the transition probabilities. For the convergence proofs, it is useful to have the chains for each h and all possible controls defined on the same probability space.[9] This is done as follows. Let $\{\chi_n\}$ be a sequence of i.i.d random variables, uniformly distributed on $[0, 1]$ with $\{\chi_l, l \geq n\}$ being independent of $\{\xi_l^h, u_l^h, l \leq n\}$. For each value of the current state $x = \xi_n^h$ and control $\alpha = u_n^h$, order the finite number of possible next states y, and divide $[0, 1]$ into subintervals whose lengths are $p^h(x, y|\alpha)$. Then with $x = \xi_n^h, \alpha = u_n^h$ given, the next state value is selected according to where the random choice for χ_n falls in $[0, 1]$. The same random variables $\{\chi_n\}$ will be used for all h.

[9]This representation of the chain is not needed and was not used for the single player problem nor for the zero-sum game. For the non-zero-sum game, it helps with the difficulties in the convergence proof that are unique to the problem of this paper.

The continuous-time interpolation $\xi^h(\cdot)$ of the chain. The discrete-time chain ξ_n^h is used for the numerical computations. The proofs of convergence are based on continuous-time interpolation of the $\{\xi_n^h\}$. These interpolations will approximate the controlled diffusion process $x(\cdot)$. Two types of interpolations were used in [18]. The one that was of most use in [18] is called $\psi^h(\cdot)$ and will be briefly described below, so that the results of the reference can be appealed to where needed. In the present paper it is simpler to work mostly with the other interpolation $\xi^h(\cdot)$, which is constructed as follows. Define $\Delta t_n^h = \Delta t^h(\xi_n^h, u_n^h)$, and $t_n^h = \sum_{i=0}^{n-1} \Delta t_i^h$. Define $\xi^h(t) = \xi_n^h$ and $u_i^h(\cdot) = u_n^h$ on $[t_n^h, t_{n+1}^h)$. Let the (continuous time) relaxed control representation be denoted by $r_i^h(\cdot)$. Define $r^h(\cdot) = (r_1^h(\cdot), r_2^h(\cdot))$, with time derivative $r^{h,\prime}(\alpha, t) = r_1^{h,\prime}(\alpha_1, t) r_2^{h,\prime}(\alpha_2, t)$. We use $\mathcal{U}_i^h$ for the set of continuous time interpolations of the control (ordinary or relaxed) for Player i.

By (6.1), we can write

$$\xi_{n+1}^h = \xi_n^h + b^h(\xi_n^h, u_n^h)\Delta t_n^h + \beta_n^h \tag{6.4}$$

where β_n^h is a martingale difference with $E_n^h[\beta_n^h][\beta_n^h]' = a^h(\xi_n^h, u_n^h)\Delta t_n^h$. There are martingale differences[10] δw_n^h with conditional (given $\mathcal{F}_n^h$) covariance $\Delta t_n^h I$ such that (see [18, Section 10.4.1] and [13, Section 6.6]) $\beta_n^h = \sigma^h(\xi_n^h, u_n^h)\delta w_n^h$. Let $w^h(\cdot)$ denote the continuous time interpolation of $\sum_{i=0}^{n-1} \delta w_n^h$ with intervals Δt_n^h. Then we can write[11]

$$\begin{aligned} \xi^h(t) &= x(0) + \int_0^t b^h(\xi^h(s), u^h(s))ds + \int_0^t \sigma^h(\xi^h(s))dw^h(s) + \epsilon^h(t), \\ \int_0^t \sigma^h(\xi^h(s), u^h(s))dw^h(s) &= \int_0^t \sigma(\xi^h(s))dw^h(s) + \epsilon^h(t), \end{aligned} \tag{6.5}$$

where, for each $T < \infty$, $\epsilon^h(\cdot)$ satisfies

$$\lim_{h\to 0} \sup_{u^h} E \sup_{s\le T} |\epsilon^h(s)|^2 = 0. \tag{6.6}$$

The $\epsilon^h(t)$ is due to the $O(\Delta t^h)$ approximation of $a^h(x,\alpha)$ by $\sigma(x)\sigma(x)'$.

The interpolation $\psi^h(\cdot)$. In [18] an alternative continuous-time interpolation $\psi^h(\cdot)$ was used primarily since it simplified the proofs there. For each h, let $\nu_n^h, n = 0, 1, \ldots,$ be mutually independent and exponentially distributed random variables with unit mean, and that are independent of $\{\xi_n^h, u_n^h, n \ge 0\}$. Define $\Delta\tau_n^h = \nu_n^h \Delta t_n^h$ and $\tau_n^h = \sum_{i=0}^{n-1} \Delta\tau_i^h$. The $\delta\tau_n^h$ is a new interpolation interval, and τ_n^h provides a new time scale. Define the interpolations $\psi^h(\cdot)$ and $u_\psi^h(\cdot)$ by: $\psi^h(t) = \xi_n^h$ and $u_\psi^h(t) = u_n^h$ on $[\tau_n^h, \tau_{n+1}^h)$.

---

[10]Possibly requiring the augmentation of the probability space by the addition of an independent Wiener process.

[11]The equation is only valid at the times t_n^h, but the error is asymptotically negligible, in that it satisfies (6.6).

The advantage of $\psi^h(\cdot)$ is that $\psi^h(\cdot)$ is a point process. If the controls u_n^h are feedback, then $\psi^h(\cdot)$ is a continuous-time Markov chain and when in state x and with control α used the transition rate is $1/\Delta t^h(x,\alpha)$ and the probability is just $p^h(x,y|\alpha)$ that y is the next state. Given a jump, the distribution of the next state is given by the $p^h(x,y|\alpha)$, and the conditional mean change is $b^h(x,\alpha)\Delta t^h(x,\alpha)$. In general, $\psi^h(\cdot)$ can be decomposed in terms of a compensator and martingale as

$$\psi^h(t) = x(0) + \int_0^t b^h(\psi^h(s), u_\psi^h(s))ds + M^h(t), \tag{6.7}$$

where the martingale $M^h(t)$ has quadratic variation process

$$\int_0^t a^h(\psi^h(s), u_\psi^h(s))ds.$$

It can be shown that ([18, Sections 5.7.3 and 10.4.1]) there is a martingale $\hat{w}^h(\cdot)$ (with respect to the filtration generated by the state and control processes, possibly augmented by an "independent" Wiener process) such that

$$M^h(t) = \int_0^t \sigma^h(\psi^h(s), u_\psi^h(s))d\hat{w}^h(s) = \int_0^t \sigma(\psi^h(s))d\hat{w}^h(s) + \epsilon^h(t),$$

where $\sigma^h(\cdot)[\sigma^h(\cdot)]' = a^h(\cdot)$. The martingale $\hat{w}^h(\cdot)$ has quadratic variation It and converges weakly to a standard Wiener process. The martingale $\epsilon^h(\cdot)$ arises due to the difference between $\sigma(x)$ and $\sigma^h(x)$ (recall the $o(\Delta t^h)$ terms in (6.1)) and satisfies (6.6). Thus,

$$\begin{aligned}\psi^h(t) = \;& x(0) + \int_0^t \int_U b^h(\psi^h(s), \alpha) r_\psi^{h,\prime}(d\alpha, s)ds \\ &+ \int_0^t \sigma(\psi^h(s))d\hat{w}^h(s) + \epsilon^h(t),\end{aligned} \tag{6.8}$$

where $r_\psi^h(\cdot)$ is the relaxed control representation of $u_\psi^h(\cdot)$.

The time scales based on τ_n^h and τ_n^h are asymptotically equivalent, since the random variations due to the ν_n^h variables averages out, as seen in the following theorem. Consequently, the interpolations $\xi^h(\cdot)$ and $\psi^h(\cdot)$ are asymptotically equivalent, so that any asymptotic results for one are also asymptotic results for the other.

THEOREM 6.1. *Assume the last line of* (6.1). *Then the time scales with intervals* Δt_n^h *and* $\Delta \tau_n^h$ *are asymptotically equivalent.*

Proof. Write $\Delta\tau_n^h - \Delta t_n^h = (\nu_n^h - 1)\Delta t_n^h$, a martingale difference, and define $f^h(t) = \min\{n : t_n^h \geq t\}$. By the martingale property

$$
\begin{aligned}
E \sup_{n<f^h(t)} |t_n^h - \tau_n^h|^2 &= E \sup_{n<f^h(t)} \left| \sum_{i=0}^{n} \Delta t_i^h (v_n^h - 1) \right| \\
&\leq 4E \sum_{i=0}^{f^h(t)} [\Delta t_i^h]^2 E(v_i^h - 1)^2.
\end{aligned}
$$

This goes to zero as $h \to 0$ by the last line of (6.1). The result is the same if we define $f^h(t) = \min\{n : \tau_n^h \geq t\}$. □

A comment on weak convergence. Weak convergence methods play an important role in the proofs in [18]. They will play a central role in Theorem 8.1, which shows that a near-equilibrium strategy for the diffusion is also one for the chain For the continuous-time processes we use the path spaces of functions that are right-continuous and with left-hand limits, and with the Skorokhod topology. The sequence $(\xi^h(\cdot), r_1^h(\cdot), r_2^h(\cdot), w^h(\cdot), \tau^h)$ is tight. For any subsequence of $(\xi^h(\cdot), r_1^h(\cdot), r_2^h(\cdot), w^h(\cdot), \tau^h)$, there is a further subsequence which converges weakly to random processes $(x(\cdot), r_1(\cdot), r_2(\cdot), w(\cdot), \tau)$, where $r_i(\cdot)$ is a relaxed control for Player i, $(x(\cdot), r_1(\cdot), r_2(\cdot), w(\cdot), I_{\{\tau^h \leq \cdot\}})$ is nonanticipative with respect to the standard vector-valued Wiener process $w(\cdot)$ and, writing $r(\cdot) = (r_1(\cdot), r_2(\cdot))$, the set satisfies

$$
x(t) = x(0) + \int_0^t \int_U b(x(s), \alpha) r'(d\alpha, s) ds + \int_0^t \sigma(x(s)) dw(s).
$$

Also, along the convergent subsequence, $W_i^h(x, r_1^h, r_2^h) \to W_i(x, r_1, r_2)$. The proofs are the same as for the one-player control case in [18, Chapter 10].

Example of the construction of $\delta w^h(\cdot)$. We will give a simple example of the construction of the random variable δw_n^h defined below (6.4). More complete details for the general case are in [18, Section 10.4.1], [13, Section 6.6]. Suppose that the matrix $\sigma(\cdot) = \sigma$ is a constant with the form $\sigma = \begin{bmatrix} \sigma_1 & 0 \\ 0 & 0 \end{bmatrix}$, where σ^1 is a square invertible matrix of dimension d_1. Partition the components of x as $x = (x_1, x_2)$, where the dimension of x_1 is d_1, and write the analogous partition $w^h(\cdot) = (w_1^h(\cdot), w_2^h(\cdot))$. Partition the $a^h(\cdot)$ in the second line of (6.1) as

$$
a^h(x, \alpha) = \begin{bmatrix} a_1^h(x, \alpha) & a_{1,2}^h(x, \alpha) \\ a_{2,1}^h(x, \alpha) & a_2^h(x, \alpha) \end{bmatrix}.
$$

As $h \to 0$, $a_1^h(\cdot) \to \sigma_1[\sigma_1]'$ and all other components of $a^h(\cdot)$ go to zero, all uniformly in (x, α). For any Wiener process $w_2(\cdot)$ that is independent of the other random variables, we can let $w_2^h(\cdot) = w_2(\cdot)$ and $\delta w_{2,n}^h = w_2(t_{n+1}^h) - w(t_n^h)$. The only component of $w^h(\cdot)$ that contributes to the limit processes is $w_1^h(\cdot)$ and the increments are defined by

$$\begin{aligned}\delta w^h_{1,n} &\equiv w^h_1(t^h_{n+1}) - w^h_1(t^h_n) \\ &= [a^h_1(\xi^h_n, u^h_n)]^{-1/2}\left[\xi^h_{1,n+1} - \xi^h_{1,n} - \int_{t^h_n}^{t^h_{n+1}}\int_U b^h_1(\xi^h_n,\alpha) r^{h,\prime}(s,d\alpha)ds\right] \\ &= [\sigma_1]^{-1}\left[\xi^h_{1,n+1} - \xi^h_{1,n} - \int_{t^h_n}^{t^h_{n+1}}\int_U b^h_1(\xi^h_n,\alpha) r^{h,\prime}(s,d\alpha)ds\right] + \delta\epsilon^{1,h}_n,\end{aligned} \tag{6.9}$$

where $\delta\epsilon^{1,h}_n$ is due to the approximation of $a^h_1(\cdot)$ by $\sigma_1[\sigma_1]'$ in (6.9) and its continuous time interpolation satisfies (6.6). If an ordinary control is used, then the double integral is just $b_1(\xi^h_n, u^h_n)\Delta t^h_n$.

7. Approximating the chain: I. The proof of convergence requires the use of a series of approximations to the chain that are analogous to what was done Theorems 4.1–4.3. One such approximation will be given in this section. It will be used for the proof that a near-equilibrium for the diffusion is also one for the chain, which is one half of the convergence proof and will be given in the next section. These approximations, and the ones in Section 9, have an independent interest and should be useful for other convergence analyses for numerical approximations. Theorem 7.1 concerns an approximation to (6.5) that is based on the same $w^h(\cdot)$ process, and will be used in Theorem 8.1.

Consider the interpolation of the chain as represented by (6.5). For the control $r^h(\cdot) = (r^h_1(\cdot), r^h_2(\cdot))$ used in (6.5) and the approximation parameters μ, δ, Δ as used in Theorem 4.3, define the approximation $u^{\mu,\delta,\Delta,h}_i(\cdot), i = 1, 2$, analogously to what was done above Theorem 4.3. For the process $w^h(\cdot)$ that appears in (6.5) under the original control $r^h(\cdot)$, define the process

$$\begin{aligned}\xi^{\mu,\delta,\Delta,h}(t) = x(0) &+ \int_0^t b(\xi^{\mu,\delta,\Delta,h}(s), u^{\mu,\delta,\Delta,h}(s))ds \\ &+ \int_0^t \sigma(\xi^{\mu,\delta,\Delta,h}(s))dw^h(s).\end{aligned} \tag{7.1}$$

Let $r^{\mu,\delta,\Delta,h}_i(\cdot)$ denote the relaxed control representation of $u^{\mu,\delta,\Delta,h}_i(\cdot)$. Keep in mind that the process $\xi^{\mu,\delta,\Delta,h}(\cdot)$ defined by (7.1) is not a Markov chain even if the controls are feedback, since the "driving" process $w^h(\cdot)$ is obtained from the process (6.5) under controls $r^h(\cdot)$ and not under the revised controls $r^{\mu,\delta,\Delta,h}_i(\cdot)$, $i = 1, 2$ that are used in (7.1). Let $W^{\mu,\delta,\Delta,h}_i(x, r^{\mu,\delta,\Delta,h}_1, r^{\mu,\delta,\Delta,h}_2)$ denote the cost for the process (7.1). Continuing with the approximations, next define the discrete-time system $\tilde{\xi}^{\mu,\delta,\Delta,h}(n\Delta)$ with initial condition $x(0)$ and, for $n \geq 0$,

$$\begin{aligned}\tilde{\xi}^{\mu,\delta,\Delta,h}(n\Delta+\Delta) = \; & \tilde{\xi}^{\mu,\delta,\Delta,h}(n\Delta) \\ & + \int_{n\Delta}^{n\Delta+\Delta} b(\tilde{\xi}^{\mu,\delta,\Delta,h}(n\Delta), u^{\mu,\delta,\Delta,h}(s))ds \\ & + \sigma(\tilde{\xi}^{\mu,\delta,\Delta,h}(n\Delta))[w^h(n\Delta+\Delta) - w^h(n\Delta)].\end{aligned} \tag{7.2}$$

Denote the piecewise-constant continuous-time interpolation by $\tilde{\xi}^{\mu,\delta,\Delta,h}(\cdot)$. Let $\tilde{W}_i^{\mu,\delta,\Delta,h}(x, r_1^{\mu,\delta,\Delta,h}, r_2^{\mu,\delta,\Delta,h})$ denote the associated cost. The next theorem is a partial analog of Theorem 4.3.

THEOREM 7.1. *Assume* (A2.1). *For* $(\mu,\delta,\Delta) > 0$, *let* $r_i^{\mu,\delta,\Delta,h}(\cdot)$ *approximate* $r_i^h(\cdot)$. *For each* $\epsilon > 0$ *and* $t < \infty$, *there are* $\mu_\epsilon > 0, \delta_\epsilon > 0, \Delta_\epsilon > 0$ *and* $\kappa_\epsilon > 0$, *such that for* $\mu \le \mu_\epsilon, \delta \le \delta_\epsilon, \Delta \le \Delta_\epsilon$ *and* $\delta/\Delta \le \kappa_\epsilon$,

$$\limsup_{h\to 0} \sup_{x, r_1^h, r_2^h} E \sup_{s\le t} \left|\xi^{\mu,\delta,\Delta,h}(s) - \xi^h(s)\right| \le \epsilon. \tag{7.3}$$

Under the additional condition (A2.2), *we can suppose that*

$$\limsup_{h\to 0} \sup_{x, r_1^h, r_2^h} \left|W_i^{\mu,\delta,\Delta,h}(x, r_1^{\mu,\delta,\Delta,h}, r_2^{\mu,\delta,\Delta,h}) - W_i^h(x, r_1^h, r_2^h)\right| \le \epsilon. \tag{7.4}$$

(7.3) *and* (7.4) *hold if* $\xi^{\mu,\delta,\Delta,h}(\cdot)$ *and* $W_i^{\mu,\delta,\Delta,h}(\cdot)$ *are replaced by* $\tilde{\xi}^{\mu,\delta,\Delta,h}(\cdot)$ *and* $\tilde{W}_i^{\mu,\delta,\Delta,h}(\cdot)$, *resp., or if only one of the controls is approximated.*

Proof. Define $\delta\xi^{\mu,\delta,\Delta,h}(t) = \tilde{\xi}^{\mu,\delta,\Delta,h}(t) - \xi^h(t)$. Then, proceeding as done in Theorem 4.1, write

$$\begin{aligned}\delta\xi^{\mu,\delta,\Delta,h}(t) = & \int_0^t \int_U \left[b(\xi^{\mu,\delta,\Delta,h}(s),\alpha) - b^h(\xi^h(s),\alpha)\right] r^{h,\prime}(d\alpha,s)ds \\ & + \int_0^t \left[\sigma(\xi^{\mu,\delta,\Delta,h}(s)) - \sigma(\xi^h(s))\right] dw^h(s) \\ & + \int_0^t \int_U b(\xi^{\mu,\delta,\Delta,h}(s),\alpha)\left[r^{\mu,\delta,\Delta,h,\prime}(d\alpha,s) - r^{h,\prime}(d\alpha,s)\right] ds + \epsilon_1^h(t).\end{aligned}$$

Both of the processes $w^h(\cdot), \epsilon_1^h(\cdot)$ are martingales with respect to the filtration induced by the data $(\xi^h(\cdot), r^h(\cdot), w^h(\cdot))$ and $w^h(\cdot)$ has quadratic variation[12] It and $\epsilon_1^h(\cdot)$ satisfies (6.6). Partition the last integral analogously to what was done in (4.7), with intervals λ. It can be shown that the process $\xi^{\mu,\delta,\Delta,h}(\cdot)$ satisfies the following estimate.

$$\begin{aligned}\sup_{\mu,\delta,\Delta} \sup_{r^h} \sup_{l\lambda\le t} E \sup_{s\le\lambda} \left|\xi^{\mu,\delta,\Delta,h}(l\lambda+s) - \xi^{\mu,\delta,\Delta,h}(l\lambda)\right|^2 \\ = O(\lambda) + \sup \Delta t^h(x,\alpha)).\end{aligned}$$

[12] Actually, they are martingales only when evaluated at the time points t_n^h, but the difference is unimportant, since they are constant between such times.

The martingale property and the Lipschitz condition and standard estimates imply that there is a constant K, depending only on t, such that

$$E \sup_{s \le t} \left|\delta\xi^{\mu,\delta,\Delta,h}(s)\right|^2 \le K \int_0^t E\left|\delta\xi^{\mu,\delta,\Delta,h}(s)\right|^2 ds + \kappa^{\lambda,h}(t) + O(\lambda)$$

$$+KE\left|\sum_{l=0}^{[t/\lambda]-1} \int_{l\lambda}^{(l+1)\lambda} b(\xi^{\mu,\delta,\Delta,h}(l\lambda),\alpha)\left[r^{\mu,\delta,\Delta,h,\prime}(d\alpha,s) - r^{h,\prime}(d\alpha,s)\right] ds\right|^2.$$

The error term $\kappa^{\lambda,h}(t)$ is due to the $\epsilon_1^h(\cdot)$ and to the use of $\xi^{\mu,\delta,\Delta,h}(l\lambda)$ in lieu of $\xi^{\mu,\delta,\Delta,h}(s)$ in the second line, and it satisfies

$$\lim_{h,\lambda\to 0} \sup_{s\le t} |\kappa^{\lambda,h}(s)| = 0.$$

For each small λ, the method of approximation of the controls implies that the second line goes to zero uniformly in h as $(\mu,\delta,\Delta) \to 0$. Then (7.3) follows from the resulting inequality and the Bellman-Gronwall Lemma. The inequality (7.4) follows from (7.3) and (A2.2). □

8. Convergence, Part I: An approximate equilibrium for the diffusion is also one for the chain. The convergence proof will be in two parts. In this section, we show that an ϵ-equilibrium for (2.1) is an ϵ_0-equilibrium for the approximating chain, where $\epsilon_0 \to 0$ as $\epsilon \to 0$. The converse result will be proved in Section 10, under a stronger condition, where it is shown that an approximate equilibria for the chain is also one for the diffusion.

THEOREM 8.1. *Assume* (A2.1), (A2.2), *and* (A5.1). *An ϵ-equilibrium value for* (2.1) *or* (2.1r) *is an ϵ_0-equilibrium value for the approximating Markov chain, where $\epsilon_0 \to 0$ as $\epsilon \to 0$.*

Proof. Let $\epsilon > 0$ be given. By (A5.1), there is an ϵ-equilibrium strategy pair for (2.1r). Theorem 5.1 says that for small enough μ, δ and Δ, it can be represented as in (5.2), where Δ/δ is an integer, and the $p_{i,k}(\cdot)$ are continuous in the w variables. We can suppose, without loss of generality, that for each n, k and i, the rule (5.2) is defined for the finite number of possible conditioning control values, Let $(\bar{c}_1^\Delta(\cdot), \bar{c}_2^\Delta(\cdot))$ denote this strategy pair. [13]

Recall that when a strategy defined by (5.2) is applied to an arbitrary relaxed control, (5.2) is actually applied to the space-time discretization of that relaxed control, as defined above Theorem 4.3. We will need to adapt the strategies $\bar{c}_i^\Delta(\cdot)$ for use on the chain ξ_n^h. This is done by simply replacing the $w(l\Delta)$ in (5.2) by the $w^h(l\Delta)$ process that was used in the representation (6.5), and whose construction was illustrated in (6.9). These

[13] The strategies $\bar{c}_i^\Delta(\cdot)$ depend on μ and δ as well as on Δ, but for simplicity we suppress that dependence in the notation.

controls are constructed for purposes of the proof only. For each integer k, the control value $u_i^{\mu,\delta,\Delta,h}(k\delta)$ that is obtained from the rule (5.2) with the $w^h(\cdot)$ samples used will be applied to the chain for all steps m for which $t_m^h \in [k\delta, k\delta+\delta)$. The resulting strategies are well-defined for all steps of the chain. They will be denoted by $\bar{c}_i^{\Delta,h}(\cdot)$ and are in $\mathcal{C}_i^h$.

To prove the theorem we will need to show that for small enough (μ, Δ, δ), there are $\epsilon_0 > 0$ and $h_0 > 0$, where $\epsilon_0 \to 0$ as $\epsilon \to 0$ such that for $h \le h_0$ and any sequence $r_i^h(\cdot)$ of admissible relaxed (or ordinary) controls for the chain,

$$\begin{aligned} W_1^h(x, \bar{c}_1^{\Delta,h}, \bar{c}_2^{\Delta,h}) &\ge W_1^h(x, r_1^h, \bar{c}_2^{\Delta,h}) - \epsilon_0, \\ W_2^h(x, \bar{c}_1^{\Delta,h}, \bar{c}_2^{\Delta,h}) &\ge W_2^h(x, \bar{c}_1^{\Delta,h}, r_2^h) - \epsilon_0. \end{aligned} \tag{8.1}$$

When writing $W_2^h(x, \bar{c}_1^{\Delta,h}, r_2^h)$, we mean that Player 1 uses strategy $\bar{c}_1^{\Delta,h}(\cdot)$ and Player 2 uses relaxed control $r_2^h(\cdot)$ (in continuous time interpolation notation, but for the chain model), or an ordinary control with this relaxed control representation, with an analogous interpretation for the other forms.

If the strategy pair $\bar{c}_i^{\Delta,h}(\cdot), i = 1, 2$, is used for the chain, then we write $\bar{r}_i^{\mu,\delta,\Delta,h}(\cdot), i = 1, 2$, for the relaxed control representation of the control actions, in its continuous-time interpolation. Let $\xi^h(\cdot)$, $w^h(\cdot)$, and τ^h denote the corresponding continuous-time interpolation of the chain, the "pre-Wiener" process, and the first exit time, resp. The sequence $(\xi^h(\cdot), \bar{r}_1^{\mu,\delta,\Delta,h}(\cdot), \bar{r}_2^{\mu,\delta,\Delta,h}(\cdot), w^h(\cdot), \tau^h)$, parametrized by h, is tight. Select a weakly convergent subsequence ($h \to 0$) with limit denoted by $(x(\cdot), r_1(\cdot), r_2(\cdot), w(\cdot), \tau)$. The set $(x(\cdot), r_1(\cdot), r_2(\cdot), I_{\{\tau \le \cdot\}})$ is non-anticipative with respect to the standard vector-valued Wiener process $w(\cdot)$, and the set $(x(\cdot), r_1(\cdot), r_2(\cdot), w(\cdot))$ solves (2.1r). The limit τ is the first hitting time of the boundary of G by the limit process $x(\cdot)$. All of the details concerning the tightness, the characterization of the limit processes and boundary hitting times, the fact that the limit processes solve (2.1r), and the convergence of the costs, are the same as for the control problem in [18, Chapters 10, 11].

From this point on, when weak convergent sequences are dealt with we will suppose (without loss of generality) that the Skorokhod representation is used so that all processes are defined on the same probability space and the weak convergence is equivalent to convergence with probability one in the topology of the path spaces [6, Theorem 1.8, Chapter 3]. Under Skorokhod representation the rule (5.2), with the $w^h(\cdot)$-samples used, converges w.p.1 to the same rule with the $w(\cdot)$-samples used, due to the convergence $w^h(\cdot) \to w(\cdot)$ and the continuity of the probabilities in (5.2) in the w-variables. Hence the limits $r_i(\cdot), i = 1, 2$, are just the control realizations of the original ϵ-equilibrium strategies $\bar{c}_i^{\Delta}(\cdot), i = 1, 2$. Since the solution to (2.1) or (2.1r) is unique for each admissible (control, Wiener process) pair, the probability law of any limit set $(x(\cdot), r_1(\cdot), r_2(\cdot), w(\cdot))$ does not depend on the particular convergent subsequence. Hence the original set of processes converges weakly to this unique limit set, where the control is determined by the rules $\bar{c}_i^{\Delta}(\cdot), i = 1, 2$.

By the weak convergence

$$\begin{aligned} W_1(x, \bar{c}_1^{\Delta}, \bar{c}_2^{\Delta}) &\leftarrow W_1^h(x, \bar{c}_1^{\Delta,h}, \bar{c}_2^{\Delta,h}) \\ &\leq \max_{r_1 \in \mathcal{U}_1^h} W_1^h(x, r_1, \bar{c}_2^{\Delta,h}) = W_1^h(x, \hat{r}_1^h, \bar{c}_2^{\Delta,h}), \end{aligned} \tag{8.2}$$

$$\begin{aligned} W_2(x, \bar{c}_1^{\Delta}, \bar{c}_2^{\Delta}) &\leftarrow W_2^h(x, \bar{c}_1^{\Delta,h}, \bar{c}_2^{\Delta,h}) \\ &\leq \max_{r_2 \in \mathcal{U}_2^h} W_2^h(x, \bar{c}_1^{\Delta,h}, r_2) = W_1^h(x, \bar{c}_1^{\Delta,h}, \hat{r}_2^h). \end{aligned} \tag{8.3}$$

Getting $\hat{r}_1^h(\cdot)$, with the strategy $\bar{c}_2^{\Delta,h}(\cdot)$ fixed, involves solving an optimal control for a non-Markov system. But, via a weak convergence argument using a maximizing sequence of controls, it can be shown that the maximizing controls $\hat{r}_1^h(\cdot)$ exist. In fact, we need only work with control processes whose cost values approximate the maximum values arbitrarily well, and there always is such a control.

We will show that

$$\limsup_{h\to 0} W_1^h(x, \hat{r}_1^h, \bar{c}_2^{\Delta,h}) \leq W_1(x, \bar{c}_1^{\Delta}, \bar{c}_2^{\Delta}) + \epsilon + \rho(\mu, \delta, \Delta), \tag{8.4}$$

where the error term $\rho(\mu, \delta, \Delta) \to 0$ as $(\mu, \delta, \Delta) \to 0$, with the analogous result for the indices reversed. The three inequalities (8.2), (8.3), and (8.4) imply that if Player 2 uses $\bar{c}_2^{\Delta,h}(\cdot)$, then as $h \to 0$, the best policy for Player 1 is to use $\bar{c}_1^{\Delta,h}(\cdot)$ (modulo $\rho(\mu, \delta, \Delta) + \epsilon$). The analogous result holds if the indices are reversed. Since (μ, δ, Δ) is arbitrarily small, the theorem is proved.

It remains to prove (8.4). Let $\{u_1^{\mu,\delta,\Delta,h}(l\delta)\}$ denote the control values that are obtained by the space and time discretization of $\hat{r}_1^h(\cdot)$, using the method described above Theorem 4.3. These are the values that the strategy $\bar{c}_2^{\Delta,h}(\cdot)$ uses. Let $\{u_2^{\mu,\delta,\Delta,h}(l\delta)\}$ denote the control choices for Player 2, based on the strategy rule $\bar{c}_2^{\Delta,h}(\cdot)$ and the control choices of Player 1. Let $r_i^{\mu,\delta,\Delta,h}(\cdot)$ denote the (continuous time interpolation) relaxed control representation of $\{u_i^{\mu,\delta,\Delta,h}(l\delta)\}$. The processes $\xi^h(\cdot)$ and $w^h(\cdot)$ now denote the interpolation of the chain and the pre-Wiener process, resp., under the strategy $\bar{c}_2^{\Delta,h}(\cdot)$ and control $\hat{r}_1^h(\cdot)$. The discretized control sequence $\{u_1^{\mu,\delta,\Delta,h}(l\delta)\}$ is used by Player 2 to get its controls, but Player 1 uses $\hat{r}_1^h(\cdot)$, and $w^h(\cdot)$ depends on $\hat{r}_1^h(\cdot)$ as well as on $\{u_2^{\mu,\delta,\Delta,h}(l\delta)\}$, the actual control sequence that is used by Player 2. This $w^h(\cdot)$ process will be fixed for each h and used in the rest of the proof.

Define the process $\xi^{\mu,\delta,\Delta,h}(\cdot)$ by (7.1), driven by the $\{u_i^{\mu,\delta,\Delta,h}(l\delta)\}, i = 1, 2$, and $w^h(\cdot)$. Theorem 7.1 implies that

$$\limsup_h \left| W_i^h(x, \hat{r}_1^h, \bar{c}_2^{\Delta,h}) - W_i^{\mu,\delta,\Delta,h}(x, r_1^{\mu,\delta,\Delta,h}, r_2^{\mu,\delta,\Delta,h}) \right| \leq \rho_1(\mu, \delta, \Delta), \tag{8.5}$$

where $\rho_1(\mu,\delta,\Delta) \to 0$ uniformly in $\hat{r}^h(\cdot)$, as $(\mu,\delta,\Delta,h) \to 0$. By the construction of $r_2^{\mu,\delta,\Delta,h}(\cdot)$, we have the equality

$$W_i^{\mu,\delta,\Delta,h}(x, r_1^{\mu,\delta,\Delta,h}, r_2^{\mu,\delta,\Delta,h}) = W_i^{\mu,\delta,\Delta,h}(x, r_1^{\mu,\delta,\Delta,h}, \bar{c}_2^{\Delta,h}). \tag{8.6}$$

Let $\tau^{\mu,\delta,\Delta,h}$ denote the first hitting time of the boundary for $\xi^{\mu,\delta,\Delta,h}(\cdot)$.

The set $(\xi^{\mu,\delta,\Delta,h}(\cdot), \hat{r}_1^h(\cdot), r_1^{\mu,\delta,\Delta,h}(\cdot), r_2^{\mu,\delta,\Delta,h}(\cdot), w^h(\cdot), \tau^{\mu,\delta,\Delta,h})$, parameterized by h, is tight. Extract a weakly convergent subsequence, index it by h also, and denote the limit by $(x(\cdot), \hat{r}_1(\cdot), r_1^{\mu,\delta,\Delta}(\cdot), r_2^{\mu,\delta,\Delta}(\cdot), w(\cdot), \tau)$. Then $(x(\cdot), \hat{r}_1(\cdot), r_1^{\mu,\delta,\Delta}(\cdot), r_2^{\mu,\delta,\Delta}(\cdot), w(\cdot), I_{\{\tau \le \cdot\}})$ is non-anticipative with respect to the standard Wiener process $w(\cdot)$, the set $(x(\cdot), r_1^{\mu,\delta,\Delta}(\cdot), r_2^{\mu,\delta,\Delta}(\cdot), w(\cdot))$ satisfies (2.1r), and τ is the first hitting time of the boundary. The $r_i^{\mu,\delta,\Delta}(\cdot)$ is just the relaxed control that is defined by the weak-sense limit $\{u_i^{\mu,\delta,\Delta}(l\delta)\}$ of $\{u_i^{\mu,\delta,\Delta,h}(l\delta)\}$. Thus, on any finite interval $[0,T]$, it is determined by a finite set of random variables $\{u_i^{\mu,\delta,\Delta}(l\delta), l\delta \le T\}$.

It must be shown that the limits $u_2^{\mu,\delta,\Delta}(l\delta)$ are chosen by the conditional probability law that defines $\bar{c}_2^{\Delta}(\cdot)$; i.e., that

$$\begin{aligned} &p_{2,k}\left(\alpha_2;\ w^h(l\Delta), l \le n; u_j^{\mu,\delta,\Delta,h}(l\delta), j=1,2, l\delta < n\Delta\right) \\ &\quad \to p_{2,k}\left(\alpha_2;\ w(l\Delta), l \le n; u_j^{\mu,\delta,\Delta}(l\delta), j=1,2, l\delta < n\Delta\right) \end{aligned} \tag{8.7}$$

for $k\delta \in [n\Delta, n\Delta+\Delta)$. Since there are only a finite number of values for the control, for any $t < \infty$ the limit $\{u_1^{\mu,\delta,\Delta}(l\delta), u_2^{\mu,\delta,\Delta}(l\delta), l\delta \le t\}$ will be achieved after a finite number of steps through the convergent subsequence $h \to 0$, w.p.1. Using this and the continuity of $p_{2k}(\cdot)$ in the w variables, we see that the $w^h(l\delta)$ in the left side of (8.7) can be replaced by the limit $w(l\delta)$ as $h \to 0$. [We will comment on this point at the end of the proof.] Thus the policy $\bar{c}_2^{\Delta}(\cdot)$ acting on any relaxed control with discretization $\{u_1^{\mu,\delta,\Delta}(l\delta)\}$ will yield the sequence $\{u_2^{\mu,\delta,\Delta}(l\delta)\}$ and it follows that

$$W_1^{\mu,\delta,\Delta,h}(x, r_1^{\mu,\delta,\Delta,h}(\cdot), \bar{c}_2^{\Delta,h}) \to W_1(x, r_1^{\mu,\delta,\Delta}(\cdot), \bar{c}_2^{\Delta}).$$

Furthermore, by (8.5) and (8.6), mod $\rho_1(\mu,\delta,\Delta)$,

$$W_1^h(x, \hat{r}_1^h, \bar{c}_2^{\Delta,h}) \to W_1(x, r_1^{\mu,\delta,\Delta}, \bar{c}_2^{\Delta}).$$

We can conclude that

$$\begin{aligned} \limsup_{h\to 0} W_1^h(x, \hat{r}_1^h, c_2^{\Delta,h}) &\le W_1(x, r_1^{\mu,\delta,\Delta}(\cdot), \bar{c}_2^{\Delta}) + \rho_1(\mu,\delta,\Delta) \\ &\le W_1(x, \bar{c}_1^{\Delta}, \bar{c}_2^{\Delta}) + \rho_1(\mu,\delta,\Delta) + \epsilon, \end{aligned} \tag{8.8}$$

where the ϵ is due to the fact that $(\bar{c}_1^{\Delta}(\cdot), \bar{c}_2^{\Delta}(\cdot))$ is an ϵ-equilibrium. The arbitrariness of the subsequence implies (8.4). The same argument is used when the indices 1, 2 are reversed.

Now, we comment briefly on the limit in (8.7). By Theorem 4.3, for small $\Delta > 0$ the controls that are used on the interval $[0, \Delta)$ have a negligible effect on the paths and costs. So let us use $u_1^{\mu,\delta,\Delta,h}(l\delta) = u_1(l\delta)$, $u_2^{\mu,\delta,\Delta,h}(l\delta) = u_2(l\delta)$ for arbitrary $u_i(l\delta)$ and $l\delta < \Delta$. For $k\delta \in [\Delta, 2\Delta)$, we have the rule

$$p_{2,k}\left(\alpha_2; w^h(\Delta); u_1(l\delta), u_2(l\delta), l\delta < \Delta\right) \tag{8.9}$$

and the probability of selecting any $\alpha_2 \in U_2^\mu$ converges as $w^h(\Delta) \to w(\Delta)$. Then the limit in (8.9) must be the law of $u_2^{\mu,\delta,\Delta}(l\delta)$ for $\Delta \leq l\delta < 2\Delta$. Using the method of selecting the control values in terms of the $\{\theta_l\}$ that was described after the proof of Theorem 5.1, we can suppose that the convergence $u_2^{\mu,\delta,\Delta,h}(l\delta) \to u_2^{\mu,\delta,\Delta}(l\delta), \Delta \leq l\delta < 2\Delta$, occurs in a finite number of steps w.p.1, as $h \to 0$ through the convergent subsequence, with the rule (8.9) used. Next, on $[2\Delta, 2\Delta + \Delta)$, we have the rule

$$p_{2,k}\left(\alpha_2; w^h(l\Delta), l \leq 2; u_i(l\delta), l\delta < \Delta; u_i^{\mu,\delta,\Delta,h}(l\delta), \Delta \leq l\delta < 2\Delta, i = 1, 2\right).$$

The $u_1^{\mu,\delta,\Delta,h}(l\delta), \Delta \leq l\delta < 2\Delta$, can be assumed to converge in a finite number of steps. Hence, proceeding as above, so do the selected values of $u_2^{\mu,\delta,\Delta,h}(l\delta), \Delta \leq l\delta < \Delta + 2\Delta$. Continuing in this way yields the form (8.7). □

9. Approximating the chain: II. Representations and approximations of the chain with control-independent driving noise. We are not able to prove the converse result to Theorem 8.1 for arbitrary $\sigma(\cdot)$ due to the fact that the $w^h(\cdot)$ process that is used in (6.5) depends on the control. When $\sigma(\cdot)$ is constant, this control dependence is small and can be factored out, and (uniform in the control) approximations in terms of an i.i.d. driving sequence can be developed. Once this control dependence is factored out, more convenient approximations to the chain can be obtained. This is done in this section and the results applied to the completion of the convergence proof in the next section. We will work with the model used in (6.9) where $\sigma = \begin{bmatrix} \sigma_1 & 0 \\ 0 & 0 \end{bmatrix}$, the dimension of x_1 is d_1, and σ_1 is a square and invertible matrix of dimension d_1. If $\sigma(\cdot)$ is constant, then the results of this section will hold for any locally consistent approximation under (A2.1) and (A2.2). But to avoid annoying minor details, we will focus on two main classes of models, each of which is typical of a large class of models and numerical algorithms. Case 1 below arises when one uses the so-called central-difference approximation to get the transition probabilities. Case 2 arises when one uses a central-difference approximation for the non-degenerate part and a one-sided or "upwind" approximation for the degenerate part [18, Chapter 5]. Both forms are locally consistent. Let $b_i(\cdot)$ denote the ith component of $b(\cdot)$.

We will show that $w^h(\cdot)$ can be factored as $w^h(\cdot) = \bar{w}^h(\cdot) + \zeta^h(\cdot)$ where $\bar{w}^h(\cdot)$ does not depend on the control or state and $\zeta^h(\cdot)$ is "asymptotically negligible."

Case 1. Let $d_1 = v$, so that σ is invertible and define the matrix $a = \sigma\sigma'$. Suppose that $a_{i,i} - \sum_{j:j\neq i} |a_{i,j}| \geq 0$. The condition can be weakened if the approximation intervals can depend on the coordinate direction. [18, Chapter 5]. We can also apply a linear transformation to diagonalize the matrix a. Then the condition always holds. Note that such a diagonalization might make the coding of the numerical algorithm near the boundary more complicated. Let e_i denote the unit vector in the ith coordinate direction. A "central-difference" version of the canonical form of the transition probabilities and interpolation interval in [18, (4.15), Chapter 5] is

$$\begin{aligned} p^h(x, x \pm e_i h|\alpha) &= \frac{q_{i,i} \pm h b_i(x,\alpha)/2}{Q}, \quad \Delta t^h(x,\alpha) = \Delta t^h = \frac{h^2}{Q}, \\ p^h(x, x + e_i h + e_j h|\alpha) &= p^h(x, x - e_i h - e_j h|\alpha) = \frac{a_{i,j}^+}{2Q}, \\ p^h(x, x + e_i h - e_j h|\alpha) &= p^h(x, x - e_i h + e_j h|\alpha) = \frac{a_{i,j}^-}{2Q}, \\ Q &= \sum_i a_{i,i} - \sum_{i,j:i\neq j} |a_{i,j}|/2, \quad q_{ii} = a_{i,i}/2 - \sum_{j:j\neq i} |a_{i,j}|/2. \end{aligned} \tag{9.1}$$

We also assume that h is small enough so that $q_{i,i} - h|b_i(x,\alpha)| \geq 0$. A simple computation using (9.1) shows that $b^h(x,\alpha) = b(x,\alpha)$ and $a^h(x,\alpha) = \sigma\sigma' + O(\Delta t^h)$. By (9.1) we can write $\Delta t_n^h = \Delta t^h$. In one dimension, (9.1) reduces to (6.3), where $q_{1,1} = \sigma^2/2$.

Case 2. Now, use (6.9) where $\sigma = \begin{bmatrix} \sigma_1 & 0 \\ 0 & 0 \end{bmatrix}$ where the dimension of x_1 is d_1, and σ_1 is a square and invertible matrix of dimension d_1. The problem concerns the effect of the degenerate part. We will use a canonical model that is motivated by the general model of [18, Chapter 5]. Define $\bar{b} = \sup_{x,\alpha} \sum_{i=d_1+1}^{v} |b_i(x,\alpha)|$ and $Q^h = Q + h\bar{b}$. For this case, redefine $\Delta t^h = \Delta t^h(x,\alpha) = h^2/Q^h$. Use the form (9.1) for $i \leq d_1$, with Q replaced by Q^h. For $i = d_1 + 1, \ldots, v$, use

$$p^h(x, x \pm e_i h|\alpha) = \frac{h b_i^{\pm}(x,\alpha)}{Q^h}.$$

In this case, we might have transitions from state x to itself, with transition probability

$$p^h(x, x|\alpha) = \frac{h\bar{b} - h\sum_{i=d_1+1}^{v} |b_i(x,\alpha)|}{Q^h}.$$

It is easy to show that $a^h(x,\alpha) = \sigma\sigma' + O(\Delta t^h)$ and $b^h(x,\alpha) = b(x,\alpha)$. Let E_n^h denote the expectation given all the system data up to step n.

THEOREM 9.1. *With either of the models Case 1 or Case 2,* $\delta w_n^h = \delta \bar{w}_n^h + \delta\zeta_n^h$*, where the components are martingale differences. The* $\delta\bar{w}_n^h$ *are i.i.d.,* $\{\delta\bar{w}_l^h, l \geq n\}$ *is independent of* $\{\xi_l^h, u_l^h, l \leq n\}$*, and* $\delta\bar{w}_n^h = O(h)$*. For either case,* $E_n^h \delta\bar{w}_n^h[\delta\bar{w}_n^h]' = \Delta t^h$*, and* $E_n^h\delta\zeta_n^h[\delta\zeta_n^h]' = O(h\Delta t^h)$*,* $E_n^h\delta\zeta_n^h[\delta\bar{w}_n^h]' = O(h\Delta t^h)$*.*

Proof. The proof is just a construction of the $\delta\bar{w}_n^h$. First we define δw_n^h as though $b(\cdot) = 0$ and $a^h(x,\alpha) = \sigma\sigma'$. The result will be $\delta\bar{w}_n^h$. Then $\delta\zeta_n^h$ is defined to make up the difference, and shown to have no effect asymptotically, as $h \to 0$. The fact that the dominant terms in the transition probabilities in (9.1) do not depend on h, and that the contributions due to the drift term $b(\cdot)$ (which contains the control and state information) are proportional to h makes this possible. In order to avoid a complicated notation and to concentrate on the essential points, we start with Case 1 in one dimension. The treatment of the higher-dimensional model follows the same pattern and this is illustrated via a two-dimensional case. Then the minor modifications that are required for Case 2 are discussed. These three examples illustrate the general procedure, which will hopefully then be apparent. The point of the proof is to show that the $d\bar{w}_n^h$ and $\delta\zeta_n^h$ can be constructed so that the theorem is satisfied. We need only know that the decomposition is possible. Define $\delta\xi_n^h = \xi_{n+1}^h - \xi_n^h$.

Case 1, one dimension. Since $b^h(\cdot) = b(\cdot)$, the double integral term in (6.9) is just $b(\xi_n^h, u_n^h)\Delta t^h$. To construct the state transitions for the chain, we will use the representation described in Section 6, that defines the state transitions in terms of i.i.d random variables $\{\chi_n\}$, uniformly distributed on $[0,1]$. In one dimension (9.1) is (6.3) and $p^h(x, x \pm h|\alpha) = 0.5 \pm hb(x,\alpha)/[2\sigma^2]$ and $\Delta t^h = h^2/\sigma^2$. Let $\delta\xi_n^h = h$ if $\chi_n \in [0, .5 + hb(\xi_n^h, u_n^h)/2\sigma^2]$, and set it equal to $-h$ otherwise. The "conditional mean" change is $2h[hb(\xi_n^h, u_n^h)/2\sigma^2] = b(\xi_n^h, u_n^h)\Delta t^h$, which partially confirms that the construction satisfies the local consistency condition (6.1). It also holds that the conditional variance of $\delta\xi_n^h$ is $\sigma^2\Delta t^h + o(\Delta t^h)$, which completes the requirements for local consistency

Next define the martingale difference term $\delta\bar{w}_n^h$. Divide $[0,1]$ into the two segments $[0,.5], (.5,1]$, If $\chi_n \in [0,.5]$, then set $\delta\bar{w}_n^h = h/\sigma$, otherwise set it equal to $-h/\sigma$. It is what δw_n^h would be if $b(\cdot) = 0$ and $a^h(x,\alpha) = \sigma\sigma'$. Next, we define $\delta\zeta_n^h = w_n^h - \delta\bar{w}_n^h$ to make up for the errors. There are two components to $\delta\zeta_n^h$. One component is due to the use of σ in lieu of $[a^h(\xi_n^h, u_n^h)]^{1/2}$ as in the last line of (6.9). Since $a^h(x,\alpha) - \sigma^2 = O(h^2)$, we have $[a^h(x,\alpha)]^{1/2} - \sigma = O(h^2)$ and the corresponding error component is $O(h^3)$. The associated interpolated error process clearly satisfies (6.6).

The second component of the error $\delta\zeta_n^h$ is due to the neglect of the term $b(\cdot)$ in constructing $\delta\bar{w}_n^h$. This is handled as follows. Suppose that

$b(\xi_n^h, u_n^h) \geq 0$ (the computation is analogous if $b(\xi_n^h, u_n^h) < 0$). Then the second component of the error is[14]

$$\delta\zeta_n^h = (2h - b(\xi_n^h, u_n^h)\Delta t^h)/\sigma, \quad \text{if } \chi_n \in [.5, .5 + hb(\xi_n^h, u_n^h)/2\sigma^2]$$

and it equals $-b(\xi_n^h, u_n^h)\Delta t^h/\sigma$ otherwise. The conditional variance of $\delta\zeta_n^h$ is

$$E_n^h \left[2h - b(\xi_n^h, u_n^h)\Delta t^h/\sigma\right]^2 \frac{hb(\xi_n^h, u_n^h)}{2\sigma^2} + E_n^h \left[b(\xi_n^h, u_n^h)\Delta t^h/\sigma\right]^2 \left(1 - \frac{hb(\xi_n^h, u_n^h)}{2\sigma^2}\right) = O(h)\Delta t^h.$$

Clearly the term $\delta\zeta_n^h$ depends on the control u_n^h and state ξ_n^h. But $\delta\bar{w}_n^h$ does not. It is just a Bernoulli sequence with values $\pm h$, with $\{\delta\bar{w}_l^h, l \geq n\}$ independent of the data up to step n. Also, $E_n^h[\delta\bar{w}_n^h]^2 = \Delta t^h$, $E_n^h \delta\bar{w}_n^h \delta\zeta_n^h = O(h)\Delta t^h$ and $E_n^h[\delta\zeta_n^h]^2 = O(h)\Delta t^h$, all uniformly in the controls.

Define the continuous-time martingales $\bar{w}^h(t)$ and $\zeta^h(t)$ by interpolating $\sum_{i=0}^{n-1} \delta\bar{w}_i^h$ and $\sum_{i=0}^{n-1} \delta\zeta_i^h$, resp., with intervals Δt^h[15] and define $w^h(t) = \bar{w}^h(t) + \zeta^h(t)$. The martingale $\bar{w}^h(\cdot)$ has quadratic variation It, and $\bar{w}^h(s), s \geq t$, is independent of $\xi^h(s), u^h(s), s \leq t$. The quadratic variation of $\zeta^h(\cdot)$ (and its quadratic covariation with $\bar{w}^h(\cdot)$) is $O(h)$, uniformly in the controls and initial condition.

The two-dimensional model for Case 1. The method in higher dimensions follows that in one dimension, except that there are more sub intervals to deal with. The general approach will be illustrated via a two-dimensional problem, where σ is a 2×2 non-singular matrix. Let $a_{1,2} \geq 0, b_i(\xi_n^h, u_n^h) \geq 0$. The method for the other cases is analogous. As for the one-dimensional case, the goal is to realize the choices determined by the transition probabilities (9.1). Divide the unit interval into successive subintervals of lengths $q_{1,1}/Q, q_{1,1}/Q, q_{2,2}/Q, q_{2,2}/Q, a_{1,2}/2Q, a_{1,2}/2Q$. If $\chi_n \in [0, (q_{1,1} + hb_1(\xi_n^h, u_n^h))/Q]$, then set $\delta\xi_{1,n}^h = h$, and $\delta\xi_{2,n}^h = 0$. If $\chi_n \in (q_{1,1} + hb_1(\xi_n^h, u_n^h))/Q, 2q_{1,1}/Q]$, then set $\delta\xi_{1,n}^h = -h$, and $\delta\xi_{2,n}^h = 0$. Using the two intervals of length $q_{2,2}/Q$, do the analogous computation with the indices 1, 2 reversed. If χ_n falls in the next to last of the six subintervals, then set $\delta\xi_n^h = (h, h)$, and set it equal to equal to $(-h, -h)$ if χ_n falls in the last of the six subintervals. Define $\delta\bar{w}_n^h$ by repeating the above with $b(x, \alpha) = 0$ and premultiplying by σ^{-1}, as follows: For χ_n falling in the six successive subintervals, define

$$\delta\bar{w}_n^h = \sigma^{-1}\left\{\begin{pmatrix} h \\ 0 \end{pmatrix}, \begin{pmatrix} -h \\ 0 \end{pmatrix}, \begin{pmatrix} 0 \\ h \end{pmatrix}, \begin{pmatrix} 0 \\ -h \end{pmatrix}, \begin{pmatrix} h \\ h \end{pmatrix}, \begin{pmatrix} -h \\ -h \end{pmatrix}\right\}.$$

[14]The $2h$ in the formula is due to the fact that on the interval $\chi_n \in [.5, .5 + hb(\xi_n^h, u_n^h)/2\sigma^2]$, the difference $\xi_{n+1}^h - \xi_n^h$ was implicitly assigned a value $-h$ when $\delta\bar{w}_n^h$ was constructed, when it should have been assigned the value $+h$.

[15]They are martingales only when sampled at the times $n\Delta t^h = \Delta t_n^h, n = 0, 1, 2, \ldots$, but this difference is unimportant, asymptotically.

The error terms $\delta\zeta_n^h$ are computed analogously to what was done for the one-dimensional model. The same pattern is followed in higher dimensions.

Case 2. The difference between Cases 1 and 2 is that for Case 2 the matrix σ is singular. The main complication is notational, and the general procedure will be illustrated via a two-dimensional model, where only the first component of $x(\cdot)$ has a Wiener process driving term. In this case, $\bar{b} = \max_{x,\alpha} |b_2(x,\alpha)|$, $Q = 2q_{1,1} = [\sigma_1]^2$ and $Q^h = Q + h\bar{b}$. The procedure differs from that of Case 1 mainly in that we need to account for the $p^h(x, x|\alpha)$, which depends on $\bar{b}$. Divide the unit interval into three subintervals of lengths

$$\frac{q_{1,1}/2}{Q^h}, \; \frac{q_{1,1}/2}{Q^h}, \; \frac{h\bar{b}}{Q^h},$$

and divide the third subinterval into two further subintervals of lengths

$$h|b_2(x,\alpha)|/Q^h, \quad h[\bar{b} - |b_2(x,\alpha)|]/Q^h,$$

resp. There are now four subintervals. The values of $\delta\xi_n^h$ are determined as follows. Analogously to what was done in the one-dimensional example of Case 1, if $\chi_n \in [0, Q/Q^h]$, then define $\delta\xi_{2,n}^h = 0$. If $\chi_n \in (Q/Q^h, 1]$, set $\delta\xi_{1,n}^h = 0$ and $\delta\bar{w}_{1,n}^h = 0$. If the value of χ_n falls into the fourth subinterval, then define $\delta\xi_{2,n}^h = 0$. If χ_n falls into the third subinterval, then define $\delta\xi_{2,n}^h = h \operatorname{sign}(b_2(\xi_n^h, u_n^h))$. If $\chi_n \in [0, (q_{11} + hb_1(\xi_n^h, u_n^h))/Q^h]$, set $\delta\xi_{1,n}^h = h$, and set it equal to $-h$ if $\chi_n \in [(q_{11} + hb_1(\xi_n^h, u_n^h))/Q^h, 2q_{11}/Q^h]$. These constructions yield (9.1) with $\Delta t^h = h^2/Q^h$.

To complete the computation of $\delta\bar{w}_{1,n}^h$, repeat the procedure with $b(\cdot) = 0$ and divide by σ_1, as follows. Set $\delta\bar{w}_{1,n}^h = h/\sigma_1$ if $\chi_n \in [0, q_{1,1}/Q^h]$. It is $-h/\sigma_1$ if $\chi_n \in (q_{1,1}/Q^h, 2q_{1,1}/Q^h]$, and it is zero otherwise. The variance is h^2/Q^h, which is the current value of Δt^h. The second component of $\delta\bar{w}_n^h$, namely $\delta\bar{w}_{2,n}^h$, is unimportant since it is eventually multiplied by zero. So, let us take its values from independent Bernouilli sequence with values $\pm h/\sqrt{Q^h}$, each taken with probability 1/2.

The error terms ζ_n^h are computed using a procedure that is analogous to that in the one-dimensional Case 1 example. In all cases, the quadratic variation process of $\bar{w}^h(\cdot)$ is It. □

In the next theorem, where $\sigma(\cdot) = \sigma$, a constant, it is shown that the continuous-time interpolation $\xi^h(\cdot)$ can be written in the form

$$\xi^h(t) = x(0) + \int_0^t \int_U b(\xi^h(s), \alpha) r^{h,\prime}(d\alpha, s) ds + \int_0^t \sigma d\bar{w}^h(s) + \epsilon_2^h(t), \quad (9.2)$$

where $\epsilon_2^h(\cdot)$ equals $\epsilon_1^h(\cdot)$ plus a stochastic integral with respect to the error martingale $\zeta^h(\cdot)$, and it satisfies (6.6). Since $\bar{w}^h(\cdot)$ does not depend on the control and is the sum of i.i.d. zero-mean random variables of size $O(h)$, we can exploit the form (9.2) to get approximation results analogous to those in Theorems 4.1–4.3. In other words, the controls can be space and time discretized with arbitrarily small change in the costs for the chain, just as in Theorems 4.1–4.3. The proof of the various assertions follows the lines of arguments used in Theorem 7.1, exploiting the martingale properties and the Lipschitz condition. The details are very similar and are omitted.

THEOREM 9.2. *Assume* (A2.1) *and any of the models that were used in Theorem* 9.1. *Define the process* $\bar{\xi}^h(\cdot)$:

$$\bar{\xi}^h(t) = x(0) + \int_0^t \int_U b(\bar{\xi}^h(s), \alpha) r^{h,\prime}(d\alpha, s)ds + \int_0^t \sigma d\bar{w}^h(s). \tag{9.3}$$

Then, for each $t > 0$,

$$\lim_{h \to 0} \sup_{x(0), r^h} E \sup_{s \leq t} \left|\xi^h(s) - \bar{\xi}^h(s)\right|^2 = 0. \tag{9.4}$$

If (A2.2) *is assumed as well, then the costs for* $\xi^h(\cdot)$ *and* $\bar{\xi}^h(\cdot)$ *are arbitrarily close, uniformly in the control and initial condition.*

Now, given μ, δ, Δ, *let* $u_i^{\mu,\delta,\Delta,h}(\cdot)$ *be the delayed and discretized approximation of* $r_i^h(\cdot)$ *that is constructed by the procedure described above Theorem* 4.3, *and let* $r^{\mu,\delta,\Delta,h}(\cdot)$ *denote the relaxed control representation of the pair of approximations. Define the process* $\bar{\xi}^{\mu,\delta,\Delta,h}(\cdot)$ *by*

$$\begin{aligned}\bar{\xi}^{\mu,\delta,\Delta,h}(t) = x(0) + &\int_0^t \int_U b(\bar{\xi}^{\mu,\delta,\Delta,h}(s), \alpha) r^{\mu,\delta,\Delta,h,\prime}(d\alpha, s)ds \\ &+ \int_0^t \sigma d\bar{w}^h(s).\end{aligned} \tag{9.5}$$

Then for $t > 0$ *and* $\gamma > 0$ *there are positive numbers* $\mu_\gamma, \delta_\gamma, \Delta_\gamma, h_\gamma, \kappa_\gamma$, *such that for* $\mu \leq \mu_\gamma$, $\delta \leq \delta_\gamma$, $\Delta \leq \Delta_\gamma$, $h \leq h_\gamma$, $\delta/\Delta \leq \kappa_\gamma$,

$$\sup_{r^h, x(0)} E \sup_{s \leq t} \left|\bar{\xi}^{\mu,\delta,\Delta,h}(s) - \bar{\xi}^h(s)\right|^2 \leq \gamma. \tag{9.6}$$

If the condition (A2.2) *is also assumed, then for small* (μ, δ, Δ, h) *the costs for* $\bar{\xi}^{\mu,\delta,\Delta,h}(\cdot)$ *and* $\bar{\xi}^h(\cdot)$ *are arbitrarily close, uniformly in the control and initial condition.*

Approximating $\bar{w}^h(\cdot)$ by a Wiener process. By the method of construction, the terms $[\bar{w}^h(n\Delta+\Delta) - \bar{w}^h(n\Delta)], n = 0, 1, \ldots$ are i.i.d. and have orthogonal components. The covariance is ΔI, where I is the identity matrix, and the set converge to (i.i.d) normally distributed random variables as $h \to 0$. It will be useful to quantify this closeness to normally distributed random variables for use in Theorems 9.4 and 10.1 via the following strong approximation theorem for i.i.d. random variables.

LEMMA 9.3. [2, Theorem 3.]*Let* $\{\phi_n\}$ *be a sequence of* $I\!R^d$*-valued i.i.d. random variables with zero mean and bounded* $(2+\delta)$*th moment, where* $0<\delta\leq 1$. *Suppose that the covariance matrix* Γ *is non-singular. Then without changing the distribution, one can redefine the sequence on a richer probability space together with a Wiener process* $B(\cdot)$ *with covariance matrix* Γ *such that*

$$\left|\sum_{i\leq n}\phi_i - B(n)\right| = o(n^{0.5-c}) \tag{9.7}$$

w.p.1 for large n, *for some* $0<c<0.5$.

The next theorem exploits Lemma 9.3 to prove that the process defined by (9.5) can be well approximated by the discrete time system (3.1), whose solution we now write as $x^{\mu,\delta,\Delta,h}(\cdot)$, since the discretized controls $r_i^{\mu,\delta,\Delta,h}(\cdot), i=1,2$, are used. The controls that are used in (9.5) are obtained from the discretization of the relaxed control representation of the interpolation of $\{u_n^h\}$, the original control sequence for the chain.

THEOREM 9.4. *Assume* (A2.1) *and* (A2.2) *and any of the models that were used in Theorem* 9.1 *Then the probability space can be defined such that* $\bar{w}^h(t)=w(t)+\rho^h(t)$, *where* $w(\cdot)$ *is a vector-valued standard Wiener process. For each* $t>0$, $\lim_{h\to 0} E\sup_{s\leq t}|\rho^h(s)|^2\to 0$. *Let* $x^{\mu,\delta,\Delta,h}(\cdot)$ *denote the solution to* (3.1) *with the same Wiener process* $w(\cdot)$ *but with the controls that are used in* (9.5). *Then, for any* $t>0$,

$$\lim_{h\to 0}\sup_{r^h,x(0)} E\sup_{s\leq t}\left|x^{\mu,\delta,\Delta,h}(s)-\bar{\xi}^{\mu,\delta,\Delta,h}(s)\right|^2 = 0. \tag{9.8}$$

Proof. The claim that we can apply the controls for the chain to (3.1) requires some explanation and a careful choice of the probability space. Consider Case 1 and define random variables $\phi_n=\delta\bar{w}_n^h/\sqrt{h^2/Q}=\delta\bar{w}_n^h/\Delta t^h$. (For Case 2, the development is the same, except that Q^h replaces Q.) Then $\{\phi_n,\chi_n\}$ satisfies the conditions of Lemma 9.3 and, without loss of generality, we can suppose that the probability space is constructed such that (9.7) holds for a standard vector-valued Wiener process $B(\cdot)$, which will be used only to approximated the sums of the ϕ_n. We will not have use for the component of the Wiener process that approximates the sums of the χ_n. Then, on this probability space define $\delta\bar{w}_n^h$ in terms of the ϕ_n, inverting the above definition. We can suppose that ξ_0^h is also defined on the same space. Now construct u_0^h, which is just a function of ξ_0^h. Then construct the random variable ξ_1^h, which is just a function of (χ_0,ξ_0^h,u_0^h), as done for Case 1 in Theorem 9.1. Continuing in this way, we can construct $\{\xi_n^h,u_n^h,\delta\bar{w}_n^h\}$ on the probability space, with the same law as originally. Given the controls u_n^h, they can then be time and space discretized and delayed. [16] as in Theorem 9.2.

[16]Actually, it is only required that the controls be approximated and delayed such

From Lemma 9.3, we for large n and w.p.1,

$$\left|\sum_{i=0}^{n} h\phi_i - hB(n)\right| = h\,o(n^{0.5-c}). \tag{9.9}$$

and $w(\cdot) = hB(\cdot/\Delta t^h)/\sqrt{Q}$ is a standard vector-valued Wiener process. It follows that there is a $c > 0$ and a $t_h \to 0$ as $h \to 0$ such that

$$\left|\sum_{i=0}^{t/\Delta t^h} \delta\bar{w}_i^h - w(t)\right| = o(t[\Delta t^h]^c) \tag{9.10}$$

w.p.1 for $t \geq t_h$ and small h. We can now write $\bar{w}^h(t) = w(t) + \rho^h(t)$, where $\rho^h(\cdot)$ has independent increments, and $\lim_{h\to 0} E\sup_{s\leq t}|\rho^h(s)|^2 = 0$. Rewrite (9.5) as

$$\begin{aligned}\bar{\xi}^{\mu,\delta,\Delta,h}(t) &= x(0)\\ &+\int_0^t\int_U b(\bar{\xi}^{\mu,\delta,\Delta,h}(s),\alpha)r^{\mu,\delta,\Delta,h,\prime}(d\alpha,s)ds + \sigma dw(t) + \sigma\rho^h(t),\end{aligned}$$

and write (3.1) with the controls $r^{\mu,\delta,\Delta,h}(\cdot)$ as (in interpolated form)

$$x^{\mu,\delta,\Delta,h}(t) = x(0) + \int_0^t\int_U b(x^{\mu,\delta,\Delta,h}(s),\alpha)r^{\mu,\delta,\Delta,h,\prime}(d\alpha,s)ds + \sigma dw(t). \tag{9.11}$$

From here on the proof is completed by using the Lipschitz condition, the martingale properties, and standard estimates and the rest of the details are omitted. □

10. Convergence: II. An approximate equilibrium for the chain is an approximate equilibrium for the diffusion. Theorem 8.1 showed that an approximate equilibrium for the diffusion is also one for the approximating chain. It follows from this that if the ϵ-equilibrium value for the chain is unique for arbitrarily small ϵ, then the converse result is true; namely that ϵ-equilibrium values for the chain are ϵ_1-equilibrium values for (2.1r), where $\epsilon_1 \to 0$ as $\epsilon \to 0$, and we are done, since it would then be implied by Theorem 8.1 that the ϵ-equilibrium values for the diffusion are also unique for small ϵ. If the ϵ-equilibrium value for the chain is not unique for small ϵ, then we will show that this "converse" assertion is true for the model used in Theorem 9.1. We are not able to show the converse result when $\sigma(\cdot)$ depends on x, so $\sigma(\cdot)$ is constant in the next theorem. Condition (A5.1) is not needed for the proof, although the problem makes little sense if (A5.1) does not hold.

---

that the control applied on $[n\Delta, n\Delta + \Delta)$ is $\mathcal{F}^h_{n\Delta-}$-measurable. The other aspects of the discretization are not needed.

THEOREM 10.1. *Assume* (A2.1) *and* (A2.2) *and the model used in Theorem* 9.1, *where* $\sigma(\cdot)$ *is constant. Then for any* $\epsilon > 0$ *there is* $\epsilon_1 > 0$ *which goes to zero as* $\epsilon \to 0$ *such that an* ϵ*-equilibrium value for the chain* ξ_n^h *for small* h *is an* ϵ_1*-equilibrium value for* (2.1r).

Proof. Theorem 9.2 says that the paths and cost functions for (9.2) (which is $\xi^h(\cdot)$ under an arbitrary control), (9.3) (where the control is as in (9.2) but the driving process is $\bar{w}^h(\cdot)$), and (9.5) (which is (9.3) with discretized controls) are arbitrarily close, uniformly in the controls, for small (μ, δ, Δ, h). Theorem 9.4 gives the same result for (9.5) and $x^{\mu,\delta,\Delta,h}(\cdot)$ given by (9.11), which is (3.1) with discretized controls. Theorem 4.3 implies the same thing for $x^{\mu,\delta,\Delta}(\cdot)$ and (2.1r). These "uniform closeness" results imply the theorem. □

REFERENCES

[1] A. BENSOUSSAN AND A. FRIEDMAN. Nonzero-sum stochastic differential games with stopping times and free boundaries. *Trans. Amer. Math Soc.*, **231**: 275–327, 1977.

[2] I. BERKES AND W. PHILIPP. Approximation theorems for independent and weakly dependent random vectors. *Ann. Probab.*, **7**: 29–54, 1979.

[3] R. BUCKDAHN, P. CARDALIAGUET, AND C. RAINIER. Nash equilibrium payoffs for nonzero-sum stochastic differential games. *SIAM J. Control and Optimization*, **43**: 624–642, 2002.

[4] E.ALTMAN, O.POURTALLIER, A. HAURIE, AND F. MORESINO. Approximating Nash equilibria in nonzero-sum games. *International Game Theory Review*, **2**: 155–172, 2000.

[5] R.J. ELLIOTT AND N.J. KALTON. *Existence of value in differential games*, Mem. AMS, **126**. Amer. Math. Soc, Providence, RI, 1974.

[6] S.N. ETHIER AND T.G. KURTZ. *Markov Processes: Characterization and Convergence.* Wiley, New York, 1986.

[7] W.F. FLEMING. Generalized solutions in optimal stochastic control. In P.T. Liu, E. Roxin, and R. Sternberg, editors, *Differential Games and Control Theory: III*, pp. 147–165. Marcel Dekker, 1977.

[8] W.H. FLEMING AND P.E. SOUGANIDIS. On the existence of value functions for two-player zero-sum differential games. *Indiana Univ. Math. J.*, **38**: 293–314, 1989.

[9] A. FRIEDMAN. Stochastic differential games. *J. Differential Equations*, **11**: 79–108, 1972.

[10] S. HAMADENE. Points d'équilibre dans les jeux stochastiques de somme non nulle. *C. R. Acad. Sci. Paris Sér. I Math*, **318**: 251–256, 1994.

[11] A.B. HAURIE, J.B. KRAWCZYK, AND M. ROCHE. Monitoring cooperative equilibria in a stochastic differential game. *J. of Optimiz. Theory and Applic.*, **81**: 73–95, 1994.

[12] A.B. HAURIE AND F. MORESENO. Computing equilibria in stochastic games of intergenerational equity. In *International Conference on Dynamic Games.* International Society of Dynamic Games, 2004.

[13] H.J. KUSHNER. *Probability Methods for Approximations in Stochastic Control and for Elliptic Equations.* Academic Press, New York, 1977.

[14] H.J. KUSHNER. Numerical methods for stochastic control problems in continuous time. *SIAM J. Control Optim.*, **28**: 999–1048, 1990.

[15] H.J. KUSHNER. Numerical methods for stochastic differential games. *SIAM J. Control Optim.*, pp. 457–486, 2002.

[16] H.J. KUSHNER. Numerical approximations for stochastic differential games: The ergodic case. *SIAM J. Control Optim.*, **42**: 1911–1933, 2004.

[17] H.J. KUSHNER. Numerical methods for stochastic differential games: The ergodic cost criterion. In S. Jorgensen, M. Quincampoix, and T. Vincent, editors, *Annals of the International Society of Dynamic Games*, 2005. To appear.

[18] H.J. KUSHNER AND P. DUPUIS. *Numerical Methods for Stochastic Control Problems in Continuous Time.* Springer-Verlag, Berlin and New York, 1992. Second edition, 2001.

[19] K.M. RAMACHANDRAN. Stochastic differential games and applications. In *Handbook of Stochastic Analysis and Applications*, pp. 473–532. Marcel Dekker, New York, 2002.

Part II: Nonparametric Estimation

ON THE ESTIMATION OF AN ANALYTIC SPECTRAL DENSITY OUTSIDE OF THE OBSERVATION BAND*

ILDAR A. IBRAGIMOV†

Abstract. We consider a Gaussian stationary process $X(t)$ with an integer analytic spectral density $f(\lambda)$ and study a problem of its estimation. The process $X(t)$ is non-observable. Instead of it we observe a linear transformation $Y(t)$, $0 \le t \le T$, of $X(t)$ with a transfer function $a(\lambda)$, $|a(\lambda)| = 1$ if λ belongs to an interval I. We study how far from I the consistent estimation of $f(\lambda)$ is possible, $T \to \infty$.

Key words. Stationary processes, band limited processes, spectral density estimation.

1. Introduction. The paper continues author's investigations devoted to the statistical approach to the unicity theorem for analytic functions. The unicity theorem says that if two functions f and g are holomorphic in a region G and $f(z) = g(z)$ for all z in some sequence of distinct points with limit point in G, then $f(z) = g(z)$ everywhere in G. The theorem means in particular that if an integer (or entire, " an entire function is an integral function in British usage" [3]) analytic function is observed on an interval I, it can be restored immediately in the whole complex plane. Of course, this problem of restoration is an ill posed one and small perturbations of the observations may drastically change the solution on large distance from the region of observation. We are interested in knowing how far from the region of observation a consistent restoration is possible under small stochastic perturbations. There are a few possible statements of the problem. One of them which has been considered in [8] is the following one.

Problem 1. We are observing an integer analytic function f on an interval $[x, y]$ in the white noise of a small intensity ε, i.e. the observation $X(t)$, $x \le t \le y$, satisfies the relation

$$dX(t) = f(t) + \varepsilon dw(t), \tag{1.1}$$

$w(t)$ is a standard Wiener process.

Denote $\mathbf{F}(M, \sigma, \rho)$ the class of integer analytic functions which satisfy the following restrictions on their growth

$$\sup_{|z| \le R} |f(z)| \le M \exp\{\sigma R^\rho\}. \tag{1.2}$$

These are well known classes of integer functions of order ρ and type σ, see [3], [12]. It has been shown in [8] that if the function f in (1.1) belongs

*Supported by the RFBR (05-01-00911), RFBR-DFG (04-01-04000), NSh-4222.2006.1, INTAS (03-51-5018), CRC-701.

†Petersburg Department of Steklov Institute of Mathematics, St. Petersburg 191023, Russia.

to the class $\mathbf{F}(M, \sigma, \rho)$, the consistent estimation of $f(z)$ when $\varepsilon \to 0$ is possible approximately in $(\ln 1/\varepsilon)^{\frac{1}{\rho}}$ vicinities of $[x, y]$ (see details in [8]).

An other scheme of observation has been considered in [9], [10]. Call it **Problem 2.** Let

$$X_1, X_2, \dots X_n \tag{1.3}$$

be a sample from a distribution with an integer analytic density function $f(x)$. It is supposed that the sample (1.3) is censored by an interval $[x, y]$: only observations $X_j \in [x, y]$ can be used to construct estimators for f, all other observations disappear. It is shown in [9], [10] that if the density function f belongs to a class $\mathbf{F}(M, \sigma, \rho)$, the consistent estimation is possible approximately in $(\ln n)^{\frac{1}{\rho}}$ vicinities of $[x, y]$ (see details in [9],[10]).

In this paper we consider one more variant of the last problem. Let $X(t)$ be a Gaussian stationary process with an integer analytic spectral density $f(\lambda)$. The process $X(t)$ is the Fourier transform of a Gaussian process $Z(\lambda)$ with independent increments and $\mathbf{E}|dZ(\lambda)|^2 = f(\lambda)d\lambda$,

$$X(t) = \int_{-\infty}^{\infty} e^{it\lambda} dZ(\lambda), \tag{1.4}$$

see, for example, [5], ch. 11.

The process $X(t)$ is unobservable. Instead of it we are observing for $0 \le t \le T$ a linear transformation $Y(t)$ of $X(t)$,

$$Y(t) = \int_{-\infty}^{\infty} e^{it\lambda} a(\lambda) dZ(\lambda) \tag{1.5}$$

with a transfer function $a(\lambda)$. The process $Y(t)$ is also a stationary Gaussian process with the spectral density $g(\lambda) = |a(\lambda)|^2 f(\lambda)$. If $|a(\lambda)| = 1$ for $\lambda \in [x, y]$, then $f(\lambda) = g(\lambda)$ for $x \le \lambda \le y$. There are many ways to construct a consistent estimator when $T \to \infty$ estimators $\hat{g}(\lambda)$ for $g(\lambda)$. Any such estimator will estimate consistently $f(\lambda)$ inside $[x, y]$. How shall we set problems on the estimation $f(\lambda)$ outside $[x, y]$? On the first glance a direct analogue of the problem 2 looks as follows: the process $Y(t)$ is bandlimited to $[x, y]$, i.e. the transfer function $a(\lambda) = 0$ for $\lambda \notin [x, y]$. But this approach will not work. Indeed, in the case the observation process

$$Y(t) = \int_x^y e^{it\lambda} a(\lambda) dZ(\lambda)$$

is analytic in the whole complex plane. Thus the observation $Y(t)$ admits the analytic continuation on $\mathbb{R}^1$. By Maruyama's theorem the Gaussian process $Y(t)$ with continuous spectrum is ergodic (see [13] or [6]) and hence by ergodic theorem the correlation function

$$R_Y(t) = \mathbf{E}Y(t)\overline{Y(0)} = \lim_{S \to \infty} \frac{1}{S} \int_0^S Y(t+s)\overline{Y(s)} ds$$

can be determined by the observation $Y(t)$, $0 \le t \le T$. The correlation function $R_Y(t)$ determines the spectral density $g(\lambda), \lambda \in \mathbb{R}^1$, and hence $f(\lambda)$ for $x \le \lambda \le y$. But then the analytic function $f(\lambda)$ is determined for all λ. (Note that linear filters with $a(\lambda) = 0$, $\lambda \notin [x, y]$ are physically non-realizable.)

Therefore we take an engineering conception of bandlimited processes. Namely, we consider linear transformations $Y(t)$ with transfer functions $a(\lambda)$ such that

$$|a(\lambda)| = 1,\ x \le \lambda \le y, \qquad |a(\lambda)| < 1,\ \lambda \notin [x, y].$$

We denote the class of such transfer functions $\mathcal{A}$. Of course it is supposed that the transfer function $a(\lambda)$ is unknown, otherwise we could estimate $f(\lambda)$ as $\hat{g}(\lambda)|a(\lambda)|^{-2}$. Thus we consider below the following estimation problem.

Problem 3. A stationary Gaussian process $X(t)$ with the spectral representation (1.4) has the spectral density $f \in \mathbf{F}(M, \sigma, \rho)$. Suppose also that

$$\int_{-\infty}^{\infty} f(\lambda) d\lambda \le A. \tag{1.6}$$

Denote the class of spectral densities f which belongs $\mathbf{F}(M, \sigma, \varrho)$ and satisfies (1.6) $\mathbf{G}(M, \sigma, \rho, A) = \mathbf{G}$. The observation process $Y(t)$, $0 \le t \le T$, is determined by the relation (1.4). The unknown transfer function $a \in \mathcal{A}$. The problem is to estimate the spectral density $f(\lambda)$, $f \in \mathbf{G}$, of $X(t)$ in and outside of the observation interval $[a, b]$.

Let us formulate the basic results.

THEOREM 1.1. *There exists an estimate $\hat{f}(\lambda)$ of $f(\lambda)$ such that*

$$\mathbf{E} \int_x^y \left| f(\lambda) - \hat{f}(\lambda) \right|^2 d\lambda \le \frac{C}{T} \frac{\ln T}{\ln \ln T} \tag{1.7}$$

where the constant C depends on M, σ, ρ, A, x, y only.

Denote $G_T(\alpha)$ the $(\ln T)^{\alpha/\rho}$-vicinity of the interval $[x, y]$.

THEOREM 1.2. *There exists an estimate $\hat{f}(\lambda)$ of $f(\lambda)$ such that for all fixed $\alpha < 1$ and $\beta < \alpha$ the following inequality holds*

$$\sup_{f \in \mathbf{G}, a \in \mathcal{A}} \mathbf{E}_f \left\{ \sup_{G_T(\beta)} \left| f(\lambda) - \hat{f}(\lambda) \right| \right\} \le C_{\alpha,\beta} T^{\frac{1-\alpha}{2}}. \tag{1.8}$$

The constant $C_{\alpha,\beta}$ depends also on M, σ, ρ, A, x, y.

The next theorem shows that the length of consistency intervals can not be essentially increased.

THEOREM 1.3. *Let L be an exterior point of the interval $G_T(\alpha)$, $\alpha > 1$. Then for all sufficiently large T*

$$\inf_{\hat{f}} \sup_{f \in \mathbf{G}, a \in \mathcal{A}} \mathbf{E}_f |\hat{f} - f(L)|^2 > c_0 > 0. \tag{1.9}$$

The constant c_0 depends on $M, \sigma, \rho, x, y, \alpha$.

2. Construction of estimators. Proof of Theorems 1, 2. The method of the proof does not depend on the observation's interval and for the sake of simplicity we suppose below that $[x, y] = [-1, 1]$.

We consider the following general scheme of estimator's construction. Let $\{\varphi_n\}$ be a complete orthonormal system in $L_2(-1, 1)$. Then we can expand the restriction of $f(\lambda)$ on $[-1, 1]$ into the Fourier series with respect to $\{\varphi_n\}$,

$$f(\lambda) = \sum_n a_n \varphi_n(\lambda), \quad a_n = \int_{-1}^{1} \varphi_n(\lambda) f(\lambda) d\lambda.$$

If $\hat{a}_n$ are estimators for a_n, estimate $f(\lambda)$ by a sum $\sum_{n \le N} \hat{a}_n \varphi_n(\lambda)$, the choice of N depends on the class $\mathbf{G}$. This method has been initiated by N.N. Chentsov [4].

To realize the scheme take as a complete orthonormal system in $L_2(-1, 1)$ the Legendre polynomials $P_n(\lambda)$, $n = 0, 1, \ldots$: P_n is a polynomial of degree n and $\int_{-1}^{1} P_n(\lambda) P_m(\lambda) d\lambda = \delta_{mn}$. See, for example, [14].

LEMMA 2.1. *Let $f \in \mathbf{F}(M, \sigma, \rho)$. Let*

$$f(\lambda) = \sum_n a_n P_n(\lambda), \quad a_n = \int_{-1}^{1} P_n(\lambda) f(\lambda) d\lambda, \tag{2.1}$$

be its expansion into the Fourier series with respect to $\{P_n\}$. Then

$$|a_k| \le cM \exp\left\{ -\frac{k}{\rho} \ln \frac{k}{e\rho\sigma} \right\} \tag{2.2}$$

where c depends on σ, ρ. If $\rho = 1$,

$$|a_k| \le M \exp\{-k \ln(k/e\sigma)\}. \tag{2.3}$$

The result should be well known. A proof of it see in [10].

LEMMA 2.2 (see[2], p. 74, or [15], p. 17). *Let $Q(z)$ be a polynomial of degree n. Then for all complex z*

$$|Q(z)| \le \max_{|x| \le 1} Q(x) \cdot \left| z + \sqrt{z^2 - 1} \right|^n. \tag{2.4}$$

LEMMA 2.3 (see [14]). *The following inequalities hold*

$$\max_{|x| \le 1} |P_n(x)| \le |P_n(1)| = \sqrt{(2n+1)/2}. \tag{2.5}$$

It follows from the lemmas above that the series (2.1) converges in the whole complex plane and determines there an analytic function. Hence the equality

$$f(z) = \sum_n a_n P_n(z) \tag{2.6}$$

holds in the whole complex plane.

Take as a basic statistics the periodogram $I_T(\lambda)$ of the observation,

$$I_T(\lambda) = \frac{1}{2\pi T}\left|\int_0^T e^{-it\lambda} Y(t)dt\right|^2 \tag{2.7}$$

and consider as estimators of linear functionals

$$(\varphi, g) = \int_{-\infty}^{\infty} \varphi(\lambda) g(\lambda) d\lambda$$

the statistics $\int_{-\infty}^{\infty} \varphi(\lambda) I_T(\lambda) d\lambda$. In particular we estimate the coefficients

$$a_n = \int_{-1}^{1} P_n(\lambda) f(\lambda) d\lambda = \int_{-1}^{1} P_n(\lambda) g(\lambda) d\lambda$$

by the statistics

$$\hat{a}_n = \int_{-1}^{1} P_n(\lambda) I_T(\lambda) d\lambda. \tag{2.8}$$

LEMMA 2.4. *The mathematical expectation*

$$\mathbf{E} I_T(\lambda) = \frac{2}{\pi T}\int_{-\infty}^{\infty} D_T^2(\lambda - \mu) g(\mu) d\mu \tag{2.9}$$

where

$$D_T(\lambda) = \frac{\sin(T\lambda/2)}{\lambda}.$$

The covariance function

$$\begin{aligned}\mathbf{E} I_T(\lambda_1) I_T(\lambda_2) = \frac{4}{\pi^2 T^2}\Big\{&\int_{-\infty}^{\infty} f(l) D_T^2(l-\lambda_1) dl \int_{-\infty}^{\infty} f(l) D_T^2(l-\lambda_2) dl \\ &+\left(\int_{-\infty}^{\infty} f(l) D_T^2(l-\lambda_1) D_T(l-\lambda_2) dl\right)^2 \\ &+\left(\int_{-\infty}^{\infty} f(l) D_T^2(l-\lambda_1) D_T(l-\lambda_2) dl\right)^2\Big\}.\end{aligned} \tag{2.10}$$

Proof for processes with the discrete time see, for example, in [1], Theorems 8.2.7, 8.2.8. The formulas (2.8), (2.9) can be proved in the same way.

LEMMA 2.5. *The bias* $\mathbf{E}\hat{a}_n - a_n$ *satisfies the inequality*

$$|\mathbf{E}\hat{a}_n - a_n| \le C T^{-1/2}. \tag{2.11}$$

The constant C *depends on the class* $\mathbf{G}$.

Proof. Since the equality

$$\frac{1}{\pi}\int_{-\infty}^{\infty} D_T^2(\lambda-\mu)d\mu = 1$$

and (2.8), we can write

$$\begin{aligned}|\mathbf{E}\hat{a}_n - a_n| &= \frac{1}{\pi}\bigg|\int_{-1}^{1} P_n(\lambda)\int_{-\infty}^{\infty} D_2^2(\mu)g(\lambda+2\mu/T)d\mu \\ &\quad -\frac{1}{\pi}\int_{-1}^{1} P_n(\lambda)d\lambda\int_{-\infty}^{\infty} D_2^2(\mu)g(\lambda)d\mu\bigg| \\ &\le \frac{1}{\pi}\int_{|\mu|>T/2}\mu^{-2}d\mu\int_{-1}^{1}|P_n(\lambda)||g(\lambda)-g(\lambda+2\mu/T)|d\lambda \\ &\quad +\frac{1}{\pi}\int_{|\mu|\le T/2} D_2^2(\mu)d\mu\int_{-1}^{1}|P_n(\lambda)||g(\lambda)-g(\lambda+2\mu/T)|d\lambda.\end{aligned} \tag{2.12}$$

The first summand on the right side of (2.11) does not exceed

$$T^{-1}\int_{-1}^{1}|P_n(\lambda)g(\lambda)d\lambda + T^{-1}\int_{-1}^{1}|P_n(\lambda)d\lambda\int_{-\infty}^{\infty} g(\mu)d\mu \le CT^{-1}.$$

If $|\lambda|\le 1$ and $|\lambda+2\mu/T|\le 1$, then $|g(\lambda)-g(\lambda+2\mu/T)|\le C|\mu|T^{-1}$. Hence the second summand in (2.11) does not exceed

$$\frac{C}{T}\int_{-T/2}^{T/2}|\mu|D_2^2(\mu)d\mu + C\int_{-T/2}^{T/2} D_2^2(\mu)d\mu\int_{\{1-2|\mu|/T\le\lambda\le 1\}}|P_n(\lambda)|d\lambda \le \frac{C}{\sqrt{T}}.$$

The lemma is proved. ☐

LEMMA 2.6. *The variance of* $\hat{a}_n$

$$\mathbf{E}|\hat{a}_n - \mathbf{E}\hat{a}_n|^2 \le \frac{C}{T}. \tag{2.13}$$

Proof. It follows from (2.9) that

$$\begin{aligned}&\mathbf{E}|\hat{a}_n - \mathbf{E}\hat{a}_n|^2 \\ &= \frac{4}{\pi^2T^2}\int_{-1}^{1}\int_{-1}^{1} P_n(\lambda)P_n(\mu)d\lambda d\mu\Big(\int_{-\infty}^{\infty} D_T(\lambda-l)D_T(\mu-l)g(l)dl\Big)^2 \\ &\quad+\frac{4}{\pi^2T^2}\int_{-1}^{1}\int_{-1}^{1} P_n(\lambda)P_n(\mu)d\lambda d\mu\Big(\int_{-\infty}^{\infty} D_T(\lambda-l)D_T(\mu+l)g(l)dl\Big)^2 \\ &= J_1 + J_2.\end{aligned} \tag{2.14}$$

We bound the first summand on the right, the second one can be bounded in the same way. We have

$$J_1 \leq \frac{8}{\pi^2 T^2}\left\{\int_{-1}^{1}\int_{-1}^{1} P_n(\lambda)P_n(\mu)d\lambda d\mu\Big(\int_{\{|l|\geq 2\}} D_T(\lambda-l)D_T(\mu-l)g(l)dl\Big)^2 + \Big(\int_{\{|l|\leq 2\}} D_T(\lambda-l)D_T(\mu-l)g(l)dl\Big)^2\right\} = J_{11}+J_{12}.$$

Further

$$J_{11} \leq CT^{-2}\int_{-1}^{1}|P_n(\lambda)|^2 d\lambda\Big(\int_{-\infty}^{\infty} g(l)dl\Big)^2 \leq CT^{-2}$$

and

$$\begin{aligned} J_{12} &\leq \frac{C}{T^2}\int_{-2}^{2}\int_{-2}^{2} g(l_1)g(l_2)\Big(\int_{-1}^{1} P_n(\lambda)P_n(\mu)D_T(\lambda-l_1)D_T(\lambda-l_2)d\lambda\Big)^2 \\ &\leq \frac{C(\max_{|l|\leq 2} f(l))^2}{T^2} \\ &\quad\times\int_{-1}^{1}\int_{-1}^{1} P_n(\lambda)P_n(\mu)d\lambda d\mu\Big(\int_{-\infty}^{\infty} D_T(\lambda-l)D_T(\mu-l)dl\Big)^2. \end{aligned}$$

But

$$\int_{-\infty}^{\infty} D_T(x-l)D_T(y-l)dl = \pi D_T(x-y).$$

Thus

$$\begin{aligned} J_{12} &\leq CT^{-1}\int_{-1}^{1}|P_n(\lambda)|d\lambda\int_{-1}^{1}|P_n(\mu)|T^{-1}D_T^2(\lambda-\mu)d\mu \\ &\leq CT^{-1}||P_n||^2 = CT^{-1} \end{aligned}$$

and the lemma follows.

Prove now theorem 1.1. Consider estimators

$$f_N(\lambda) = \sum_0^N \hat{a}_n P_n(\lambda), \quad N = 0,1,\dots. \tag{2.15}$$

The sequence $\{P_n\}$ is orthonormal in $L_2[-1,1]$ and

$$\begin{aligned} \mathbf{E}\{||f-f_N||^2\} &= \sum_0^N \mathbf{E}|\hat{a}_n - a_n|^2 + \sum_{N+1}^{\infty}|a_n|^2 \\ &\leq 2\sum_0^N |\mathbf{E}\hat{a}_n - a_n|^2 + 2\sum_0^N \mathbf{E}|\hat{a}_n - \mathbf{E}\hat{a}_n|^2 + \sum_{N+1}^{\infty}|a_n|^2. \end{aligned}$$

Lemma 2.2 implies that

$$\sum_{N+1}^{\infty} |a_n|^2 \le C \exp\Big\{ -\frac{2}{\rho} N \ln N \Big\}.$$

It follows from this inequality and Lemmas 2.5 and 2.6 that

$$\mathbf{E}\{||f - f_N||^2\} \le C\Big(NT^{-1} + \exp\Big\{ -\frac{N \ln N}{\rho} \Big\}\Big).$$

Take $N \sim \frac{\rho \ln T}{\ln \ln T}$ and put $\hat{f} = f_N$. We find that

$$\mathbf{E}\{||\hat{f} - f||^2\} \le \frac{C}{T} \frac{\ln T}{\ln \ln T}.$$

The theorem is proved. □

Remark. Suppose that the observation process is not band limited.We may conjecture (again because of the unicity theorem) that then the bound can be ameliorated. Indeed this is the case. For example suppose that $\rho = 1$ and $\sigma > 0$ is known.. For $|t| \le \sigma$ consider the following estimator $\hat{R}(t)$ for the correlation function $R(t)$ of the observation process

$$\hat{R}(t) = \frac{1}{T-t} \int_0^{T-t} X(t+s)X(s)ds,\ 0 \le t \le \sigma, \quad \hat{R}(t) = \hat{R}(-t),\ -\sigma \le t \le 0.$$

If $|t| > \sigma$, we set $\hat{R}(t) = 0$. Consider as an estimator for $f(\lambda)$

$$\hat{f}(\lambda) = \frac{1}{2\pi} \int_{-\sigma}^{\sigma} e^{-it\lambda} \hat{R}(t) dt.$$

It is rather easy to show that

$$\mathbf{E}||\hat{f} - f||_{L_2(-1,1)} \asymp \mathbf{E}||\hat{f} - f||_\infty \asymp CT^{-1}.$$

Comparing the last formula with (1.7) we see how large are losses due to band limitation.

Proof of Theorem 1.2. Consider at first the estimators f_n defined in (2.14). Let $L \ge 1$. Because of Lemmas 2.1, 2.2 and 2.6 we will have that

$$\begin{aligned} \Delta(L) &= \mathbf{E} \sup_{|\lambda \le L} |f_N(\lambda) - f(\lambda)| \\ &\le \frac{C}{\sqrt{T}} \sum_0^N \sqrt{n} |L + \sqrt{|L^2 - 1|}|^n \\ &\quad + \sum_{N+1}^{\infty} \exp\Big\{ -\frac{n}{\rho} \ln \frac{n}{e\rho\sigma} \Big\} |L + \sqrt{|L^2 - 1|}|^n. \end{aligned} \tag{2.16}$$

It follows that

$$\Delta(L) \le (CL)^N \Big(T^{-1/2} + \exp\Big\{ -\frac{N}{\rho} \ln N \Big\} \Big).$$

Take $N \sim \frac{\varrho}{2} \frac{\ln T}{\ln \ln T}$ and set $\hat{f} = f_N$. Then if $L \le C(\ln T)^{\frac{\alpha}{\rho}}$, $0 < \alpha < 1$,

$$\mathbf{E} \sup_{|\lambda \le L} \big| \hat{f}(\lambda) - f(\lambda) \big| \le C \exp\Big\{ -\frac{\ln T}{2} \Big(1 - \rho \frac{\ln LC}{\ln \ln T} \Big) \Big\} \le C T^{\frac{1-\alpha}{2}}. \quad (2.17)$$

The theorem is proved. □

3. Lower bounds. Proof of Theorem 1.3. Below we restrict ourself to the observations (1.1) with the transfer functions of the following form

$$a(\lambda) = \begin{cases} 1, & \text{if } |\lambda| \le 1, \\ \varepsilon \Big(\dfrac{\sin \lambda}{\lambda} \Big), \ 0 < \varepsilon < 1, & \text{if } |\lambda| > 1. \end{cases} \quad (3.1)$$

Denote $\mathcal{A}_0$ the set of these transfer functions.

Take a number L, $|L| > 1$. Our initial problem is the problem of estimation of the value $f(L)$ based on the observations (1.5) with $a \in \mathcal{A}_0$. Consider together with the initial problem an additional one-parametric estimation problem of the following type. Define the following one-parametric family $\{f_\theta\}$ of spectral densities

$$f_\theta(\lambda) = \varphi_L(\lambda)(1 + \theta \psi_L(\lambda)), \quad \theta \in [0, 1], \quad (3.2)$$

where $\varphi_L \in \mathbf{G}$ and $\psi_L \in \mathbf{G}$ are two given functions such that $\varphi_L(L)\psi_L(L) \ge 1/2$. The final choice of these functions will be made later. Below we usually omit the index L and write φ, ψ instead of φ_L, ψ_L. Suppose that the initial process $X(t)$ has a spectral density f_θ and consider the problem of estimation of the parameter θ on the base of observations (1.5) with a *known* transfer function $a \in \mathcal{A}_0$.

Let $\hat{f}$ be an estimator of the value $f(L)$, $f \in \mathbf{G}$, in the initial problem. Because of

$$\theta = \frac{f_\theta(L) - \varphi(L)}{\varphi(L)\psi(L)}$$

we may consider as an estimator for θ in the additional problem

$$\hat{\theta} = \frac{\hat{f} - \varphi(L)}{\varphi(L)\psi(L)}.$$

Evidently

$$\mathbf{E}_\theta\{|\hat{\theta} - \theta|^2\} = \frac{\mathbf{F}_{f_\theta} |\hat{f} - f_\theta(L)|^2}{\varphi^2(L)\psi^2(L)} \quad (3.3)$$

and hence

$$\inf_{\hat{f}} \sup_{f \in \mathbf{G}, a \in \mathcal{A}} \mathbf{E}^{\theta}|\hat{f} - f_{\theta}(L)|^2 \geq \varphi(L)\psi(L) \inf_{\hat{\theta}} \sup_{\theta, a \in \mathcal{A}_0} \mathbf{E}|\hat{\theta} - \theta|^2. \tag{3.4}$$

It follows that we prove the theorem 1.3 if we can establish the inequality

$$\sup_{a \in \mathcal{A}_0} \sup_{\theta} \mathbf{E}_{\theta}|\hat{\theta} - \theta|^2 \geq c_0 > 0. \tag{3.5}$$

We see later that the additional estimation problem corresponds to a regular statistical experiment in the sense of [11], sect. 1.7. Hence the Fisher information $I(\theta)$ of the additional problem is well defined and by Cramér – Rao's inequality (see [11], Sect. 1.7): for any estimator $\hat{\theta}$ with $\sup_{\theta} \mathbf{E}_{\theta}|\hat{\theta} - \theta|^2 < \infty$

$$\mathbf{E}_{\theta}|\hat{\theta} - \theta|^2 \geq \frac{(1 + d'(\theta))^2}{I(\theta)} + d^2(\theta). \tag{3.6}$$

Here $d(\theta) = \mathbf{E}_{\theta}\hat{\theta} - \theta$ is the bias of $\hat{\theta}$. It follows from (3.6) that if

$$\sup_{\theta} I(\theta) \leq 1, \tag{3.7}$$

then

$$\sup_{\theta} \mathbf{E}_{\theta}|\hat{\theta} - \theta|^2 \geq c_1 > 1/16. \tag{3.8}$$

In the next part of the section we prove the inequality (3.7). Because of (3.4), (3.5), (3.8) it will prove the theorem.

Because the interval $G_T(\alpha)$ of theorem 1.3 depends on ρ only and to simplify the proof we will not try to construct spectral densities with given σ, ρ, M, A but satisfy ourselves by constructing $f \in \mathbf{G}$ with given ρ and *some* σ, M, A. In other words we prove the following weakened version of Theorem 1.3: for a given ρ one can find σ, M, A such that for $f \in \mathbf{G}(M, \sigma, \rho, A)$ the inequality (1.9) will be satisfied (of course these σ, M, A will not depend on the interval $G_T(\alpha)$).

Return to the definition of the functions φ, ψ determining the family (3.2). Let $\varphi_0(\lambda) = (\sin(\lambda + i)(\lambda + i)^{-1})(\sin(\lambda - i)(\lambda - i)^{-1})$. Set

$$\varphi_L(\lambda) = \varphi(\lambda) = \varphi_0(\lambda - L) + \varphi_0(\lambda + L). \tag{3.9}$$

The function $\varphi(z)$ is an integer function. It is easy to see that

$$|\varphi_L(z)| \leq Ce^{2|z|}$$

where the constant C does not depend on L.

Let now $\psi_0(z)$ be a function from $\mathbf{F}(\sigma, M, \rho)$ with some σ, M and let the function ψ_0 satisfy the following conditions:

on the real line

$$0 \leq \psi_0(\lambda) \leq e^{-|\lambda|^\rho}, \ \psi_0(\lambda) = \psi_0(-\lambda), \ \psi_0(0) = 1; \tag{3.10}$$

for all $L \in R^1$

$$|\psi_0(z - L)| \leq C \exp\{c|z|^\rho\} \tag{3.11}$$

where the constants C, c do not depend on L.

The existence of such functions will be proved at the end of the section. We define now ψ_L as

$$\psi_L(z) = \psi_0(z + L) + \psi_0(z - L). \tag{3.12}$$

Under such choice of φ, ψ all the functions f_θ will belong to $\mathbf{G}$ (with a given ρ and some σ, M, A determined by φ_0, ψ_0).

The observation process $Y(t)$ of the additional problem determines on the Hilbert space $L_2(0,T) = L_T$ the family of Gaussian measures P_θ. We show that all these measures are absolutely continuous one with respect to other. To do this we apply some general results on absolute continuity of Gaussian measures in Hilbert spaces.

LEMMA 3.1 (see [7], Th. 2, Sect. 4, Ch. VII). *Let P_1 and P_2 be two Gaussian measures with mean values zero and correlation operators $\mathbf{R}_1$ and $\mathbf{R}_2$ correspondingly defined in a Hilbert space H. In order that these measures be equivalent, it is necessary and sufficient that the operator $D = \mathbf{R}_2^{-1/2}\mathbf{R}_1\mathbf{R}_2^{-1/2} - \mathbf{I}$, $\mathbf{I}$ be the identity operator , be a Hilbert - Schmidt operator and its eigenvalues δ_k satisfy the inequality $\delta_k > -1$. If the measures P_1, P_2 are equivalent, the Radon - Nykodim derivative*

$$\frac{dP_2}{dP_1}(x) = \exp\left\{ -\frac{1}{2}\sum_k \left[(\mathbf{R}_2^{-1/2}x, e_k)^2 \frac{\delta_k}{1+\delta_k} - \ln(1+\delta_k) \right] \right\} \tag{3.13}$$

where e_k are the eigenfunctions of the operator $\mathbf{D}$ corresponding to the eigenvalues δ_k.

In the case when P is a Gaussian measure on $L_T = L_2(0, T)$ generated by a stationary process $X(t)$ with the correlation function $B(t)$ and the spectral density $b(\lambda)$ the correlation operator is defined on L_T by the formula

$$(\mathbf{B}u)(t) = \int_0^T B(t-s)u(s)ds.$$

To the operator $\mathbf{B}$ corresponds the bilinear form

$$(\mathbf{B}u, v)_T = \int_0^T \int_0^T B(t-s)u(t)\overline{u(s)}dtds. \tag{3.14}$$

Continue a function $u \in L_T$ on the whole real line setting $u(t) = 0$, $t \notin [0, T]$ and denote $\tilde{u}(\lambda)$ the Fourier transform of the continued function,

$$\tilde{u}(\lambda) = \int_0^T e^{it\lambda} u(t) dt.$$

We can rewrite then the form $(\mathbf{B}u, v)_T$ in the terms of $b(\lambda)$, namely

$$(\mathbf{B}u, v)_T = \int_{-\infty}^{\infty} b(\lambda) \tilde{u}(\lambda) \overline{\tilde{v}(\lambda)} d\lambda. \tag{3.15}$$

Denote E_T the class of integer analytic functions $\varphi(z)$ square summable on the real line and satisfying the inequality

$$|\varphi(z)| \leq C e^{T|z|}$$

(integer functions which satisfy the last inequality are called integer functions of exponential type T, see their theory in [3] or [12].)

The famous Paley–Wiener theorem (see, for example, [3], Th. 6.8.1) asserts that the class E_T coincides with the class of functions u which can be represented as

$$u(\lambda) = \int_{-T}^{T} e^{it\lambda} v(t) dt, \quad v \in L_2(-T, T).$$

In particular, the unit ball determined by the form $(\mathbf{B}u, u)_T : \{u \in L_T : (\mathbf{B}u, u) \leq 1\}$ corresponds to the set $\{u \in E_T : \int_{-\infty}^{\infty} b(\lambda) |u(\lambda)|^2 d\lambda \leq 1\}$.

LEMMA 3.2 (see [7], Th. 3, Sect. 5, Ch. VII). *Let P_1 and P_2 be two Gaussian measures generated by two stationary Gaussian processes ξ_1, ξ_2 possessing spectral densities $f_1(\lambda), f_2(\lambda)$ correspondingly. Suppose that the spectral densities f_1, f_2 satisfy the following conditions:*

1. there exists a function $\varphi_0 \in E_\sigma$ and positive constants c_1, c_2 such that

$$c_1 |\varphi_0(\lambda)|^2 \leq f_1(\lambda) \leq c_2 |\varphi_0(\lambda)|^2, \ \lambda \in R^1;$$

2.

$$\int_{-\infty}^{\infty} \left| \frac{f_2(\lambda) - f_1(\lambda)}{f_1(\lambda)} \right| d\lambda < \infty.$$

Then the restrictions P_{1T}, P_{2T} of the measures P_1, P_2 on L_T are equivalent for any positive T.

The observation process of the additional problem has as the possible spectral densities the functions

$$g_\theta(\lambda) = |a(\lambda)|^2 \varphi(\lambda)(1 + \theta\psi(\lambda))$$

and it follows from the last lemma that for all θ the measures P_0, P_θ are equivalent for all $T > 0$.

Denote $R_\theta(t)$ the correlation function corresponding to the spectral density $g_\theta(\lambda)$ and let $\mathbf{R}_\theta$ denote the corresponding correlation operator. Let $h(\lambda) = \varphi(\lambda)|a(\lambda)|^2\psi(\lambda)$. The Fourier transform of this function

$$H(t) = \int_{-\infty}^{\infty} e^{it\lambda} h(\lambda)d\lambda$$

is a correlation function with the corresponding correlation operator $\mathbf{H}$ on L_T,

$$\mathbf{H}u(t) = \int_0^T H(t-s)u(s)ds.$$

Thus the operators

$$\mathbf{R}_\theta = \mathbf{R}_0 + \theta\mathbf{H}$$

and the operators $\mathbf{D}_\theta$ corresponding to the operator $\mathbf{D}$ of Lemma 6 are

$$\mathbf{D}_\theta = \mathbf{R}_0^{-1/2}\mathbf{R}_\theta\mathbf{R}_0^{-1/2} - \mathbf{I} = \theta\mathbf{R}_0^{-1/2}\mathbf{H}\mathbf{R}_0^{-1/2} = \theta\mathbf{D}.$$

In particular, the eigenvalues of $\mathbf{D}_\theta$ are equal to $\theta\delta_k$ where δ_k are the eigenvalues of $\mathbf{D}$. It follows then from Lemma 3.1 that the density function

$$\begin{aligned} p_\theta(x) &= \frac{dP_\theta}{dP_0}(x) \\ &= \exp\left\{\frac{1}{2}\sum\left\{(\mathbf{R}_0^{-1/2}x, e_k)_T^2\frac{\theta\delta_k}{1+\theta\delta_k} - \ln(1+\theta\delta_k)\right\}\right\} \end{aligned} \tag{3.16}$$

where e_k are the eigenfunctions of the operator $\mathbf{D}$.

LEMMA 3.3. *The Fisher information*

$$I(\theta) = \mathbf{E}_\theta\left\{\left(\frac{d}{d\theta}\ln p_\theta(Y)\right)^2\right\} = \frac{1}{2}\sum_k \frac{\delta_k^2}{(1+\theta\delta_k)^2}. \tag{3.17}$$

Proof. Apply (3.16). We get that

$$\frac{d}{d\theta}\ln p_\theta(x) = \frac{1}{2}\sum_k \frac{\delta_k}{1+\theta\delta_k}\left\{\frac{(\mathbf{R}_0^{-1/2}x, e_k)^2}{1+\theta\delta_k} - 1\right\}$$

and hence

$$I(\theta) = \mathbf{E}_\theta\left|\frac{d}{d\theta}\ln p_\theta(Y)\right|^2 = \frac{1}{4}\mathbf{E}_\theta\left|\sum_k \frac{\delta_k}{1+\theta\delta_k}\left[\frac{(\mathbf{R}_0^{-1/2}Y, e_k)^2}{1+\theta\delta_k} - 1\right]\right|^2. \tag{3.18}$$

To compute the last expectation we apply the following evident result:

if Z is a Gaussian stationary process with the correlation operator $\mathbf{B}$, then for all $v \in L_T$ random variables $(Z, v)_T$ are Gaussian and

$$\mathbf{E}(Z,v)_T = \mathbf{E}Z(0)\cdot(1,v)_T,\ \mathbf{E}(Z,u)_T(Z,v)_T = (\mathbf{B}u,v)_T.$$

It follows that under P_θ the random variables $\xi_k = (\mathbf{R}_0^{-1/2}Y, e_k)_T = (Y, \mathbf{R}_0^{-1/2}e_k)_T$ are Gaussian with means zero and correlations

$$\mathbf{E}_\theta(\xi_k\xi_l) = \big((\mathbf{R}_0+\theta\mathbf{H})\mathbf{R}_0^{-1/2}e_k, \mathbf{R}_0^{-1/2}e_l\big) = \big((\mathbf{I}+\mathbf{D}_\theta)e_k, e_l\big) = (1+\theta\delta_k)\delta_{kl}$$

where the last δ_{kl} is the Kroneker symbol. Thus the Gaussian random variables ξ_k are independent with means zero and variances $\mathbf{E}\xi_k^2 = 1+\theta\delta_k$. The equality (3.18) gives then that

$$I(\theta) = \frac{1}{4}\sum_k \frac{\delta_k}{(1+\theta\delta_k)^2}\mathbf{E}_\theta\left|\frac{(\mathbf{R}_0^{-1/2}Y, e_k)^2}{1+\theta\delta_k} - 1\right|^2 = \frac{1}{2}\sum_k \frac{\delta_k}{(1+\theta\delta_k)^2}.$$

The lemma is proved. It follows from the lemma that

$$I(\theta) \le \frac{1}{2}\sum_k \delta_k^2 \le \frac{1}{2}\max_k \delta_k \sum_k \delta_k.$$

□

Below we study two last factors.

LEMMA 3.4. *The eigenvalues δ_k of the operator $\mathbf{D}$ satisfy the inequality*

$$\max_k \delta_k \le C\big(e^{-(L/2)^\rho} + \varepsilon^2 e^{36LT\ln 9L} + e^{-tL}\big) \tag{3.19}$$

where ε is defined by (3.1). The constant C does not depend on L and T.

Proof. We have

$$\max\delta_k = \sup_{\|u\|_T=1}(\mathbf{D}u,u)_T = \sup_{\|u\|_T=1}(\mathbf{R}_0^{-1/2}\mathbf{H}\mathbf{R}_0^{-1/2}u,u)_T = \sup_v(\mathbf{H}v,v)_T$$

where the last upper bound is taken over all $v \in L_T$ such that $(\mathbf{R}_0 v, v)_T = 1$. Taking into account the relation (3.15) we find that

$$\begin{aligned}\max\delta_k &= \sup_v \int_{-\infty}^{\infty} h(\lambda)|v(\lambda)|^2 d\lambda \\ &= \sup_v \int_{-\infty}^{\infty} |a(\lambda)|^2\psi(\lambda)\varphi(\lambda)|v(\lambda)|^2 d\lambda\end{aligned} \tag{3.20}$$

where the upper bound is taken over all $v \in E_T$ such that

$$\int_{-\infty}^{\infty} |a(\lambda)|^2\varphi(\lambda)|v(\lambda)|^2 = 1. \tag{3.21}$$

Further a non-negative function $u \in E_\sigma$ can be represented as $|v(\lambda)|^2$, $v \in E_{\sigma/2}$ (see [12]). Thus

$$\max \delta_k \le \sup_v \int_{-\infty}^{\infty} \psi(\lambda)|a(\lambda)|^2|v(\lambda)|^2 d\lambda$$

and the upper bound is taken over all $v \in E_{T+1}$ such that

$$\int_{-\infty}^{\infty} |a(\lambda)|^2|v(\lambda)|^2 d\lambda = \int_{-1}^{1} |v(\lambda)|^2 d\lambda + \varepsilon^2 \int_{|\lambda|>1} \left|\frac{\sin\lambda}{\lambda} v(\lambda)\right|^2 d\lambda \le 1. \quad (3.22)$$

The properties (3.10) (3.11) of the function ψ imply that

$$\max \delta_k \le C\big(\exp\{-(L/2)^\rho\}\big) + \varepsilon^2 \sup_v \max_{3L/2\ge|\lambda|\ge L/2} \left|\frac{\sin\lambda}{\lambda} v(\lambda)\right|^2. \quad (3.23)$$

To estimate the last summand on the right hand side, take a function v satisfying (3.22) and set $v_1(z) = v(z)\frac{\sin z}{z}$. Then $v_1 \in E_{T+2}$ and satisfies the inequalities

$$\int_{-1}^{1} |v_1(\lambda)|^2 d\lambda \le 1, \int_{-\infty}^{\infty} |v_1(\lambda)|^2 d\lambda \le \varepsilon^{-2}. \quad (3.24)$$

The second inequality together with Paley – Wiener theorem implies that

$$|v_1(z)| \le \varepsilon^{-1} e^{|\Im z|(T+2)} \le \varepsilon^{-1} e^{(T+2)|z|}. \quad (3.25)$$

Represent the function $v_1(\lambda)$ by its expansion with respect to the Legendre polynomials (see Sect. 2). We get that $v_1(\lambda) = \sum_n a_n P_n(\lambda)$ and (see Lemma 2.1)

$$\sum_n |a_n|^2 \le 1, \quad |a_n| \le C\varepsilon^{-1} \exp\{-n \ln n/eT\}.$$

Hence (see Lemma 2.2) for $L/2 \le |\lambda| \le 3L/2$

$$\begin{aligned}|v(\lambda)| &\le \left(\sum_0^N |P_n(\lambda)|^2\right)^{1/2} + C\varepsilon^{-1} \sum_{N+1}^{\infty} |a_n||P_n(\lambda)|^2 \\ &\le (9L)^N + C\varepsilon^{-1}(9L)^N e^{-N \ln N/eT}.\end{aligned}$$

Take here $N \asymp 18TL$. We find that

$$\varepsilon \max_{L/2\le|\lambda|\le 3L/2} |v(\lambda)(\sin\lambda)\lambda^{-1}| \le C\big(\varepsilon \exp\{18TL \ln 9L\} + e^{-TL}\big). \quad (3.26)$$

□

LEMMA 3.5. *The following inequality holds*

$$tr\mathbf{D} = \sum_k \delta_k \le C(T + \ln^2 \varepsilon^{-1}). \quad (3.27)$$

Proof. The trace of the operator $\mathbf{D} = \mathbf{R}_0^{-1/2}\mathbf{H}\mathbf{R}_0^{-1/2}$ is

$$tr\mathbf{D} = \sum_k (\mathbf{D}u_k, u_k)_T$$

where $\{u_k\}$ is an orthonormal basis in L_T. If we consider such basis for which $v_k = \mathbf{R}_0^{1/2}u_k$ are defined, we find that

$$tr\mathbf{D} = \sum_k (\mathbf{H}v_k, v_k)_T$$

where $\{v_k\}$ are orthonormal with respect to the bilinear form corresponding to the operator $\mathbf{R}_0$, $(\mathbf{R}_0 v_k, v_l)_T = \delta_{kl}$. Arguing as above, rewrite the forms $(\mathbf{H}v, v)_T, (\mathbf{R}_0 v, v)_T$ in the terms of the corresponding spectral densities. We find that

$$tr\mathbf{D} = \sum_k \int_{-\infty}^{\infty} \psi(\lambda)|a(\lambda)|^2|v(\lambda)|^2 d\lambda \tag{3.28}$$

where the functions $v_k \in E_{T+1}$ and satisfy the conditions

$$\int_{-\infty}^{\infty} |a(\lambda)|^2 v_k(\lambda)v_l(\lambda)d\lambda = \delta_{kl}. \tag{3.29}$$

Setting $a(\lambda)v_k(\lambda) = r_k(\lambda)$ we find that

$$tr\mathbf{D} = \int_{-\infty}^{\infty} \psi(\lambda)\sum_k r_k^2(\lambda)d\lambda \tag{3.30}$$

and the set of the functions $\{r_k\}$ is an orthonormal system in the subspace $a\dot{E}_{T+1}$ of $L_2(-\infty,\infty)$

We estimate now the sum $\sum_k r_k^2(\lambda)$. Let

$$r_j(\lambda) = \int_{-\infty}^{\infty} e^{i\lambda x}\tilde{r}_j(x)dx.$$

Then

$$\sum_k r_k^2(\lambda) = \frac{1}{2\pi}\sum_k\left(\int_{-\infty}^{\infty} e^{i\lambda x}\sqrt{2\pi}\tilde{r}_k(x)dx\right)^2$$

and the functions $\{\sqrt{2\pi}\tilde{r}_k\}$ constitute an orthonormal system in $L_2(-\infty,\infty)$. It follows that

$$\sum_k r_k^2(\lambda) \le \frac{1}{2\pi}\sup_r\left|\int_{-\infty}^{\infty} e^{i\lambda x}r(x)dx\right|^2 \tag{3.31}$$

and the upper bound is taken over all $r, ||r|| = 1$, which are the Fourier transforms of functions represented as av, $v \in E_{T+1}$. Hence

$$\sum_k r_k^2(\lambda) \le |a(\lambda)|^2 \sup_v |v(\lambda)|^2 \tag{3.32}$$

where upper bound is taken over all $v \in E_{T+1}$ under the condition

$$\int_{-\infty}^{\infty} |a(\lambda)|^2 |v(\lambda)|^2 d\lambda = 1. \tag{3.33}$$

Consider now separately two cases: $|\lambda| > 1$ and $|\lambda| \le 1$.

Let $|\lambda| > 1$. It follows from (3.33) that

$$\int_{-\infty}^{\infty} \frac{\sin^2 \lambda}{\lambda^2} |v(\lambda)|^2 d\lambda \le \varepsilon^{-2}.$$

Hence

$$\left| \frac{\sin \lambda}{\lambda} v(\lambda) \right| \le \frac{\sqrt{T+2}}{\varepsilon}$$

and

$$\sum_k r_k^2(\lambda) \le (|a(\lambda)| \sup_v |v(\lambda)|)^2 \le T+2.$$

Let now $|\lambda| \le 1$. Functions $v_1(\lambda) \frac{\sin \lambda}{\lambda}$ satisfy the inequalities

$$\int_{-1}^{1} |v_1(\lambda)|^2 d\lambda \le 1, \quad \int_{-\infty}^{\infty} |v_1(\lambda)|^2 d\lambda \le \varepsilon^{-2}.$$

Hence expanding the function $v_1(\lambda)$ with respect to the Legendre polynomials (see Sect.2) we find that for $N > eT$

$$|v_1(\lambda)| \le \left(\sum_0^N |P_n(1)|^2 \right)^{1/2} + \frac{C}{\varepsilon} \sum_{N+1} e^{-n \ln n/eT} \le C(N + \varepsilon^{-1} e^{-N}). \tag{3.34}$$

Take here $N \sim \ln 1/\varepsilon$. We get that

$$|v(\lambda)| \le C \ln \varepsilon^{-1}. \tag{3.35}$$

□

Now we are ready to prove the inequality (3.7).

LEMMA 3.6. *Let the numbers ε, L in the additional problem satisfy the relation*

$$\ln \varepsilon^{-1} = 18LT \ln 9L - LT. \tag{3.36}$$

Then for sufficiently large T and $L > (\ln T)^{\alpha/\rho}$, $\alpha > 1$, the Fisher information of the additional problem satisfies the inequality

$$I(\theta) \leq C\big(Te^{(-L/4)^{\rho}} + e^{-LT/2}\big) < 1. \tag{3.37}$$

The lemma is an immediate corollary of the inequalities (3.19) and (3.27). We have noticed above that the inequality (3.7) proves the theorem. Thus to finish the proof we have only to construct the function ψ_0 satisfying the conditions (3.10), (3.11). The construction coincides in principle with analogous constructions in papers [9], [10] and we omit the details. We will distinguish three cases.

The first and the simplest case is when ρ is an even integer, $\rho = 2k$. We set in this case $\psi_0(z) = \exp\{-z^{2k}\}$. Evidently this function satisfies (3.10). The function $\psi_L(z) = \psi_0(z - L)$ is an integral function and for $|z| < L/4k$ $\Re(z-L)^2k > 0$ and hence $|\psi_L(z)| < 1$. If $|z| > L/4k$, then $|z - L|^{2k} \leq 2^{2k-1}(|z|^{2k} + L^{2k}) \leq (C|z|)^{2k}$.

If $\rho = 2k+1$, $k \geq 0$, is an odd integer, we set

$$\psi_0(z) = \left(\Pi_{n=1}^{\infty} \frac{\sin(z^{\rho} n^{-(1+\delta)})}{z^{\rho} n^{-(1+\delta)}}\right)^2, \quad \delta > 0.$$

Applying the Stirling formula we get that

$$|\psi_0(\lambda)| \leq \left(\Pi_{n<|\lambda|^{\rho/(1+\epsilon)}}\big(|\lambda| n^{-(1+\delta)}\big)^{\rho}\right)^{-2} \leq Ce^{-c|\lambda|^{\rho}}.$$

Further

$$\left|\frac{\sin z}{z}\right| \leq |z|^{-1} e^{|\Im z|}$$

and hence for $|z| \leq \alpha|L|$ with α being a small positive number,

$$|\psi_L(z)| \leq C\exp\{-c_1|z-L|^{\rho} + c_2|\Im(z-L)^{\rho}|\} \leq C.$$

For $|z| > \alpha|L|$ evidently $|\psi_L(z)| \leq Ce^{c|z|^{\rho}}$.

The case of non integer ρ is more complicated. Roughly speaking the construction is the following one. Define the integer p from the relations $p < \rho < p+1$. Consider at first the function

$$V(z) = \Pi_{r=1}^{\infty} G\big(\frac{z}{r^{1/\rho}}, p\big)$$

where

$$G(u,p) = (1-u)\exp\left\{u + \frac{u^2}{2} + \cdots + \frac{u^p}{p}\right\}.$$

Take $\alpha = \frac{3\pi}{2\rho}$ and set $V_\alpha(z) = V(ze^{-i\alpha})$. The function ψ_0 is then defined as follows

$$\psi_0(z) = V_\alpha(z)V_{-\alpha}(z)V_\alpha(-z)V_{-\alpha}(-z).$$

For the proof that this function satisfies all the necessary conditions, see [10].

This work has been started in summer of 2006 when I visited the Bielefeld University and finished January of 2007 again in Bielefeld. I am very thankful to Prof. F. Goetze for his hospitality during my visit. I would like to thank the secretaries Mrs. A.L. Cole and Mrs. N. Epp for their help during the visits.

REFERENCES

[1] ANDERSON T.W., The statistical analysis of time series, J. Wiley, 1971.
[2] BERNSTEIN S.N., Extremal properties of polynomials, ONTI, Moscow, 1937 (in Russian).
[3] BOAS R., Entire functions, Academic Press, 1954.
[4] CHENTSOV N.N., Estimation of unknown probability density based on observations, *Dokl. Akad. Nauk SSSR* (1962), **147**: 45–48 (in Russian).
[5] DOOB J.L., Stochastic processes, J. Wiley, 1953.
[6] GRENANDER U., Abstract inference, J. Wiley, 1981.
[7] GIHMAN I. AND SKOROHOD A., The theory of stochastic processes I, Springer, 1974.
[8] IBRAGIMOV I., On the extrapolation of entire functions observed in a Gaussian white noise, *Ukrainian Math. J.* (2000), **52**: 1383–1395.
[9] IBRAGIMOV I., An estimate for the analytic density of distribution, based on a censored sample, *J. Math. Sci.* (2006), **133**: 1290–1297.
[10] IBRAGIMOV I., On censored sample estimation of a multivariate analytic probability density, *Theory Probab. Appl.* (2007), **51**: 1–13.
[11] IBRAGIMOV I.A. AND KHASMINSKII R.Z., Statistical estimation, Asymptotic theory, Springer, 1981.
[12] LEVIN B., Distribution of zeros of entire functions, Amer. Math. Soc., Providence, RI, 1964.
[13] MARUYAMA G., The harmonic analysis of stationary stochastic processes, *Mem. Fac. Sci. Kyusyn Univ. A*, **4**: 45–106.
[14] SZEGO G, Orthogonal polynomials, Amer. Math. Soc. Colloq. Publ., N.Y., 1967.
[15] WALSH J.L., Approximation by bounded analytic functions, Gauthier–Villars, Paris, 1960.

ON ORACLE INEQUALITIES RELATED TO HIGH DIMENSIONAL LINEAR MODELS

YURI GOLUBEV*

Abstract. We consider the problem of estimating an unknown vector θ from the noisy data $Y = A\theta + \epsilon$, where A is a known $m \times n$ matrix and ϵ is a white Gaussian noise. It is assumed that n is large and A is ill-posed. Therefore in order to estimate θ, a spectral regularization method is used and our goal is to choose a spectral regularization parameter with the help of the data Y. We study data-driven regularization methods based on the empirical risk minimization principle and provide some new oracle inequalities related to this approach.

Key words. Spectral regularization, excess risk, ordered smoothers, empirical risk minimization principle, oracle inequality.

AMS(MOS) subject classifications. Primary 62G05, 62G20.

1. Introduction and main results. In this paper, we deal with a classical problem of recovering an unknown vector $\theta = (\theta(1), \dots, \theta(n))^\top \in \mathbb{R}^n$ from the noisy data

$$Y = A\theta + \epsilon, \tag{1.1}$$

where A is a known $m \times n$ - matrix and $\epsilon \in \mathbb{R}^n$ is a white Gaussian noise with a known variance $\sigma^2 = \mathbf{E}\epsilon^2(k),\ k = 1, \dots, m$.

The standard way to estimate θ is based on the maximum likelihood estimator

$$\hat{\theta}_0 = \arg\min_{\theta \in \mathbb{R}^n} \|Y - A\theta\|^2,$$

where $\|x\|^2 = \sum_{k=1}^m x^2(k)$. It is easy to see that $\hat{\theta}_0 = (A^\top A)^{-1}A^\top Y$ and the mean square risk of this estimator is computed as follows

$$\begin{aligned}\mathbf{E}\|\hat{\theta}_0 - \theta\|^2 &= \mathbf{E}\|(A^\top A)^{-1}A^\top \epsilon\|^2 = \sigma^2 \text{trace}\big[(A^\top A)^{-1}\big] \qquad (1.2)\\ &= \sigma^2 \sum_{k=1}^n \lambda_k,\end{aligned}$$

where λ_k and $\phi_k \in \mathbb{R}^n$ are the eigenvalues and the eigenfunctions of $(A^\top A)^{-1}$:

$$\lambda_k A^\top A \phi_k = \phi_k.$$

In what follows, it is assumed that A is ill-posed i.e., $\lambda_1 \le \lambda_2 \le \cdots \le \lambda_n$. The equation (1.2) reveals the principal difficulty in $\hat{\theta}_0$: *its risk may be*

*CNRS and Institute for Problems of Information Transmission CMI, 39 rue F. Joliot-Curie, 13453 Marseille, France.

very large when n is large or when A has a large condition number. In this paper, we suppose that n is large (it may be infinity), so the risk of $\hat{\theta}_0$ is also large.

The basic method to improve $\hat{\theta}_0$ is to make the variance $\sigma^2 \sum_{k=1}^n \lambda_k$ smaller by suppressing large λ_k. The simplest way to implement this idea is to smooth $\hat{\theta}_0$ with the help of a properly chosen $n \times n$ - matrix H i.e., using a new estimator $H\hat{\theta}_0$. In this paper, we focus on with the following family of linear estimators

$$\hat{\theta}_\alpha = H_\alpha \hat{\theta}_0 = H_\alpha\big[(A^\top A)^{-1}\big](A^\top A)^{-1}A^\top Y,$$

where $H_\alpha(z)$ is an analytic function $H_\alpha(z) = \sum_{k=0}^\infty h_\alpha(k)z^k$ such that $\lim_{\alpha\to 0} H_\alpha(z) = 1$, $\lim_{z\to\infty} H_\alpha(z) = 0$. This method is called *spectral regularization* (see [3]). The regularization parameter α controls the quality of $\hat{\theta}_\alpha$. Indeed, with an elementary algebra we get the standard bias-variance decomposition

$$\mathbf{E}\|\hat{\theta}_\alpha - \theta\|^2 = \sum_{k=1}^n \big[1 - H_\alpha(\lambda_k)\big]^2 \langle \theta, \psi_k\rangle^2 + \sigma^2 \sum_{k=1}^n \lambda_k H_\alpha^2(\lambda_k), \quad (1.3)$$

where $\psi_k = A\phi_n/\|A\phi_k\|$ and $\langle \theta, \psi_k\rangle = \sum_{l=1}^n \theta(l)\psi_k(l)$. It is clear that the spectral regularization may substantially improve $\hat{\theta}_0$ when $\langle\theta,\psi_k\rangle^2$ is small for large k.

In practice, a good choice of $H_\alpha(\cdot)$ is a delicate problem related to the numerical complexity $\hat{\theta}_\alpha$. For instance, to make use of the spectral cut-off regularization with $H_\alpha(z) = \mathbf{1}\{\alpha z \le 1\}$, one has to compute the singular value decomposition (SVD) of A. For large n this numerical problem may be difficult or even infeasible.

The very popular Tikhonov's [7] regularization is defined by

$$\hat{\theta}_\alpha = \arg\min_\theta \Big\{ \|Y - A\theta\|^2 + \alpha\|\theta\|^2 \Big\}.$$

For this method we have $H_\alpha(z) = 1/(1+\alpha z)$. Notice here that this regularization technique is good if A is really ill-posed. Indeed in view of (1.3), $\hat{\theta}_\alpha$ may improve the banal estimator $\hat{\theta}_0$ if

$$\sum_{k=1}^n \frac{\lambda_k}{[1+\alpha\lambda_k]^2} \ll \sum_{k=1}^n \lambda_k.$$

This means for instance that for inverse problems with $\lambda_k \approx 1$ Tikhonov's regularization makes no sense.

Another widespread regularization technique is due to Landweber. This method is based on a very simple idea: to find recursively a root of equation

$$A^\top Y = A^\top A\theta.$$

Notice that for positive a we can write $A^\top Y = [A^\top A - aI]\theta + a\theta$, or equivalently $\theta = [I - a^{-1}A^\top A]\theta + a^{-1}A^\top Y$. This formula motivates Landweber's iterations defined by

$$\theta_i = [I - a^{-1}A^\top A]\theta_{i-1} + a^{-1}A^\top Y.$$

It is easy to see that these iterations converge if $a\lambda_1 < 1$. It is also easy to check that

$$H_i(z) = 1 - \left(1 - \frac{1}{az}\right)^{i+1} \tag{1.4}$$

In spite of its iterative character, the numerical complexity of Landweber's iterations may be very hight. Indeed, when the noise is small, $H_i(z)$ should be 1, and (1.4) results in

$$i > \operatorname{cond}(A) \stackrel{\text{def}}{=} \frac{\lambda_n}{\lambda_1}$$

So, if A is severely ill-posed, the number of iterations may be very large, thus making the method infeasible. A substantial improvement of Landweber's iterations is provided with the ν-method (see e.g., [3]).

Whatever an inversion method is used, the principal question is how to choose its regularization parameter. Intuitively, (see (1.3)), this parameter should minimize in some sense the risk

$$L[\alpha, \theta] = \sum_{k=1}^{n} \bigl[1 - H_\alpha(\lambda_k)\bigr]^2 \langle \theta, \psi_k \rangle^2 + \sigma^2 \sum_{k=1}^{n} \lambda_k H_\alpha^2(\lambda_k).$$

In statistics, there are two main ideas to formalize this optimization problem:

- to assume that θ belongs to a known set $\Theta \in \mathbb{R}^n$ and to take

$$\alpha^* = \arg\min_{\alpha} \sup_{\theta \in \Theta} L[\alpha, \theta]$$

- to construct based on the data an "estimate" $\hat{L}[\alpha, Y]$ of $L[\alpha, \theta]$ and to compute

$$\hat{\alpha} = \arg\min_{\alpha} \hat{L}[\alpha, Y]$$

Statistical literature related to these approaches is so vast that it would impractical to cite it here. We refer interested reader to [6] and [2] as typical representatives of its. Notice that the first approach is related to the theory of minimax estimation [4].

This paper focuses on the second approach, namely, it deals with data-driven regularization parameters computed with help of the empirical risk minimization principle. This method says that the regularization parameter should be computed as follows

$$\hat{\alpha} = \arg\min_{\alpha} R_{Pen}[Y, \alpha],$$

where

$$R_{Pen}[Y, \alpha] = \|\hat{\theta}_0 - \hat{\theta}_\alpha\|^2 + Pen(\alpha)\sigma^2 - \sigma^2 \sum_{k=1}^{n} \lambda_k,$$

and $Pen(\cdot)$ is a given function $\mathbb{R}^+ \to \mathbb{R}^+$. A heuristic motivation behind this method is rather transparent. Indeed, the best regularization parameter is obviously given by

$$\alpha^* = \arg\min_{\alpha} \|\theta - \hat{\theta}_\alpha\|^2. \tag{1.5}$$

Evidently, α^* cannot be used since it depends on θ which is unknown. So, the first idea is replace θ in (1.5) by $\hat{\theta}_0$. It is clear, that directly this idea doesn't work because $\min_\alpha \|\hat{\theta}_0 - \hat{\theta}_\alpha\|^2 = 0$. Therefore we need to correct $\|\hat{\theta}_0 - \hat{\theta}_\alpha\|^2$ by an additional term, thus arriving at $R_{Pen}[Y, \hat{\alpha}]$. Intuitively, this idea assumes that the best $Pen(\alpha)$ should be a minimal function such that uniformly in $\theta \in \mathbb{R}^n$

$$\mathbf{E}\|\theta - \hat{\theta}_{\hat{\alpha}}\|^2 \lesssim \mathbf{E} R_{Pen}[Y, \hat{\alpha}]. \tag{1.6}$$

Unfortunately, the mathematical formalization of "the best penalty" and (1.6) is a very delicate problem. We refer interested readers to [1], which provides a reasonable approach to this formalization.

In this paper, we assume that the penalty is given and our goal is to bound from above $\mathbf{E}_\theta\|\hat{\theta}_{\hat{\alpha}} - \theta\|^2$. The simplest way to analyze this risk is to use SVD. Let λ_k and ϕ_k be the eigenfunctions and eigenvectors of $(A^\top A)^{-1}$. Denoting $\psi_k = A\phi_n/\|A\phi_k\|$, one checks easily

$$y(k) \stackrel{\text{def}}{=} \langle Y, \psi_k \rangle \sqrt{\lambda_k} = \langle \theta, \psi_k \rangle + \sigma\xi(k)\sqrt{\lambda_k}, \tag{1.7}$$

where $\xi(k)$ are i.i.d. $\mathcal{N}(0,1)$. Notice that $\hat{\theta}_\alpha$ admits the following representation

$$\langle \hat{\theta}_\alpha, \psi_k \rangle = H_\alpha(\lambda_k) y(k)$$

and

$$\|\hat{\theta}_0 - \hat{\theta}_\alpha\|^2 = \sum_{k=1}^{n} [1 - H_\alpha(\lambda_k)]^2 y^2(k) \tag{1.8}$$

$$\|\theta - \hat{\theta}_\alpha\|^2 = \sum_{k=1}^{n} [\theta(k) - H_\alpha(\lambda_k) y(k)]^2. \tag{1.9}$$

We have already mentioned that the main idea in the empirical risk minimization is to control $\mathbf{E}_\theta\|\hat\theta_{\hat\alpha}-\theta\|^2$ with the help of $\mathbf{E}_\theta R_{Pen}[Y,\hat\alpha]$. Mathematically this idea can be expressed as follows. For any $\mu>0$ and a given penalty $Pen(\cdot)$ define the excess risk by

$$\Delta_{Pen}(\mu)=\sup_{\theta\in\mathbb{R}^n}\Big\{\mathbf{E}_\theta\|\theta-\hat\theta_{\hat\alpha}\|^2-(1+\mu)\mathbf{E}_\theta R_{Pen}[Y,\hat\alpha]\Big\} \tag{1.10}$$

and we will say that the penalty is admissible if $\Delta_{Pen}(\mu)<\infty$ for any $\mu>0$.

If $Pen(\alpha)$ is admissible, then we obtain immediately

$$\mathbf{E}_\theta\|\theta-\hat\theta_{\hat\alpha}\|^2\le(1+\mu)\inf_\alpha\bar R_{Pen}[\theta,\alpha]+\Delta_{Pen}(\mu), \tag{1.11}$$

where

$$\bar R_{Pen}[\theta,\alpha]=\mathbf{E}_\theta R_{Pen}[Y,\alpha]=\sum_{k=1}^n[1-H_\alpha(\lambda_k)]^2\langle\theta,\psi_k\rangle^2 + \sigma^2\sum_{k=1}^n\lambda_kH_\alpha^2(\lambda_k)+\sigma^2 Pen(\alpha)-2\sigma^2\sum_{k=1}^n\lambda_kH_\alpha(\lambda_k).$$

This equation can be interpreted as a bias-variance decomposition of $\hat\theta_\alpha$ related to the empirical risk. The empirical bias term is given by $\sum_{k=1}^n[1-H_\alpha(\lambda_k)]^2\langle\theta,\psi_k\rangle^2$ and it coincides with the standard one. However the variance term differs from $\sigma^2\sum_{k=1}^n\lambda_kH_\alpha^2(\lambda_k)$ and it is computed as follows

$$\Sigma_{Pen}(\alpha)=\sigma^2\sum_{k=1}^n\lambda_kH_\alpha^2(\lambda_k)+\sigma^2Pen(\alpha)-2\sigma^2\sum_{k=1}^n\lambda_kH_\alpha(\lambda_k).$$

The inequality (1.11) can be rewritten in the form of an oracle inequality

$$\mathbf{E}_\theta\|\theta-\hat\theta_{\hat\alpha}\|^2\le r_{Pen}[\theta]+\inf_\mu\Big\{\mu r_{Pen}[\theta]+\Delta_{Pen}(\mu)\Big\},$$

where $r_{Pen}[\theta]=\inf_\alpha\bar R_{Pen}[\theta,\alpha]$ is the oracle risk. Thus, to control the risk of our data-driven method we need to compute the excess risk. Notice that when n is large the exact computation of the excess risk is infeasible: indeed, for given θ with the Monte-Carlo method we can compute

$$D(\theta,\mu)=\mathbf{E}_\theta\|\theta-\hat\theta_{\hat\alpha}\|^2-(1+\mu)\mathbf{E}_\theta R_{Pen}[Y,\hat\alpha]$$

but we cannot maximize this function numerically over $\mathbb{R}^n$ for large n.

In order to overcome this difficulty, let us introduce

$$\Delta^C_{Pen}(\mu)\stackrel{\text{def}}{=}\mathbf{E}_0\sup_\alpha\Big\{2(1+\mu)\sum_{k=1}^n\xi^2(k)\lambda_kH_\alpha(\lambda_k) -\mu\sum_{k=1}^n\lambda_kH_\alpha^2(\lambda_k)\xi^2(k)+\frac{C\max_k\lambda_kH_\alpha^2(\lambda_k)}{\mu}-(1+\mu)Pen(\alpha)\Big\}. \tag{1.12}$$

Notice that in contrast to the excess risk, $\Delta^C_{Pen}(\mu)$ can be computed by the Monte-Carlo method, and we will show that for a sufficiently large class of spectral regularization methods $\Delta_{Pen}(\mu) \leq \sigma^2 \Delta^C_{Pen}(\mu)$. These regularization methods (smoothers) are called *ordered smoothers.* They were firstly introduced in [5].

DEFINITION 1.1. *The family of smoothers* $\{H_\alpha(\cdot),\ \alpha \geq 0\}$ *is called ordered if:*

1. for all $\alpha \geq 0$ *and* $\lambda \geq 0$, $0 \leq H_\alpha(\lambda) \leq 1$

2. $H_{\alpha_1}(\lambda) \geq H_{\alpha_2}(\lambda)$, *for all* $\alpha_1 \leq \alpha_2$ *and all* $\lambda > 0$.

Typical examples of ordered smoothers are provided the Tikhonov regularization, the spectral cut-off method, the Landweber iterations.

The main result of this paper is given by the following theorem

THEOREM 1.1. *Let* $\{H_\alpha(\cdot),\ \alpha \geq 0\}$ *be a family of ordered smoothers. Then for some* $C > 0$ *and for all* $\mu > 0$

$$\Delta_{Pen}(\mu) \leq \sigma^2 \Delta^C_{Pen}(\mu).$$

Let us illustrate how this theorem works. Suppose

$$Pen(\alpha) = \underline{Pen}(\alpha) = 2\sum_{k=1}^{n} H_\alpha(\lambda_k)\lambda_k.$$

It is well known that this penalty is related to the unbiased risk estimation, since for given α

$$\mathbf{E}_\theta\|\theta - \hat{\theta}_\alpha\|^2 = \bar{R}_{\underline{Pen}}[\theta,\alpha] = \sum_{k=1}^{n}[1 - H_\alpha(\lambda_k)]^2\langle\theta,\psi_k\rangle^2 + \sigma^2\sum_{k=1}^{n}\lambda_k H^2_\alpha(\lambda_k).$$

Assume also that A is not severely ill-posed. More precisely, suppose there exists $\kappa < 1$ such that for all $\alpha \geq 0$

$$\max_k \lambda_k H^2_\alpha(\lambda_k) \leq \lambda_1\left[\frac{1}{\lambda_1}\sum_{k=1}^{n}\lambda_k H^2_\alpha(\lambda_k)\right]^\kappa, \tag{1.13}$$

$$\sum_{k=1}^{n}\lambda_k^2 H^2_\alpha(\lambda_k) \leq \frac{\lambda_1^2}{1-\kappa}\left[\frac{1}{\lambda_1}\sum_{k=1}^{n}\lambda_k H^2_\alpha(\lambda_k)\right]^{1+\kappa}. \tag{1.14}$$

It is easy to see that these conditions allow only a polynomial growth of λ_k. Indeed, if

$$\frac{\lambda_k}{\lambda_1} = k^m \quad \text{for some } m \in [0,\infty),$$

then for the spectral cut-off with $H_\alpha(\lambda) = \mathbf{1}\{\alpha\lambda \leq 1\}$ it is easy to check that $\kappa = m/(m+1)$.

THEOREM 1.2. *Let $\{H_\alpha(\cdot),\ \alpha \geq 0\}$ be a family of ordered smoothers and (1.13-1.14) hold true, then for some $C > 1$ and any $\mu \in (0,1)$*

$$\Delta_{\underline{Pen}}(\mu) \leq \frac{C^{1/(1-\kappa)}\lambda_1\sigma^2}{(1-\kappa)^{1/(1-\kappa)}\mu^{(1+\kappa)/(1-\kappa)}}.$$

This theorem improves oracle inequalities from [2] over the classes of ordered smoothers. It says that the upper bound for the excess risk doesn't depend on n. Moreover, it allows the minimization of the empirical risk over all possible regularization parameters, thus showing that a priory restrictions on α are not essential. On the other hand, it reveals some serious difficulties related to the penalty $\underline{Pen}(\alpha)$: the upper bound for excess risk explodes as $\kappa \to 1$. It means that this penalty is not good for severely ill-posed inverse problems. More details about this effect along with an improved penalty can be found in [1].

2. Proofs.

2.1. Ordered processes and their properties. Our method of deriving oracle inequalities is related to a special class of random processes. Let $\xi(t),\ t \geq 0$ be a random process with $\mathbf{E}\xi(t) = 0$ and a finite variance $\mathbf{E}\xi^2(t) = \sigma^2(t)$ which is assumed to be monotone

$$\sigma^2(t_2) \geq \sigma^2(t_1), \quad t_2 \geq t_1.$$

The process $\xi(t),\ t \geq 0$, is called *ordered* if it is separable and for all $t_2 \geq t_1$

$$\mathbf{E}[\xi(t_2) - \xi(t_1)]^2 \leq \sigma^2(t_2) - \sigma^2(t_1). \tag{2.1}$$

Obviously, this condition can be rewritten as

$$\mathbf{E}\xi(t_2)\xi(t_1) \geq \min\{\mathbf{E}\xi^2(t_2), \mathbf{E}\xi^2(t_1)\}.$$

So, an ordered process can be viewed as a natural generalization of the Wiener process $W(t)$ for which $\mathbf{E}W(t_1)W(t_2) = \min\{\mathbf{E}W^2(t_1), \mathbf{E}W^2(t_2)\}$.

Denote for brevity

$$\Delta_\xi(t_1,t_2) = \frac{\xi(t_1) - \xi(t_2)}{\sqrt{\mathbf{E}[\xi(t_1) - \xi(t_2)]^2}}.$$

The main property of ordered processes is given by the following lemma.

LEMMA 2.1. *Suppose there exists $\lambda > 0$ such that*

$$\varphi(\lambda) \stackrel{\text{def}}{=} \sup_{t_1,t_2} \mathbf{E}\cosh\big[\lambda\Delta_\xi(t_1,t_2)\big] < \infty. \tag{2.2}$$

Then there exists a constant C depending on λ such that for all $T > 0$ and all $p \geq 1$

$$\Big[\mathbf{E}\sup_{t,s\in[0,T]} |\xi(t) - \xi(s)|^p\Big]^{1/p} \leq Cp\sigma(T), \tag{2.3}$$

where C is a generic constant.

Proof. We provide the proof of (2.3) only for reader's convenience. In fact, its proof is standard and it is based on the classical chaining arguments [8]. For simplicity, we assume that $\sigma^2(t)$ is a continuous function. Then for a given integer $s \geq 0$ we can find points t_k^s on $[0, T]$ such that

$$\sigma^2(t_k^s) = 2^{-s} k \sigma^2(T), \quad k = 0, \dots, 2^s - 1$$

and denote by $\mathcal{T}^s$ the set of these points. Let u be an arbitrary point in $\mathcal{T}^s$. Then we can find a chain, i.e. the points $\tau_j(u) \in \mathcal{T}^j$, $j = 0, \dots, s$ such that

1. $u = \tau_s(u)$, $0 = \tau_0(u)$
2. $|\sigma^2(\tau_j(u)) - \sigma^2(\tau_{j-1}(u))| \leq 2^{-j+1}\sigma^2(T)$.

To verify that such points exist, one can imagine the standard binary tree with the nodes at the level j associated with the points t_l^j, $l = 0, \dots, 2^j - 1$. It is clear that there exists a unique way connecting $u \in \mathcal{T}^s$ and 0 (top of the tree). This way passes via nodes, which are denoted by $\tau_k(u)$. So, we can write

$$u = \sum_{k=0}^{s-1} [\tau_{k+1}(u) - \tau_k(u)],$$

and for arbitrary points u, v in $\mathcal{T}^s$ we get

$$u - v = \sum_{k=0}^{s-1} [\tau_{k+1}(u) - \tau_k(u)] - \sum_{k=0}^{s-1} [\tau_{k+1}(v) - \tau_k(v)].$$

Therefore in view of (2.2), we have

$$\begin{aligned}
&\mathbf{E}^{1/p} \sup_{u,v \in \mathcal{T}^s} |\xi(u) - \xi(v)|^p \\
&\leq 2 \sum_{k=0}^{s-1} \Big[\mathbf{E} \sup_{u \in \mathcal{T}^{k+1}} |\xi(\tau_{k+1}(u)) - \xi(\tau_k(u))|^p \Big]^{1/p} \\
&= 2 \sum_{k=0}^{s-1} \Big[\mathbf{E} \sup_{u \in \mathcal{T}^{k+1}} \big|\Delta_\xi(\tau_{k+1}(u), \tau_k(u))\big|^p \qquad\qquad (2.4)\\
&\times \big[\mathbf{E}[\xi(\tau_{k+1}(u)) - \xi(\tau_k(u))]^2\big]^{p/2} \Big]^{1/p} \\
&\leq 2\sigma(T) \sum_{k=0}^{s-1} 2^{-k/2} \Big[\mathbf{E} \sup_{u \in \mathcal{T}^{k+1}} \big|\Delta_\xi(\tau_{k+1}(u), \tau_k(u))\big|^p \Big]^{1/p}.
\end{aligned}$$

Let $L(x) = \log^p(x + \mathrm{e}^{p-1})$. Notice that for any $p \geq 1$ this function is convex on $(0, \infty)$, since

$$L''(x) = \frac{p \log^{p-2}(x + \mathrm{e}^{p-1})}{(x + \mathrm{e}^{p-1})^2} \Big[p - 1 - \log(x + \mathrm{e}^{p-1}) \Big] \leq 0.$$

Therefore, using convexity of $L(x)$ and (2.2), we obtain

$$\begin{aligned}\mathbf{E}^{1/p} &\sup_{u\in\mathcal{T}^{k+1}} \left|\Delta_\xi(\tau_{k+1}(u),\tau_k(u))\right|^p \\ &\le \frac{1}{\lambda} L\Bigg[\sum_{u\in\mathcal{T}^{k+1}} \mathbf{E}e^{\lambda|\Delta_\xi(\tau_{k+1}(u),\tau_k(u))|}\Bigg] \\ &\le \frac{1}{\lambda}\log\left[2^{k+2}\varphi(\lambda)+e^{p-1}\right] = \frac{(k+2)\log(2)+p-1}{\lambda} + \frac{\log[\varphi(\lambda)]}{\lambda}.\end{aligned}$$

Substituting this in (2.4), we arrive at the inequality

$$\Big[\mathbf{E}\sup_{u,v\in\mathcal{T}^s} |\xi(u)-\xi(v)|^p\Big]^{1/p} \le C\sigma(T)$$

which proves the lemma, since by separability of $\xi(t)$

$$\Big[\mathbf{E}\sup_{u,v\in[0,T]} |\xi(u)-\xi(v)|^p\Big]^{1/p} = \limsup_{s\to\infty}\Big[\mathbf{E}\sup_{u,v\in\mathcal{T}^s} |\xi(u)-\xi(v)|^p\Big]^{1/p}.$$

□

Lemma 2.1 almost immediately results in the following fact:

LEMMA 2.2. *Let $\xi(t)$ be an ordered process satisfying (2.1, 2.2) and such that $\xi(0)=0$. Then there exists a constant C depending on λ such that for all $\gamma>0$*

$$\mathbf{E}\sup_{t\ge 0}\left[\xi(t)-\gamma\sigma^q(t)\right]_+^p \le \frac{C[2q(p+2)-4]^{q(p+2)-2}}{\gamma^{p/(q-1)}}, \tag{2.5}$$

where $[x]_+ = \max(0,x)$.

Proof. We will use the following form of the Markov inequality

$$\mathbf{E}\eta^p \mathbf{1}\{\eta > x\} \le \frac{\mathbf{E}|\eta|^{p+d}}{x^d} \tag{2.6}$$

which immediately results from the banal inequality

$$\eta^p\mathbf{1}\{\eta>x\} \le |\eta|^p\left|\eta/x\right|^d.$$

Without loss of generality, we may assume that $\sigma^2(t)$ is continuous and such that $\lim_{t\to\infty}\sigma^2(t)=\infty$. Then for any integer $k\ge 0$ we can find $t_k(\gamma)$ such that

$$\sigma^{q-1}\big(t_k(\gamma)\big) = \frac{k}{\gamma}.$$

Using that $f(x) = x^p \mathbf{1}\{x > x_0\}$ is monotone in $x > 0$, we have

$$
\begin{aligned}
&\mathbf{E}\sup_{t\ge 0}\bigl[\xi(t) - \gamma\sigma^q(t)\bigr]_+^p \\
&\le \sum_{k=0}^{\infty}\mathbf{E}\sup_{t\in[t_k(\gamma),t_{k+1}(\gamma)]}\xi^p(t)\mathbf{1}\bigl\{\xi(t)\ge\gamma\sigma^q(t)\bigr\} \\
&\le \sum_{k=0}^{\infty}\mathbf{E}\sup_{t\in[t_k(\gamma),t_{k+1}(\gamma)]}\xi^p(t)\mathbf{1}\bigl\{\xi(t)\ge\gamma\sigma^q(t_k(\gamma))\bigr\} \\
&\le \sum_{k=0}^{\infty}\mathbf{E}\sup_{t\in[t_k(\gamma),t_{k+1}(\gamma)]}\xi^p(t)\mathbf{1}\Bigl\{\sup_{t\in[t_k(\gamma),t_{k+1}(\gamma)]}\xi(t)\ge\gamma\sigma^q(t_k(\gamma))\Bigr\} \\
&\le \mathbf{E}\sup_{0\le t\le t_1(\gamma)}|\xi(t)|^p \\
&\quad + \sum_{k=1}^{\infty}\mathbf{E}\sup_{0\le t\le t_{k+1}(\gamma)}\xi^p(t)\mathbf{1}\Bigl\{\sup_{0\le t\le t_{k+1}(\gamma)}\xi(t)\ge\gamma\sigma^q(t_k(\gamma))\Bigr\}.
\end{aligned}
\tag{2.7}
$$

By Lemma 2.1, the first term at the right-hand side of the above inequality is bounded as follows

$$
\mathbf{E}\sup_{0\le t\le t_1(\gamma)}|\xi(t)|^p \le Cp^p, \sigma^p(t_1(\gamma)) = \frac{Cp^p}{\gamma^{p/(q-1)}} \tag{2.8}
$$

whereas the second one, in view of (2.6), is controlled by

$$
\begin{aligned}
&\sum_{k=1}^{\infty}\mathbf{E}\sup_{0\le t\le t_{k+1}(\gamma)}\xi^p(t)\mathbf{1}\Bigl\{\sup_{0\le t\le t_{k+1}(\gamma)}\xi(t)\ge\gamma\sigma^q(t_k(\gamma))\Bigr\} \\
&\le C(p+d)^{p+d}\sum_{k=1}^{\infty}\frac{\sigma^{p+d}(t_{k+1}(\gamma))}{[\gamma\sigma^q(t_k(\gamma))]^d} = \frac{C(p+d)^{p+d}}{\gamma^{p/(q-1)}}\sum_{k=1}^{\infty}\frac{(k+1)^{p+d}}{k^{qd/(q-1)}} \\
&\le \frac{C[2(p+d)]^{p+d}}{\gamma^{p/(q-1)}}\sum_{k=1}^{\infty}\frac{1}{k^{d/(q-1)-p}}.
\end{aligned}
$$

Setting $d = (q-1)(p+2)$ in the above inequality and using (2.7) together with (2.8), we prove (2.5). □

2.2. Some examples of ordered processes. The simplest example of an ordered process is $\xi(t) = \xi t$, where ξ is a zero mean random variable with a finite exponential moment $\mathbf{E}\cosh(\lambda\xi) < \infty$ for some $\lambda > 0$. As we have already mentioned, the Wiener process $W(t)$ is an ordered process. At the first glance, ξt and $W(t)$ are quite different, but from the viewpoint of Lemma 2.2 they are equivalent. Of course, the distribution of $\max_{t\ge 0}[W(t) - \gamma t]$ is well-known

$$
\mathbf{P}\bigl\{\max_{t\ge 0}[W(t) - \gamma t] \ge x\bigr\} = \exp(-2\gamma x).
$$

The next two examples play an essential role in adaptive estimation. Let $H_t(\cdot)$ be a family of ordered smoothers (see Definition 1.1). Consider the following Gaussian processes

$$\xi_+(t) = \sum_{k=1}^{n} [H_{t_0}(\lambda_k) - H_{t_0+t}(\lambda_k)] b_k \xi(k), \quad t \geq 0$$

$$\xi_-(t) = \sum_{k=1}^{n} [H_{t_0}(\lambda_k) - H_{t_0-t}(\lambda_k)] b_k \xi(k), \quad 0 \leq t \leq t_0,$$

where $\xi(k)$ are i.i.d. $\mathcal{N}(0,1)$ and $\sum_{i=1}^{\infty} b_i^2 < \infty$. It is easy to see that $\xi_+(t)$ and $\xi_-(t)$ are ordered processes. Indeed, in view of (1.12) we have for $t_2 \geq t_1$

$$\begin{aligned}
\mathbf{E}\xi_+^2(t_1) &= \sum_{k=1}^{n} [H_{t_0}(\lambda_k) - H_{t_0+t_1}(\lambda_k)][H_{t_0}(\lambda_k) - H_{t_0+t_1}(\lambda_k)] b_k^2 \\
&\leq \sum_{k=1}^{n} [H_{t_0}(\lambda_k) - H_{t_0+t_1}(\lambda_k)][H_{t_0}(\lambda_k) - H_{t_0+t_2}(\lambda_k)] b_k^2 \\
&= \mathbf{E}\xi_+(t_1)\xi_+(t_2),
\end{aligned}$$

and similarly,

$$\begin{aligned}
\mathbf{E}\xi_+^2(t_1) &= \sum_{k=1}^{n} [H_{t_0-t_1}(\lambda_k) - H_{t_0}(\lambda_k)][H_{t_0-t_1}(\lambda_k) - H_{t_0}(\lambda_k)] b_k^2 \\
&\leq \sum_{k=1}^{n} [H_{t_0-t_1}(\lambda_k) - H_{t_0}(\lambda_k)][H_{t_0-t_2}(\lambda_k) - H_{t_0}(\lambda_k)] b_k^2 \\
&= \mathbf{E}\xi_+(t_1)\xi_+(t_2).
\end{aligned}$$

Therefore with Lemma 2.2 we get

$$\mathbf{E} \sup_{\alpha \geq \alpha_0} \left[\xi_+(\alpha) - \gamma \sum_{k=1}^{n} [H_{\alpha_0}(\lambda_k) - H_\alpha(\lambda_k)]^2 b_k^2 \right]_+^p \leq \frac{C(p)}{\gamma^p},$$

$$\mathbf{E} \sup_{\alpha \leq \alpha_0} \left[\xi_-(\alpha) - \gamma \sum_{k=1}^{n} [H_{\alpha_0}(\lambda_k) - H_\alpha(\lambda_k)]^2 b_k^2 \right]_+^p \leq \frac{C(p)}{\gamma^p},$$

thus arriving at

LEMMA 2.3. *Let $\{H_\alpha(\cdot),\ \alpha \geq 0\}$ be a family of ordered smoothers, then for any $\gamma > 0$*

$$\begin{aligned}
\mathbf{E} \sup_{\alpha \geq 0} \Bigg[&\sum_{k=1}^{n} [H_{\alpha_0}(\lambda_k) - H_\alpha(\lambda_k)] b_k \xi(k) \\
&- \gamma \sum_{k=1}^{n} [H_{\alpha_0}(\lambda_k) - H_\alpha(\lambda_k)]^2 b_k^2 \Bigg]_+^p \leq \frac{C(p)}{\gamma^p}.
\end{aligned} \tag{2.9}$$

The next important ordered process is defined by

$$\eta(t) = \sum_{k=1}^{n} H_{1/t}(\lambda_k)(\xi^2(k) - 1),$$

where $\xi(k)$ are i.i.d. $\mathcal{N}(0,1)$ and $\{H_t(\cdot),\ t \geq 0\}$ is a family of ordered smoothers. It is easy to check that

$$\mathbf{E}\eta_2^2(t_1) \leq \mathbf{E}\eta_2(t_2)\eta_2(t_1), \quad t_1 \leq t_2.$$

So, in order to apply Lemma 2.2, it remains to check (2.2). Denoting for brevity

$$\|H_{\alpha_2} - H_{\alpha_1}\|^2 = \sum_{k=1}^{n} [H_{\alpha_2}(\lambda_k) - H_{\alpha_1}(\lambda_k)]^2,$$

we have

$$\begin{aligned} &\mathbf{E}\exp\big[\lambda\Delta_\xi(\alpha_2,\alpha_1)\big] \\ &= \exp\Bigg[-\frac{\lambda}{\sqrt{2}\|H_{\alpha_2} - H_{\alpha_1}\|}\sum_{k=1}^{n}[H_{\alpha_2}(\lambda_k) - H_{\alpha_1}(\lambda_k)] \qquad (2.10) \\ &-\frac{1}{2}\sum_{k=1}^{n}\log\bigg(1 - \sqrt{2}\lambda\frac{H_{\alpha_2}(\lambda_k) - H_{\alpha_1}(\lambda_k)}{\|H_{\alpha_2} - H_{\alpha_1}\|}\bigg)\Bigg]. \end{aligned}$$

Since obviously

$$\max_k [H_{\alpha_2}(\lambda_k) - H_{\alpha_1}(\lambda_k)] \leq \|H_{\alpha_2} - H_{\alpha_1}\|,$$

then using the Taylor expansion for $\log(1 - \cdot)$ at the right-hand side of (2.10), we get for $\lambda \leq 1/2$

$$\mathbf{E}\exp\big[\lambda\Delta_\xi(\alpha_2,\alpha_1)\big] \leq \exp(C\lambda^2),$$

thus proving (2.2). Therefore using Lemma 2.2, we obtain the following fact.

LEMMA 2.4. *Let* $\{H_\alpha(\cdot),\ \alpha \geq 0\}$ *be a family of ordered smoothers, then for all* $\gamma > 0$

$$\mathbf{E}\sup_{\alpha>0}\Bigg[\sum_{k=1}^{\infty} H_\alpha(\lambda_k)[\xi^2(k) - 1] - \gamma\bigg(2\sum_{k=1}^{\infty} H_\alpha^2(\lambda_k)\bigg)^q\Bigg]_+^p \leq \frac{C(p)}{\gamma^{p/(1-q)}}. \qquad (2.11)$$

2.3. Proof of Theorem 1.1. Denote for brevity $\theta_k = \langle\theta, \phi_k\rangle$. We begin with a simple auxiliary lemma that is cornerstone for the proof.

LEMMA 2.5. *Let $\{H_\alpha(\cdot),\ \alpha \geq 0\}$ be a family of ordered smoothers. Then there exists a constant C such that for any data-driven smoothing parameter $\hat\alpha$*

$$\begin{aligned}\Big[\mathbf{E}_\theta \sum_{k=1}^{\infty}[1 - H_{\hat\alpha}(\lambda_k)]\sqrt{\lambda_k}\theta_k\xi(k)\Big]^2 \\ \leq C\mathbf{E}_\theta \max_k \lambda_k H^2_{\hat\alpha}(\lambda_k)\mathbf{E}_\theta \sum_{k=1}^{\infty}[1 - H_{\hat\alpha}(\lambda_k)]^2\theta_k^2.\end{aligned} \tag{2.12}$$

and

$$\begin{aligned}\Big[\mathbf{E}_\theta \sum_{k=1}^{\infty}[1 - H_{\hat\alpha}(\lambda_k)]^2\sqrt{\lambda_k}\theta_k\xi(k)\Big]^2 \\ \leq C\mathbf{E}_\theta \max_k \lambda_k H^2_{\hat\alpha}(\lambda_k)\mathbf{E}_\theta \sum_{k=1}^{\infty}[1 - H_{\hat\alpha}(\lambda_k)]^2\theta_k^2.\end{aligned} \tag{2.13}$$

Proof. Let α_0 be a given smoothing parameter. We obviously have

$$\begin{aligned}&\mathbf{E}_\theta \sum_{k=1}^{\infty}[1 - H_{\hat\alpha}(\lambda_k)]\sqrt{\lambda_k}\theta_k\xi(k) \\ &\quad = \mathbf{E}_\theta \sum_{k=1}^{\infty}[H_{\alpha_0}(\lambda_k) - H_{\hat\alpha}(\lambda_k)]\sqrt{\lambda_k}\theta_k\xi(k).\end{aligned} \tag{2.14}$$

It follows immediately from (2.9) that

$$\begin{aligned}&\Big|\mathbf{E}_\theta \sum_{k=1}^{\infty}[H_{\alpha_0}(\lambda_k) - H_{\hat\alpha}(\lambda_k)]\sqrt{\lambda_k}\theta_k\xi(k)\Big| \\ &\quad \leq \gamma\mathbf{E}_\theta \sum_{k=1}^{\infty}[H_{\alpha_0}(\lambda_k) - H_{\hat\alpha}(\lambda_k)]^2\lambda_k\theta_k^2 + \frac{C}{\gamma}.\end{aligned}$$

Therefore minimizing the right-hand side in γ, we obtain

$$\begin{aligned}&\Big|\mathbf{E}_\theta \sum_{k=1}^{\infty}[H_{\alpha_0}(\lambda_k) - H_{\hat\alpha}(\lambda_k)]\lambda_k\theta_k\xi(k)\Big| \\ &\quad \leq C\Big\{\mathbf{E}_\theta \sum_{k=1}^{\infty}[H_{\alpha_0}(\lambda_k) - H_{\hat\alpha}(\lambda_k)]^2\lambda_k\theta_k^2\Big\}^{1/2}.\end{aligned} \tag{2.15}$$

To bound from above the right-hand side at the above display, we use once again that $H_\alpha(\cdot)$ are ordered smoothers. So, when $\hat\alpha \leq \alpha_0$ we obtain

$$\sum_{k=1}^{\infty}[H_{\alpha_0}(\lambda_k) - H_{\hat{\alpha}}(\lambda_k)]^2\lambda_k^2\theta_k^2 = \sum_{k=1}^{\infty} H_{\alpha_0}^2(\lambda_k)\left[1 - \frac{H_{\hat{\alpha}}(\lambda_k)}{H_{\alpha_0}(\lambda_k)}\right]^2 \lambda_k^2\theta_k^2$$
$$\leq \max_k \lambda_k H_{\alpha_0}^2(\lambda_k) \sum_{k=1}^{\infty}[1-H_{\hat{\alpha}}(\lambda_k)]^2\theta_k^2$$

and similarly for $\hat{\alpha} \geq \alpha_0$

$$\sum_{k=1}^{\infty}[H_{\alpha_0}(\lambda_k) - H_{\hat{\alpha}}(\lambda_k)]^2\lambda_k\theta_k^2 \leq \max_k \lambda_k H_{\hat{\alpha}}(\lambda_k) \sum_{k=1}^{\infty}[1 - H_{\alpha_0}(\lambda_k)]^2\theta_k^2.$$

Therefore combining these inequalities with (2.14) and (2.15), we get

$$\left|\mathbf{E}_\theta \sum_{k=1}^{\infty}[1 - H_{\hat{\alpha}}(\lambda_k)]\sqrt{\lambda_k}\theta_k\xi(k)\right|$$
$$\leq C\Big[\mathbf{E}_\theta \max_k \lambda_k H_{\hat{\alpha}}^2(\lambda_k)\Big]^{1/2}\left[\sum_{k=1}^{\infty}[1 - H_{\alpha_0}(\lambda_k)]^2\theta_k^2\right]^{1/2}$$
$$+C \max_k \sqrt{\lambda_k} H_{\alpha_0}(\lambda_k)\left[\mathbf{E}_\theta \sum_{k=1}^{\infty}[1 - H_{\hat{\alpha}}(\lambda_k)]^2\theta_k^2\right]^{1/2}.$$

Using the elementary inequality $2ab \leq \mu a^2 + b^2/\mu$, we can continue the above display as follows

$$\left|\mathbf{E}_\theta \sum_{k=1}^{\infty}[1 - H_{\hat{\alpha}}(\lambda_k)]\sqrt{\lambda_k}\theta_k\xi(k)\right|$$
$$\leq \mu \sum_{k=1}^{\infty}[1 - H_{\alpha_0}(\lambda_k)]^2\theta_k^2 + \frac{C \max_k \lambda_k H_{\alpha_0}^2(\lambda_k)}{\mu}$$
$$+\mu\mathbf{E}_\theta \sum_{k=1}^{\infty}[1 - H_{\hat{\alpha}}(\lambda_k)]^2\theta_k^2 + \frac{C\mathbf{E}_\theta \max_k \lambda_k H_{\hat{\alpha}}^2(\lambda_k)}{\mu}.$$

Therefore minimizing the right-hand side in α_0, we get

$$\left|\mathbf{E}_\theta \sum_{k=1}^{\infty}[1 - H_{\hat{\alpha}}(\lambda_k)]\sqrt{\lambda_k}\theta_k\xi(k)\right|$$
$$\leq \inf_{\alpha_0}\left\{\mu \sum_{k=1}^{\infty}[1 - H_{\alpha_0}(\lambda_k)]^2\theta_k^2 + \frac{C \max_k \lambda_k H_{\alpha_0}^2(\lambda_k)}{\mu}\right\}$$
$$+\mathbf{E}_\theta\left\{\mu \sum_{k=1}^{\infty}[1 - H_{\hat{\alpha}}(\lambda_k)]^2\theta_k^2 + \frac{C \max_k \lambda_k H_{\hat{\alpha}}^2(\lambda_k)}{\mu}\right\}$$
$$\leq 2\mathbf{E}_\theta\left\{\mu \sum_{k=1}^{\infty}[1 - H_{\hat{\alpha}}(\lambda_k)]^2\theta_k^2 + \frac{C \max_k \lambda_k H_{\hat{\alpha}}^2(\lambda_k)}{\mu}\right\}.$$

To finish the proof of (2.12) it suffices to minimize the right-hand side in μ.

Inequality (2.13) follows from (2.12) since $\tilde{H}_\alpha(\lambda) = 2H_\alpha(\lambda) - H^2_\alpha(\lambda)$ are ordered smoothers and we can apply (2.12) with $H_\alpha(\cdot) = \tilde{H}_\alpha(\cdot)$. □

In view of the definition of the empirical risk and (1.9), we have

$$
\begin{aligned}
R_{Pen}[Y, \hat{\alpha}] &= \|\hat{\theta}_0 - \hat{\theta}_{\hat{\alpha}}\|^2 + \sigma^2 Pen(\hat{\alpha}) - \sigma^2 \sum_{k=1}^{n} \lambda_k \\
&= \sum_{k=1}^{n} [1 - H_{\hat{\alpha}}(\lambda_k)]^2 \theta_k^2 + \sigma^2 Pen(\hat{\alpha}) \\
&\quad + \sum_{k=1}^{n} [H^2_{\hat{\alpha}}(\lambda_k) - 2H_{\hat{\alpha}}(\lambda_k)] \lambda_k \xi^2(k) \\
&\quad + 2\sigma \sum_{k=1}^{n} [1 - H_{\hat{\alpha}}(\lambda_k)]^2 \sqrt{\lambda_k} \theta_k \xi(k)
\end{aligned}
$$

and

$$
\begin{aligned}
\|\theta - \hat{\theta}_{\hat{\alpha}}\|^2 &= \sum_{k=1}^{n} [\theta(k) - H_{\hat{\alpha}}(\lambda_k) y(k)]^2 \\
&= \sum_{k=1}^{n} [1 - H_{\hat{\alpha}}(\lambda_k)]^2 \theta_k^2 + \sigma^2 \sum_{k=1}^{\infty} \lambda_k H^2_{\hat{\alpha}}(\lambda_k) \xi^2(k) \\
&\quad - 2\sigma \sum_{k=1}^{n} [1 - H_{\hat{\alpha}}(\lambda_k)] \theta_k \sqrt{\lambda_k} H_{\hat{\alpha}}(\lambda_k) \xi(k).
\end{aligned}
$$

Therefore for the excess risk we have

$$
\begin{aligned}
\Delta_{Pen}(\mu) &= \sup_{\theta \in \mathbb{R}^n} \mathbf{E}_\theta \Big\{ \|\theta - \hat{\theta}_{\hat{\alpha}}\|^2 - (1+\mu) R_{Pen}[Y, \hat{\alpha}] \Big\} \\
&= \sup_{\theta \in \mathbb{R}^n} \mathbf{E}_\theta \bigg\{ -\mu \sum_{k=1}^{n} [1 - H_{\hat{\alpha}}(\lambda_k)]^2 \theta_k^2 - \frac{C\sigma^2 \max_k \lambda_k H^2_{\hat{\alpha}}(\lambda_k)}{\mu} \\
&\quad - 2\sigma \sum_{k=1}^{n} [1 - H_{\hat{\alpha}}(\lambda_k)] \theta_k \sqrt{\lambda_k} \xi(k) \\
&\quad - 2\mu\sigma \sum_{k=1}^{n} [1 - H_{\hat{\alpha}}(\lambda_k)]^2 \theta_k \sqrt{\lambda_k} \xi(k) \\
&\quad + \frac{C \max_k \sigma^2 \lambda_k H^2_{\hat{\alpha}}(\lambda_k)}{\mu} + 2(1+\mu)\sigma^2 \sum_{k=1}^{n} \lambda_k H_{\hat{\alpha}}(\lambda_k) \xi^2(k) \\
&\quad - (1+\mu)\sigma^2 Pen(\hat{\alpha}) - \mu\sigma^2 \sum_{k=1}^{n} \lambda_k H^2_{\hat{\alpha}}(\lambda_k) \xi^2(k) \bigg\}.
\end{aligned}
\tag{2.16}
$$

The last two lines can be bounded by $\sigma^2\Delta^C_{Pen}(\mu)$. Indeed,

$$
\begin{aligned}
&\mathbf{E}\Big\{\frac{C\sigma^2\max_k \lambda_k H^2_{\hat\alpha}(\lambda_k)}{\mu} + 2(1+\mu)\sigma^2\sum_{k=1}^n \lambda_k H_{\hat\alpha}(\lambda_k)\xi^2(k) \\
&\qquad -(1+\mu)\sigma^2 Pen(\hat\alpha) - \mu\sigma^2\sum_{k=1}^n \lambda_k H^2_{\hat\alpha}(\lambda_k)\xi^2(k)\Big\} \\
&\le \mathbf{E}\sup_\alpha\Big\{\frac{C\sigma^2\max_k \lambda_k H^2_\alpha(\lambda_k)}{\mu} \\
&\qquad +2(1+\mu)\sigma^2\sum_{k=1}^n \lambda_k H_\alpha(\lambda_k)\xi^2(k) \\
&\qquad -(1+\mu)\sigma^2 Pen(\alpha) - \mu\sigma^2\sum_{k=1}^n \lambda_k H^2_\alpha(\lambda_k)\xi^2(k)\Big\} \\
&= \sigma^2\Delta^C_{Pen}(\mu).
\end{aligned}
\tag{2.17}
$$

Finally, with Lemma 2.5 we obtain

$$
\begin{aligned}
&\mathbf{E}_\theta\Big\{-\mu\sum_{k=1}^n[1-H_{\hat\alpha}(\lambda_k)]^2\theta_k^2 - \frac{C\sigma^2\max_k \lambda_k H^2_{\hat\alpha}(\lambda_k)}{\mu} \\
&\quad -2\sigma\sum_{k=1}^n[1-H_{\hat\alpha}(\lambda_k)]\theta_k\sqrt{\lambda_k}\xi(k) - 2\mu\sigma\sum_{k=1}^n[1-H_{\hat\alpha}(\lambda_k)]^2\theta_k\sqrt{\lambda_k}\xi(k)\Big\} \\
&\le 0.
\end{aligned}
$$

This inequality together with (2.16) and (2.17) completes the proof of the theorem.

2.3.1. Proof of Theorem 1.2. In view of Theorem 1.1, it suffices to check that

$$
\Delta^C_{\underline{Pen}}(\mu) \le \frac{C^{1/(1-\kappa)}\lambda_1\sigma^2}{(1-\kappa)^{1/(1-\kappa)}\mu^{(1+\kappa)/(1-\kappa)}}, \tag{2.18}
$$

where

$$
\begin{aligned}
\Delta^C_{\underline{Pen}}(\mu) = \mathbf{E}\sup_\alpha\Big\{&\frac{C\max_k \lambda_k H^2_\alpha(\lambda_k)}{\mu} - \mu\sum_{k=1}^n \lambda_k H^2_\alpha(\lambda_k) \\
&+(2+\mu)\sum_{k=1}^n \frac{2(1+\mu)H_\alpha(\lambda_k) - \mu H^2_\alpha(\lambda_k)}{2+\mu}\lambda_k[\xi^2(k)-1]\Big\}.
\end{aligned}
$$

We begin the proof of (2.18) with the deterministic term. By (1.13) we get

$$\sup_{\alpha>0}\left\{\frac{C\max_k \lambda_k H_\alpha^2(\lambda_k)}{\mu}-\frac{\mu}{2}\sum_{k=1}^n \lambda_k H_\alpha^2(\lambda_k)\right\} \tag{2.19}$$
$$\le \sup_{\alpha>0}\left\{\frac{C\lambda_1^{1-\kappa}}{\mu}\left[\sum_{k=1}^n \lambda_k H_\alpha^2(\lambda_k)\right]^\kappa-\frac{\mu}{2}\sum_{k=1}^n \lambda_k H_\alpha^2(\lambda_k)\right\}$$
$$\le \sup_{x\ge 0}\left\{\frac{C\lambda_1^{1-\kappa}}{\mu}x^\kappa-\frac{\mu}{2}x\right\}=C^{1/(1-\kappa)}\lambda_1\mu^{(\kappa+1)/(\kappa-1)}.$$

Denote for brevity

$$\tilde{H}_\alpha(\lambda)=\frac{2(1+\mu)H_\alpha(\lambda_k)-\mu H_\alpha^2(\lambda_k)}{2+\mu}.$$

Our next step is to show that

$$\mathbf{E}\sup_{\alpha>0}\left\{(2+\mu)\sum_{k=1}^n \lambda_k \tilde{H}_\alpha(\lambda_k)[\xi^2(k)-1]-\frac{\mu}{2}\sum_{k=1}^n \lambda_k H_\alpha^2(\lambda_k)\right\} \tag{2.20}$$
$$\le C^{1/(1-\kappa)}\lambda_1(1-\kappa)^{-1/(1-\kappa)}\mu^{(\kappa+1)/(\kappa-1)}.$$

It is easy to see that in view of (1.14)

$$\sigma^2(\alpha)=\mathbf{E}\left[\sum_{k=1}^n \lambda_k \tilde{H}_\alpha(\lambda_k)[\xi^2(k)-1]\right]^2=2\sum_{k=1}^n \lambda_k^2\tilde{H}_\alpha^2(\lambda_k)$$
$$\le \frac{2(2+2\mu)^2}{(2+\mu)^2}\sum_{k=1}^n \lambda_k^2 H_\alpha^2(\lambda_k)$$
$$\le \frac{2\lambda_1^2(2+2\mu)^2}{(1-\kappa)(2+\mu)^2}\left[\frac{1}{\lambda_1}\sum_{k=1}^n \lambda_k H_\alpha^2(\lambda_k)\right]^{1+\kappa}.$$

Next notice that if $\{H_\alpha(\cdot),\ \alpha\ge 0\}$ is a family of ordered smoothers, then $\{\tilde{H}_\alpha(\cdot),\ \alpha\ge 0\}$ is also a family of ordered smoothers. Therefore by Lemma 2.4, for any $\mu>0$ we obtain

$$\mathbf{E}\sup_{\alpha}\left\{(2+\mu)\sum_{k=1}^n \lambda_k\tilde{H}_\alpha(\lambda_k)[\xi^2(k)-1]-\frac{\mu}{2}\sum_{k=1}^n \lambda_k H_\alpha^2(\lambda_k)\right\}$$
$$\le (2+\mu)\mathbf{E}\sup_{\alpha}\left\{\sum_{k=1}^n \lambda_k\tilde{H}_\alpha(\lambda_k)[\xi^2(k)-1]\right.$$
$$\left.-\frac{\mu\lambda_1}{2(2+\mu)}\left[\frac{\sigma^2(\alpha)(1-\kappa)(2+\mu)^2}{2\lambda_1^2(2+2\mu)^2}\right]^{1/(1+\kappa)}\right\}$$
$$\le C(1+\mu)^{2/(1-\kappa)}2^{(4+\kappa)/(1-\kappa)}\lambda_1(1-\kappa)^{-1/(1-\kappa)}\mu^{(\kappa+1)/(\kappa-1)}.$$

thus proving (2.20). Thus (2.18) follows obviously from (2.19) and (2.20).

REFERENCES

[1] L. Cavalier and Yu. Golubev, *Risk hull method and regularization by projections of ill-posed inverse problems,* Ann. of Stat. (2006), **34**: 1653–1677.

[2] L. Cavalier, G.K. Golubev, D. Picard, and A.B. Tsybakov, *Oracle inequalities for inverse problems*, Ann. of Stat. (2002), **30**: 843–874.

[3] H.W. Engl, M. Hanke, and A. Neubauer, *Regularization of Inverse Problems*, Kluwer Academic Publishers, 1996.

[4] I.A. Ibragimov and R.Z. Khasminskii, *Statistical Estimation. Asymptotic Theory,* Springer-Verlag, NY, 1981.

[5] A. Kneip, *Ordered linear smoothers*, Ann. Statist. (1994), **22**: 835–866.

[6] B. Mair and F.H. Ruymgaart, *Statistical estimation in Hilbert scale.* SIAM J. Appl. Math. (1996), **56**: 1424–1444.

[7] A.N. Tikhonov and V.A. Arsenin, *Solution of Ill-posed Problems*, Winston & Sons, Washington, 1977.

[8] A. van der Vaart and J. Wellner *Weak convergence and empirical processes.* Springer-Verlag, NY, 1996.

HYPOTHESIS TESTING UNDER COMPOSITE FUNCTIONS ALTERNATIVE

OLEG V. LEPSKI* AND CHRISTOPHE F. POUET*†

Abstract. In this paper, we consider the problem of the minimax hypothesis testing in the multivariate white gaussian noise model. We want to test the hypothesis about the absence of the signal against the alternative belonging to the set of smooth composite functions separated away from zero in sup-norm. We propose the test procedure and show that it is optimal in view of the minimax criterion if the smoothness parameters of the composition obey some special assumption. In this case we also present the explicit formula for minimax rate of testing. If this assumption does not hold, we give the explicit upper and lower bounds for minimax rate of testing which differ each other only by some logarithmic factor. In particular, it implies that the proposed test procedure is "almost" minimax. In both cases the minimax rate of testing as well as its upper and lower bounds are completely determined by the smoothness parameters of the composition.

Key words. nonparametric hypothesis testing, separation rate, minimax rate of testing, composite functions, structural models, metric entropy, gaussian random function, Implicit Function Theorem.

AMS(MOS) subject classifications. 62G10.

1. Introduction. In this paper, we study the problem of minimax hypothesis testing in the multidimensional gaussian white noise model

$$dX_\varepsilon(t) = g(t)dt + \varepsilon dW(t), \quad t = (t_1, \dots, t_d) \in \mathcal{D}_0 \tag{1.1}$$

where $\mathcal{D}_0$ is an open interval in $\mathbb{R}^d$, $d \geq 1$, W is the standard Brownian sheet in $\mathbb{R}^d$ and $0 < \varepsilon < 1$ is the noise level. Our goal is to test the hypothesis on the absence of the signal g, i.e.

$$\mathbf{H}: \quad g \equiv 0,$$

against the alternative written in the following form

$$\mathbf{A}: \quad g \in \mathbb{G}(\psi_\varepsilon),$$

$$\mathbb{G}(\psi_\varepsilon) = \left\{ g \in \mathbb{G}: \quad ||g||_\infty \triangleq \sup_{t \in \mathcal{D}} |g(t)| \geq \psi_\varepsilon \right\}.$$

Here $\mathbb{G}$ is a compact subset of $\mathbb{L}_2(\mathcal{D}_0)$ endowed by the metric generated by $||\cdot||_\infty$, $\mathcal{D} \subset \mathcal{D}_0$ is an open interval in $\mathbb{R}^d$ and $\psi_\varepsilon \to 0$, $\varepsilon \to 0$, is the *separation sequence.* We consider the observation set $\mathcal{D}_0$ which is larger than $\mathcal{D}$ in order to avoid the discussion of the boundary effects. Without loss of generality we will assume that $\mathcal{D}_0 = [-1, 1]^d$ and $\mathcal{D} = [-1/2, 1/2]^d$.

We define a *decision rule* to be any measurable function of the observation $\{X_\varepsilon(t), t \in \mathcal{D}_0\}$ taking the values 0 and 1.

---

*Laboratoire d'Analyse, Topologie et Probabilités, UMR CNRS 6632, Université d'Aix-Marseille 1, 39, rue F. Joliot–Curie, 13453 Marseille cedex 13, France (lepski@cmi.univ-mrs.fr).

†(pouet@cmi.univ-mrs.fr).

1.1. Minimax approach. To measure the performance of any decision rule we will use so-called *minimax criterion*. Let $\mathbb{P}_g$ be the probability measure on the Borel σ-algebra of $\mathbb{C}(\mathcal{D}_0)$ generated by the observation (1.1). For any decision rule $\tilde{\Delta}$ we introduce the *risk function*

$$\mathcal{R}(\tilde{\Delta}, \psi_\varepsilon) = \mathbb{P}_0\{\tilde{\Delta} = 1\} + \sup_{g \in \mathbb{G}(\psi_\varepsilon)} \mathbb{P}_g\{\tilde{\Delta} = 0\}, \tag{1.2}$$

which is the sum of the first and the second errors probabilities.

We say that the separation sequence φ_ε is the *minimax rate of testing* and $\hat{\Delta}$ is the *minimax decision rule* if

1. $\lim_{C\to\infty} \limsup_{\varepsilon\to 0} \mathcal{R}(\hat{\Delta}, C\varphi_\varepsilon) = 0$;
2. $\lim_{C\to 0} \liminf_{\varepsilon\to 0} \inf_{\tilde{\Delta}} \mathcal{R}(\tilde{\Delta}, C\varphi_\varepsilon) = 1$,

where inf is taken over all possible decision rules.

It is worthy of mentioning that the *minimax rate of testing* as well as the *minimax decision rule* usually depend on the parameter set $\mathbb{G}$ that makes its choice very important. In the next section we discuss the choice of $\mathbb{G}$ for a model described by a multivariate function.

1.2. Choice of parameter set. Structural assumptions. It is well known that the main difficulty in testing under alternatives being multivariate functions is the curse of dimensionality: the best attainable rate of testing decreases very fast, as the dimension grows. To illustrate this effect, suppose, for example, that the underlying function g belongs to $\mathbb{G} = \mathbb{H}_d(\nu, L)$, $\nu > 0, L > 0$, where $\mathbb{H}_d(\nu, L)$ is an isotropic Hölder class of functions. We give the exact definition of this functional class later. Here we only mention that $\mathbb{H}_d(\nu, L)$ consists, in particular, of functions g with bounded partial derivatives of order less or equal $\lfloor \nu \rfloor$ and such that, for all $x, y \in \mathcal{D}_0$,

$$|g(y) - P_g(x, y-x)| \le L||x-y||^\nu,$$

where $P_g(x, y-x)$ is the Taylor polynomial of order $\lfloor \nu \rfloor$ obtained by expansion of g around the point x, and $||\cdot||$ is the Euclidean norm in $\mathbb{R}^d$. Parameter ν characterizes the isotropic (i.e., the same in each direction) smoothness of function g.

If we use the risk (1.2), then ([11], [16]) the rate of testing is given by

$$\varphi_{\varepsilon,d}(\nu) = \left(\varepsilon\sqrt{\ln(1/\varepsilon)}\right)^{2\nu/(2\nu+d)}.$$

It is clear that if ν is fixed and d is large enough this asymptotics of the rate is too pessimistic to be used for real data: the value $\varphi_{\varepsilon,d}(\nu)$ is small only if the noise level ε is unreasonably small. On the other hand, if the noise level ε is realistically small the above asymptotics might be of no use already in dimension 2 or 3.

This problem arises because the d-dimensional Hölder class $\mathbb{H}_d(\nu, L)$ is too massive. A way to overcome the curse of dimensionality is to consider models with poorer functional classes $\mathbb{G}$. Clearly, if the class of candidate functions g is smaller, the rate of testing is faster. Note that the "poverty" of a functional class can be described in terms of restrictions on its metric entropy and there are several ways to do it.

The way we will follow in the present paper consists in imposing a *structural assumption* on the function g. The classical examples are provided by the single index additive and projection pursuit structures ([3], [4], [6], [7], [8], [9], [20] and [21] among others).

(i) *[Single–index model.]* Let e be a direction vector in $\mathbb{R}^d$, and assume that $g(x) = f(e^T x)$ for some unknown univariate function f.

(ii) *[Additive model.].* Assume that $g(x) = \sum_{i=1}^d f_i(x_i)$, where f_i are unknown univariate functions.

(iii) *[Projection pursuit regression.]* Let $e_1, \dots, e_d$ be direction vectors in $\mathbb{R}^d$, and assume that $g(x) = \sum_{i=1}^d f_i(e^T x)$, where f_i are as in (ii).

(iv) *[Multi–index model.]* Let $e_1, \dots, e_m$, $m < d$ are direction vectors and assume that $g(x) = f(e_1^T x, \dots, e_m^T x)$ for some unknown m-dimensional function f.

In general, under structural assumptions the rate of testing improves, as compared to the slow d-dimensional rate $\varphi_{\varepsilon,d}(\nu)$.

On the other hand, the assumption that the underlying function g belongs to a poor functional class can lead to inadequate model. In general, it is quite restrictive to assume that g has some *parametric* structure, i.e. the structure described by a finite dimensional parameter and remained unchangeable in the whole domain of observation. Thus, we seek a rather general (*nonparametric*) structural restriction which would admit powerful testing procedures without sacrificing flexibility of the modeling. We argue that this program can be realized if the underlying function g is a composition of two smooth functions.

1.3. Composite functions. We now define our *nonparametric structure* imposed on the model. We will assume that g is a *composite function*, i.e., that $g(t) = f(G(t))$, where $f : [0,1] \to \mathbb{R}$ and $G : \mathcal{D}_0 \to [0,1]$. All our results remain valid if we replace $[0,1]$ in the definition of f and G by some other bounded interval in $\mathbb{R}$.

We will further suppose that f and G are smooth functions such that $f \in \mathbb{H}_1(\gamma, L_1)$ and $G \in \mathbb{H}_d(\beta, L_2)$ where γ, L_1, β, L_2 are positive constants. Here and in what follows $\mathbb{H}_1(\gamma, L_1)$ and $\mathbb{H}_d(\beta, L_2)$ are the Hölder class on $[0,1]$ and the isotropic Hölder class on $\mathcal{D}_0$ respectively (see Definition 1.1 below). The class of composite functions g with such f and G will be denoted by $\mathbb{H}(\mathfrak{a}, \mathfrak{L})$, where $\mathfrak{a} = (\gamma, \beta) \in \mathbb{R}_+^2$ and $\mathfrak{L} = (L_1, L_2) \in \mathbb{R}_+^2$.

The performance of a testing procedure will be measured by the risk function (1.2) where we set $\mathbb{G} = \mathbb{H}(\mathfrak{a}, \mathfrak{L})$. We will see that the value of $\mathfrak{a}$ determines the quality of testing associated to our model, i.e., the rate of testing.

We now give the definitions of anisotropic Hölder ball and discuss some trivial cases of testing under alternatives being composite functions.

DEFINITION 1.1. *Fix $\nu > 0$ and $L > 0$ and an interval $\mathcal{D}'$ in $\mathbb{R}^d$. Let $\lfloor \nu \rfloor$ be the largest integer which is strictly less than ν, and for $\vec{k} = (k_1, \ldots, k_d) \in \mathbb{N}^d$ set $|\vec{k}| = k_1 + \cdots + k_d$. The* ISOTROPIC HÖLDER CLASS $\mathbb{H}_d(\nu, L)$ ON *$\mathcal{D}'$ is the set of all functions $G : \mathcal{D}' \to \mathbb{R}$ having on $\mathcal{D}'$ all partial derivatives of order $\lfloor \nu \rfloor$ and such that*

$$\sum_{0 \le |\vec{k}| \le \lfloor \nu \rfloor} \sup_{x \in \mathcal{D}'} \left| \frac{\partial^{|\vec{k}|} G(x)}{\partial x_1^{k_1} \cdots \partial x_d^{k_d}} \right| \le L,$$

$$\sum_{|\vec{k}| = \lfloor \nu \rfloor} \left| \frac{\partial^{|\vec{k}|} G(y)}{\partial x_1^{k_1} \cdots \partial x_d^{k_d}} - \frac{\partial^{|\vec{k}|} G(z)}{\partial x_1^{k_1} \cdots \partial x_d^{k_d}} \right| \le L ||y - z||^{\nu - \lfloor \nu \rfloor}, \quad \forall\, y, z \in \mathcal{D}'.$$

where x_j and y_j are the jth components of x and y.

REMARK 1.1. *It is easily to see if $G \in \mathbb{H}_d(\nu, L)$ then*

$$\left| G(y) - \sum_{0 \le |\vec{k}| \le \lfloor \nu \rfloor} \frac{\partial^{|\vec{k}|} G(x)}{\partial x_1^{k_1} \cdots \partial x_d^{k_d}} \prod_{j=1}^{d} \frac{(y_j - x_j)^{k_j}}{k_j!} \right| \le L ||y - x||^{\nu}, \quad \forall\, x, y \in \mathcal{D}'.$$

where x_j and y_j are the jth components of x and y.

REMARK 1.2. *It is also evident that if $G \in \mathbb{H}_d(\nu, L)$ then*

$$\frac{\partial G}{\partial x_1} \in \mathbb{H}_d(\nu - 1, L), \quad \frac{\partial G}{\partial x_2} \in \mathbb{H}_d(\nu - 1, L).$$

Trivial cases.

Zone of super-slow rate: $0 < \gamma, \beta \le 1$.

Clearly, in this zone $\mathbb{H}(\mathfrak{a}, \mathfrak{L}) \subset \mathbb{H}_d(\gamma\beta, L_3)$, where L_3 is a positive constant depending only on γ, β and $\mathfrak{L}$. Due to this inclusion, a standard testing procedure [11] converges with the rate $\varphi_{\varepsilon,d}(\gamma\beta) = \left(\varepsilon\sqrt{\ln(1/\varepsilon)}\right)^{2\gamma\beta/(2\gamma\beta+d)}$. It is not hard to show that this rate is optimal. For the proof of the lower bound, it is sufficient to consider a smaller class of functions by fixing $f(\cdot) = |\cdot|^{\gamma}$, and then to use standard techniques of deriving of lower bounds [11]. So, the minimax rate of testing equals to $\varphi_{\varepsilon,d}(\gamma\beta)$. This rate is very slow because $\gamma\beta \le 1$.

Zone of inactive structure: $\gamma \ge \beta$, $\gamma \ge 1$.

In this zone we easily get the inclusions $\mathbb{H}_d(\beta, L_4) \subset \mathbb{H}(\mathcal{A}, \mathcal{L}) \subset \mathbb{H}_d(\beta, L_5)$, where L_4 and L_5 are positive constants depending only on β and $\mathfrak{L}$. To show the left inclusion it suffices to fix a linear function f. Therefore, the rate of testing corresponding to $\mathbb{H}(\mathfrak{a}, \mathfrak{L})$ is the same as for any isotropic Hölder class $\mathbb{H}_d(\beta, \cdot)$ i.e. coincide with $\varphi_{\varepsilon,d}(\beta)$. Note that we test as if there were no structure, and the rate of testing does not depend on γ. This is the zone of *inactive structure*.

We finally remark that if $\beta \le 1$ the composite function g is rather non-smooth. The minimax rate of testing on $\mathbb{H}(\mathfrak{a}, \mathfrak{L})$ is the same as on the Hölder class $\mathbb{H}_d((1 \wedge \gamma)\beta, \cdot)$. This is a very slow rate $\varphi_{\varepsilon,d}((1 \wedge \gamma)\beta)$. Therefore, only for $\beta > 1$ we can expect to find decision rules with powerful statistical properties and later on only the case $\beta > 1$ will be studied.

Moreover, all results in the paper are proved under assumption $d = 2$. Below we discuss in detail this restriction.

2. Test procedure and main results.

2.1. Test procedure. Our test procedure uses two different constructions of the decision rules. Both of them are based on the kernel estimators of the underlying function $g = f(G)$.

Let $\mathcal{K} : \mathbb{R} \to \mathbb{R}$ be a function satisfying the following assumption.

Assumption (K).

1. $\mathcal{K}(u) = 0, \quad \forall u \notin [-1/2, 1/2]$.

2. $\int \mathcal{K}(u)du = 1$.

3. $\exists N \in \mathbb{N}^*$ *such that* $\int u^j \mathcal{K}(u)du = 0, \ \forall j = \overline{1, N}$.

4. $\mathcal{K} \in \mathbb{H}_1(1,1)$ *on* $\mathbb{R}$.

Put $K(v) = \mathcal{K}(v_1)\mathcal{K}(v_2), \ v \in \mathbb{R}^2$. We will say that K is a *kernel.*

First construction.

Put

$$h^* = \begin{cases} \varepsilon^{\frac{2\beta}{2\gamma\beta+\beta+1}}, & \text{if } \beta > 2\gamma - 1; \\ \left[\varepsilon\sqrt{\ln(1/\varepsilon)}\right]^{\frac{2\beta}{(\beta+1)^2}}, & \text{if } \beta \le 2\gamma - 1; \end{cases} \qquad \varkappa = \begin{cases} 1, & \text{if } \beta \le 2; \\ \alpha, & \text{if } \beta > 2; \end{cases}$$

where $\alpha = \beta - \lfloor \beta \rfloor$, and define $\forall \lambda > 0$

$$h[\lambda] = h^* \lambda^{\frac{\varkappa - \beta}{\beta+1}}.$$

We set for any $\tau = (q, \lambda, x)$, where $q : \mathbb{R} \to \mathbb{R}, \ \lambda > 0, \ x \in \mathcal{D}$,

$$\hat{g}_{\varepsilon,1}(\tau) = \frac{1}{\sqrt{\lambda h[\lambda]}\|K\|_2} \int K\left(\frac{t_1 - x_1 - \lambda^{\varkappa} q((t_2 - x_2)/\lambda)}{h[\lambda]}, \frac{t_2 - x_2}{\lambda}\right) dX_\varepsilon(t);$$

$$\hat{g}_{\varepsilon,2}(\tau) = \frac{1}{\sqrt{\lambda h[\lambda]}\|K\|_2} \int K\left(\frac{t_2 - x_2 - \lambda^{\varkappa} q((t_1 - x_1)/\lambda)}{h[\lambda]}, \frac{t_1 - x_1}{\lambda}\right) dX_\varepsilon(t).$$

Later on we will see that $\hat{g}_{\varepsilon,i}(\tau)\big(\|K\|_2/\sqrt{\lambda h}\big), i = 1, 2$, with the special choice of the function q and parameter λ can be viewed as the kernel estimator of the function g at the point x.

Set

$$\varsigma = \begin{cases} \varepsilon^{\frac{\beta-1}{2\gamma\beta+\beta+1}} \Big/ \sqrt{\ln(1/\varepsilon)}, & \text{if } \beta > 2\gamma - 1; \\ \left[\varepsilon\sqrt{\ln(1/\varepsilon)}\right]^{\frac{\beta-\gamma}{\gamma(\beta+1)}}, & \text{if } \beta \le 2\gamma - 1, \end{cases}$$

$$\Lambda = \left\{\lambda_j = \left(\varsigma 2^j\right)^{\kappa}, \quad j = \overline{0, J}\right\}, \quad 1/2 < \varsigma 2^J \leq 1,$$

where $\kappa = 1/(\beta - 1) \vee 1$,.

Put $\mathfrak{T} = \mathbb{H}_1(\beta, \mathcal{S}) \otimes \Lambda \otimes \mathcal{D}$, where $\mathcal{S}$ is the constant from Lemma D.1.

Introduce the random events

$$\mathcal{A}_i = \left\{\sup_{\tau \in \mathfrak{T}} \left|\hat{g}_{\varepsilon,i}(\tau)\left(h[\lambda]/\lambda^{\varkappa}\right)^{1/2\beta}\right| \geq 3\varepsilon\mathcal{L}(\beta, L_2)\right\}, \quad i = 1, 2,$$

where $\mathcal{L}(\beta, L_2) = \sqrt{C}(8\mathcal{S})^{\frac{1}{2\beta}}[1 - 1/\beta]$ and the constant C is defined in Lemma A.1.

The first decision rule is defined as $\mathbb{I}_{\mathcal{A}_1} \vee \mathbb{I}_{\mathcal{A}_2}$.

<u>*Second construction.*</u>

Put

$$\mu = \begin{cases} \varepsilon^{\frac{\beta+1}{2\gamma\beta+\beta+1}}\sqrt{\ln(1/\varepsilon)}, & \text{if} \quad \beta > 2\gamma - 1; \\ \left[\varepsilon\sqrt{\ln(1/\varepsilon)}\right]^{\frac{1}{\beta+1}}, & \text{if} \quad \beta \leq 2\gamma - 1, \end{cases}$$

and define for any $x \in \mathcal{D}$

$$\tilde{g}_\varepsilon(x) = \frac{1}{\mu||K||_2}\int K\left(\frac{t_1 - x_1}{\mu}, \frac{t_2 - x_2}{\mu}\right) dX_\varepsilon(t).$$

Clearly, that $\tilde{g}_\varepsilon(x)[||K||_2/\mu]$ is the standard kernel estimator with the bandwidth μ.

Introduce the random event

$$\mathcal{B} = \left\{\sup_{x \in \mathcal{D}} |\tilde{g}_\varepsilon(x)| \geq \left[3\sqrt{\ln(1/\mu)} + 2\right]\varepsilon\right\}.$$

The second decision rule is defined as $\mathbb{I}_{\mathcal{B}}$.

Our final decision rule is simply the maximum between the first and second one, i.e $\Delta^*_\varepsilon = \mathbb{I}_{\mathcal{A}_1} \vee \mathbb{I}_{\mathcal{A}_2} \vee \mathbb{I}_{\mathcal{B}}$.

2.2. Main results. For any separation sequence $\phi_\varepsilon \to 0, \varepsilon \to 0$ let us denote

$$\mathbb{H}(\phi_\varepsilon) = \left\{g \in \mathbb{H}(\mathfrak{a}, \mathfrak{L});\ ||g||_\infty \geq \phi_\varepsilon\right\}$$

and consider the separation sequence

$$\varphi_\varepsilon(\mathfrak{a}) = \begin{cases} \varepsilon^{\frac{2\beta\gamma}{2\gamma\beta+\beta+1}}, & \text{if} \quad \beta > 2\gamma - 1; \\ \left[\varepsilon\sqrt{\ln(1/\varepsilon)}\right]^{\frac{\beta}{\beta+1}}, & \text{if} \quad \beta \leq 2\gamma - 1. \end{cases}$$

THEOREM 2.1. *Let $\mathfrak{a} \in \mathbb{R}^2_+$ and $\mathfrak{L} \in \mathbb{R}^2_+$ be fixed. There exists $C^* = C(\mathfrak{a}, \mathfrak{L}) > 0$ such that $\forall C > C^*$*

$$\limsup_{\varepsilon \to 0}\left[\mathbb{P}_0\{\mathcal{A}_1 \cup \mathcal{A}_2 \cup \mathcal{B}\} + \sup_{g \in \mathbb{H}\left(C\varphi_\varepsilon(\mathfrak{a})\right)} \mathbb{P}_g\{\bar{\mathcal{A}}_1 \cap \bar{\mathcal{A}}_2 \cap \bar{\mathcal{B}}\}\right] = 0.$$

The explicit expression of the constant C^* can be found in the proof of the theorem.

THEOREM 2.2. *For any given* $\mathfrak{a} \in \mathbb{R}^2_+$ *and* $\mathfrak{L} \in \mathbb{R}^2_+$

$$\liminf_{C\to 0}\liminf_{\varepsilon\to 0}\inf_{\tilde{\Delta}_\varepsilon}\Big[\mathbb{P}_0\{\tilde{\Delta}_\varepsilon = 1\} + \sup_{g\in\mathbb{H}\left(C\varphi_\varepsilon(\mathfrak{a})/[\ln(1/\varepsilon)]^\eta\right)}\mathbb{P}_g\{\tilde{\Delta}_\varepsilon = 0\}\Big] = 1,$$

where inf *is taken over all possible decision rules. Here* $\eta = 0$ *if* $\beta \le 2\gamma - 1$ *and* $\eta > 0$ *is an arbitrary number if* $\beta > 2\gamma - 1$.

REMARK 2.1. **(Upper bound)**

1. *The assertions proved in Theorem 2.1 and Theorem 2.2 allow us to state that* Δ^*_ε *is minimax decision rule and* $\varphi_\varepsilon(\mathfrak{a})$ *is the minimax rate of testing in the case* $\beta \le 2\gamma - 1$.
2. *Moreover, if* $\beta \le 2\gamma - 1$ *the rate of testing* $\varphi_\varepsilon(\mathfrak{a})$ *does not depend on* γ *and is equal to the minimax rate* $\varphi_{\varepsilon,2}(\beta)$ *associated to the isotropic Hölder class* $\mathbb{H}_d(\beta,\cdot)$. *This is rather surprising: the quality of the test under alternatives written in the form the* $g = f(G)$ *is the same as for the alternatives described by function* G *independently of function* f. *Such a behavior cannot be explained in the terms of smoothness:* $f(G) \in \mathbb{H}_2(\gamma,\cdot)$ *and does not belong to* $\mathbb{H}_2(\beta,\cdot)$
3. *Also, it worths to mention that the method of obtaining of lower bound in the case* $\beta \le 2\gamma - 1$ *can be simply generalized for an arbitrary dimension* $d > 2$. *The corresponding domain would be* $\beta \le d(\gamma - 1) + 1$.

REMARK 2.2. **(Lower bound)**

1. *In the case* $\beta > 2\gamma - 1$ *the separation sequences in Theorem 2.1 and in Theorem 2.2 do not coincide. However, they differ from each other only by the factor* $[\ln(1/\varepsilon]^\eta$, *where* η *is an arbitrary positive number. It allows us to say that the decision rule* Δ^*_ε *is "almost" minimax decision rule. We are definitely sure that the separation sequence* $\varphi_\varepsilon(\mathfrak{a})$ *from Theorem 2.1 is the minimax rate of testing and, therefore, the lower bound result should be improved.*
2. *Let us also note that the lower bound construction used in the case* $\beta > 2\gamma - 1$ *is heavily based on the assumption* $d = 2$.

In the context of the last remark it is interesting to compare our results with the results recently obtained in [13] for the estimation of composite function. In [13] the authors study the *minimax rate of convergence of estimators*, i.e. the asymptotics of minimax risk which is defined as

$$\inf_{\tilde{g}_\varepsilon}\sup_{g\in\mathbb{H}(\mathfrak{a},\mathfrak{L})}\mathbb{E}_g\left(\|\tilde{g}_\varepsilon - g\|_\infty\right),$$

where inf is taken over all possible estimators.

The lower bound result proved in [13] states that for any dimension $d \ge 2$ minimax rate of convergence can not be faster than

$$\psi_\varepsilon(\mathfrak{a}) = \begin{cases} \left[\varepsilon\sqrt{\ln(1/\varepsilon)}\right]^{\frac{2\beta\gamma}{2\gamma\beta+\beta+d-1}}, & \text{if } \beta > d(\gamma-1)+1; \\ \left[\varepsilon\sqrt{\ln(1/\varepsilon)}\right]^{\frac{2\beta}{2\beta+d}}, & \text{if } \beta \le d(\gamma-1)+1. \end{cases}$$

Comparing this result with the results given by Theorem 2.1 we can conclude that the *minimax rate of convergence of estimators* and *minimax rate of testing* differ if $\beta > 2\gamma - 1$ and $d = 2$. This result is very unusual. As far as we know it is the first problem described in the literature where the *minimax rate of convergence of estimators* and the *minimax rate of testing* are different in the case of **sup-norm losses**. In particular, the upper bound result proved in [13] under additional assumption $\gamma \in (0,2), \beta \in (1,2]$ shows that $\psi_\varepsilon(\mathfrak{a})$ is the *minimax rate of convergence of estimators* if $\beta \le d(\gamma-1)+1$. As we see both rates, $\varphi_\varepsilon(\mathfrak{a})$ and $\psi_\varepsilon(\mathfrak{a})$ coincide if $\beta \le 2\gamma - 1$ and $d = 2$.

2.3. Open problems.

OTHER STATISTICAL MODELS AND PROBLEMS.

1. We considered the gaussian white noise model because it is the simplest and idealized object to study, its analysis requires a minimum of technicalities. Composition structures can be studied in the same spirit for more realistic models, such as nonparametric regression with random design, nonparametric density estimation and classification. Note that our theorems can be directly transposed to gaussian nonparametric regression model with fixed equidistant design using the equivalence of experiments argument [2], [18].
2. We restrict our study to the sup-norm loss and to the Hölder smoothness classes. A natural extension would be to consider models where the *risk function* is described by other norms and other smoothness classes, such as Sobolev and Besov classes, or, for instance, by the classes of monotone or convex functions. The case of functional classes with anisotropic smoothness is of interest as well.
3. We consider only the simplest composition $f(G)$, where $f : \mathbb{R} \to \mathbb{R}$ and $G : \mathbb{R}^d \to \mathbb{R}$. A more general description could be $f(G_1, \dots, G_k)$, where $f : \mathbb{R}^k \to \mathbb{R}$ and $G_s : \mathbb{R}^{d_s} \to \mathbb{R}$, $s = \overline{1,k}$, and $d_1 + \dots + d_k = d$.
4. A related more complex modeling can be based on Kolmogorov's theorem of representation of a continuous function of several variables by compositions and sums of functions of one variable addition [14], [19].

REFINEMENT OF ASSUMPTIONS.

1. In the present paper we are able to treat only the case $d = 2$. We believe that our upper bound result (Theorem 2.1), including testing procedure, with minor changes can be directly applied in any dimension in the case $\beta \le d(\gamma-1)+1$. It will require only

to prove smoothness properties of the implicit function of $d-1$ variables analogously to result obtained in Lemma D.1.
This conjecture is partially confirmed by the results from [13] for the case $\gamma \in (0,2), \beta \in (1,2]$. To construct the decision rule one can apply the estimation procedure proposed in [13] which is quite different from the constructions used in the present paper. In view of Remark 2.2 this approach would bring a minimax decision rule.

2. However, the lower bound result given by Theorem 2.2 if $\beta > 2\gamma-1$ uses rather sophisticated construction of random walk on $\mathbb{R}$ recently proposed in [5]. As far as we know such result does not exist in the dimension larger than 1, that restricts the use of our lower bound construction in the dimension larger that 2. Moreover, as it has been already mentioned in Remark 2.2, even if $d=2$ the lower bound is not exact.
As a consequence, in the case $\beta > d(\gamma-1)+1$ the extension to an arbitrary dimension as well as the exact lower bound for $d=2$ for the risk function (1.2) remain open problems.

3. Proofs.

3.1. Proof of Theorem 1. I. Let us find the upper bound for the first error probability, i.e. for $\mathbb{P}_0\{\mathcal{A}_1 \cup \mathcal{A}_2 \cup \mathcal{B}\}$.
For given $\lambda \in \Lambda$ let us denote $\mathfrak{T}_\lambda = \Big\{\tau = (q,\lambda,x),\ q \in \mathbb{H}(\beta,\mathcal{S}),\ x \in \mathcal{D}\Big\}$ and put

$$\mathcal{A}_{\lambda,i} = \left\{\sup_{\tau \in \mathfrak{T}_\lambda} \left|\hat{g}_{\varepsilon,i}(\tau)\Big(h[\lambda]/\lambda^{\varkappa}\Big)^{1/2\beta}\right| \geq \varepsilon\mathcal{K}_1\right\},\ i=1,2.$$

Note that

$$\mathbb{P}_0\{\mathcal{A}_1 \cup \mathcal{A}_2 \cup \mathcal{B}\} \leq \mathbb{P}_0\{\mathcal{B}\} + 2\sum_{\lambda \in \Lambda} \mathbb{P}_0\{\mathcal{A}_{\lambda,1}\}, \tag{3.1}$$

since the distributions of the gaussian random functions $\hat{g}_{\varepsilon,1}$ and $\hat{g}_{\varepsilon,2}$ coincide under $\mathbb{P}_0$.
Put also

$$\hat{\xi}(\tau) = \frac{1}{\sqrt{\lambda h[\lambda]}\|K\|_2} \int K\left(\frac{t_1-x_1-\lambda^{\varkappa} q((t_2-x_2)/\lambda)}{h[\lambda]}, \frac{t_2-x_2}{\lambda}\right) dW(t), \quad \tau \in \mathfrak{T};$$

$$\tilde{\xi}(x) = \frac{1}{\mu\|K\|_2} \int K\left(\frac{t_1-x_1}{\mu}, \frac{t_2-x_2}{\mu}\right) dW(t), \quad x \in \mathcal{D}.$$

First we find the upper bounds for $\mathbb{E}\sup_{\tau \in \mathfrak{T}_\lambda} \hat{\xi}(\tau), \lambda \in \Lambda$, and for $\mathbb{E}\sup_{x \in \mathcal{D}} \tilde{\xi}(x)$.

I.1.a. *Upper bound for* $\mathbb{E}\sup_{\tau \in \mathfrak{T}} \hat{\xi}(\tau)$.

Let us fix $\lambda \in \Lambda$, $\tau = (q, \lambda, x) \in \mathfrak{T}_\lambda$, $\tau' = (q', \lambda, x') \in \mathfrak{T}_\lambda$ and consider the intrinsic semi-metric generated by gaussian random function $\hat{\xi}(\cdot)$:

$$\hat{\rho}_\lambda(\tau, \tau') = \sqrt{\mathbb{E}\Big(\hat{\xi}(\tau) - \hat{\xi}(\tau')\Big)^2}.$$

Putting $\forall t \in \mathbb{R}$

$$\frac{1}{\sqrt{\lambda h[\lambda]}||K||_2} K\left(\frac{t_1 - x_1 - \lambda^\varkappa q\big((t_2 - x_2)/\lambda\big)}{h[\lambda]}, \frac{t_2 - x_2}{\lambda}\right) = K^\lambda_{q,x}(t)$$

we have

$$\hat{\rho}_\lambda(\tau, \tau') = \left\| K^\lambda_{q,x} - K^\lambda_{q',x'} \right\|_2$$

and, therefore,

$$\begin{aligned} \hat{\rho}_\lambda(\tau, \tau') &\le \left\| K^\lambda_{q,x} - K^\lambda_{q',x} \right\|_2 + \left\| K^\lambda_{q',x} - K^\lambda_{q',x'} \right\|_2 \\ &\triangleq \sqrt{I_1} + \sqrt{I_2}. \end{aligned} \tag{3.2}$$

Taking into account that $\left\| K^\lambda_{q,x} \right\|_2 = 1$, $\forall \tau = (q, \lambda, x) \in \mathfrak{T}$ and the definition of the kernel K we obtain by direct calculations

$$\begin{aligned} I_1 &= 2\Big[1 - \langle K^\lambda_{q,x}, K^\lambda_{q',x}\rangle\Big] = -2\langle K^\lambda_{q,x}, K^\lambda_{q',x} - K^\lambda_{q,x}\rangle \\ &= -\frac{2}{||K||_2^2} \int \mathcal{K}^2(v_2)\mathcal{K}(v_1) \Big[\mathcal{K}\Big(v_1 + (\lambda^\varkappa/h[\lambda])\,[q(v_2) - q'(v_2)]\Big) - \mathcal{K}(v_1)\Big]\, dv_1 dv_2. \end{aligned}$$

Applying the assumption **(K1)**, **(K2)** and **(K4)** we get

$$I_1 \le 2\left[\frac{||\mathcal{K}||_1}{||\mathcal{K}||_2^2}\right]\left[\frac{\lambda^\varkappa}{h[\lambda]}\right] ||q - q'||_\infty \le 2\left[\frac{\lambda^\varkappa}{h[\lambda]}\right] ||q - q'||_\infty \triangleq \mathbb{Q}(\lambda)||q - q'||_\infty. \tag{3.3}$$

Putting $q'_{x,x'}(\cdot) = q'(\cdot) - q'\big(\cdot + (x_2 - x'_2)/\lambda\big)$ we have similarly to (3.3)

$$\begin{aligned} I_2 &= -2\langle K^\lambda_{q',x}, K^\lambda_{q',x'} - K^\lambda_{q',x}\rangle \\ &= -\frac{2}{||K||_2^2} \int K(v)\left[K\left(v_1 + \frac{x_1 - x'_1 - \lambda^\varkappa q'_{x,x'}(v_2)}{h[\lambda]}, v_2 + \frac{x_2 - x'_2}{\lambda}\right) - K(v)\right] dv. \end{aligned} \tag{3.4}$$

Taking into account that $q' \in \mathbb{H}(\beta, \mathcal{S})$ and, therefore,

$$||q'_{x,x'}||_\infty \le \mathcal{S}\frac{|x_2 - x'_2|}{\lambda} \tag{3.5}$$

applying assumptions **(K1)**, **(K2)** and **(K4)** we obtain from (3.4) and (3.5)

$$\begin{aligned}
I_2 &\le 2\Bigg[\frac{|x_2-x_2'|}{\lambda}\left(1+\frac{\mathcal{S}\lambda^{\varkappa}}{h[\lambda]}\right)+\frac{|x_1-x_1'|}{h[\lambda]} \\
&\quad + \mathrm{diam}(\mathcal{D})\left(\frac{|x_1-x_1'|}{\lambda h[\lambda]}+\frac{\mathcal{S}\lambda^{\varkappa}|x_2-x_2'|}{\lambda^2 h[\lambda]}\right)\Bigg] \\
&\le \frac{2\left[1+\ \mathrm{diam}(\mathcal{D})+[\mathcal{S}\vee 1]\ \mathrm{diam}(\mathcal{D})\right]}{\lambda^{2-\varkappa}h[\lambda]}\left\{|x_1-x_1'|+|x_2-x_2'|\right\} \\
&\le 2\sqrt{2}\left[h^*\lambda_0^{3/2}\right]^{-1}\left[1+\ \mathrm{diam}(\mathcal{D})+[\mathcal{S}\vee 1]\ \mathrm{diam}(\mathcal{D})\right]||x-x'||_1,
\end{aligned}$$

where $||\cdot||_1$ is $\mathbb{L}_1$-norm on $\mathbb{R}^2$. Thus, we have

$$\begin{aligned}
I_2 &\le 2\sqrt{2}\left[h^*\lambda_0^{3/2}\right]^{-1}\left[1+\ \mathrm{diam}(\mathcal{D})+[\mathcal{S}\vee 1]\ \mathrm{diam}(\mathcal{D})\right]||x-x'||_1 \\
&\triangleq \mathbb{C}_\varepsilon||x-x'||_1.
\end{aligned} \tag{3.6}$$

Finally, we obtain from (3.2), (3.3) and (3.6)

$$\hat{\rho}_\lambda(\tau,\tau') \le \sqrt{\mathbb{Q}(\lambda)||q-q'||_\infty}+\sqrt{\mathbb{C}_\varepsilon||x-x'||_1}. \tag{3.7}$$

For any $\delta>0$ let us denote by $\mathcal{E}_\delta^{(\infty)}\left(\mathbb{H}_1(\beta,\mathcal{S})\right)$, $\mathcal{E}_\delta^{(1)}(\mathcal{D})$ and $\mathcal{E}_\delta^{(\hat{\rho}_\lambda)}(\mathfrak{T}_\lambda)$, $\lambda\in\Lambda$, the δ-entropy of $\mathbb{H}_1(\beta,\mathcal{S})$ w.r.t uniform norm, δ-entropy of $\mathcal{D}$ w.r.t. $||\cdot||_1$-norm, and δ-entropy of $\mathfrak{T}_\lambda$ w.r.t. intrinsic semi-metric $\hat{\rho}_\lambda$ respectively.

Clearly, that $\forall\delta>0$

$$\mathcal{E}_\delta^{(1)}(\mathcal{D}) \le 2\ln\left(\frac{\mathrm{diam}(\mathcal{D})}{\delta}\right) \tag{3.8}$$

Note also that (3.7) implies $\forall u>0$ and $\forall\lambda\in\Lambda$

$$\mathcal{E}_u^{(\hat{\rho}_\lambda)}(\mathfrak{T}_\lambda) \le \mathcal{E}^{(\infty)}_{\frac{u^2}{4\mathbb{Q}(\lambda)}}\left(\mathbb{H}(\beta,\mathcal{S})\right)+\mathcal{E}^{(1)}_{\frac{u^2}{4\mathbb{C}_\varepsilon}}(\mathcal{D}). \tag{3.9}$$

Therefore, we deduce from (3.8), Lemma A.1 and (3.9) that $\forall u>0$ and $\forall\lambda\in\Lambda$

$$\mathcal{E}_u^{(\hat{\rho}_\lambda)}(\mathfrak{T}_\lambda) \le \mathcal{C}\left[4\mathcal{S}\mathbb{Q}(\lambda)\right]^{\frac{1}{\beta}}u^{-\frac{2}{\beta}}+4\ln(1/u)+2\ln\left(4\mathbb{C}_\varepsilon\ \ \mathrm{diam}(\mathcal{D})\right) \tag{3.10}$$

Noting that $\mathbb{E}\hat{\xi}^2(\tau)=1$, $\forall\tau\in\mathfrak{T}$, and applying Lemma B.1 with $\sigma=1$ we obtain from (3.10) $\forall\lambda\in\Lambda$

$$\mathbb{E}\sup_{\tau\in\mathfrak{T}_\lambda}\hat{\xi}(\tau) \le \mathcal{L}(\beta,L_2)\left[\frac{\lambda^{\varkappa}}{h[\lambda]}\right]^{\frac{1}{2\beta}}+\sqrt{2\ln\mathbb{C}_\varepsilon}+\mathcal{L},$$

where $\mathcal{L}(\beta,L_2)=\sqrt{\mathcal{C}}(8\mathcal{S})^{\frac{1}{2\beta}}[1-1/\beta]$ and $\mathcal{L}=\sqrt{2\ln\left(4\ \ \mathrm{diam}(\mathcal{D})\right)}+2$.

Simple calculations show that $\ln \mathbb{C}_\varepsilon \asymp \ln(1/\varepsilon)$ and

$$\left[\inf_{\lambda\in\Lambda} \frac{\lambda^\varkappa}{h[\lambda]}\right]^{\frac{1}{2\beta}} = \left[\frac{\lambda_0^\varkappa}{h[\lambda_0]}\right]^{\frac{1}{2\beta}} \geq \varepsilon^{-a}, \tag{3.11}$$

where $a > 0$ depends on β only. Thus, we have for all $\varepsilon > 0$ small enough and $\forall \lambda \in \Lambda$

$$\mathbb{E} \sup_{\tau\in\mathfrak{T}_\lambda} \hat{\xi}(\tau) \leq 2\mathcal{L}(\beta, L_2)\left[\frac{\lambda^\varkappa}{h[\lambda]}\right]^{\frac{1}{2\beta}}. \tag{3.12}$$

I.1.b. *Upper bound for* $\mathbb{E}\sup_{x\in\mathcal{D}} \tilde{\xi}(x)$.

Putting $\forall x, x' \in \mathcal{D}$

$$\tilde{\rho}(x, x') = \sqrt{\mathbb{E}\Big(\tilde{\xi}(x) - \hat{\xi}(x')\Big)^2}.$$

we obtain similarly to (3.6) $\forall x, x' \in \mathcal{D}$

$$\tilde{\rho}(x, x') \leq \mu^{-1}\sqrt{\operatorname{diam}(\mathcal{D})\|x - x'\|_1}.$$

This together with $\mathbb{E}\tilde{\xi}^2(x) = 1,\ \forall x \in \mathcal{D}$, (3.8) and Lemma B.1 gives

$$\mathbb{E}\sup_{x\in\mathcal{D}} \tilde{\xi}(x) \leq 2\sqrt{\ln(1/\mu)} + 2 \tag{3.13}$$

Note that $\forall \lambda \in \Lambda$ in view of (3.12) and (3.11) we have

$$\begin{aligned}
\mathbb{P}_0\{\mathcal{A}_{\lambda,1}\} &= \mathbb{P}\left\{\sup_{\tau\in\mathfrak{T}_\lambda} |\hat{\xi}(\tau)| \geq 3\mathcal{L}(\beta, L_2)\Big(\lambda^\varkappa/h[\lambda]\Big)^{1/2\beta}\right\} \\
&= 2\mathbb{P}\left\{\sup_{\tau\in\mathfrak{T}_\lambda} \hat{\xi}(\tau) \geq 3\mathcal{L}(\beta, L_2)\Big(\lambda^\varkappa/h[\lambda]\Big)^{1/2\beta}\right\} \\
&\leq 2\mathbb{P}\left\{\sup_{\tau\in\mathfrak{T}_\lambda} \hat{\xi}(\tau) - \mathbb{E}\sup_{\tau\in\mathfrak{T}_\lambda} \hat{\xi}(\tau) \geq \mathcal{L}(\beta, L_2)\Big(\lambda^\varkappa/h[\lambda]\Big)^{1/2\beta}\right\} \\
&\leq 2\mathbb{P}\left\{\sup_{\tau\in\mathfrak{T}_\lambda} \hat{\xi}(\tau) - \mathbb{E}\sup_{\tau\in\mathfrak{T}_\lambda} \hat{\xi}(\tau) \geq \mathcal{L}(\beta, L_2)\varepsilon^{-a}\right\}.
\end{aligned}$$

Applying Lemma B.2 with $\sigma = 1$ we get $\forall \lambda \in \Lambda$

$$\mathbb{P}_0\{\mathcal{A}_{\lambda,1}\} \leq 2\exp\left\{-\frac{1}{2}\mathcal{L}^2(\beta, L_2)\varepsilon^{-2a}\right\}$$

and, therefore

$$\sum_{\lambda\in\Lambda} \mathbb{P}_0\{\mathcal{A}_{\lambda,1}\} \leq 2\log_2(1/\varsigma)\exp\left\{-\frac{1}{2}\mathcal{L}^2(\beta, L_2)\varepsilon^{-2a}\right\}. \tag{3.14}$$

Using the same arguments we have from (3.13)

$$\mathbb{P}_0\{\mathcal{B}\} \le 2\mathbb{P}\left\{\sup_{x\in\mathcal{D}} \tilde{\xi}(x) - \mathbb{E}\sup_{x\in\mathcal{D}} \tilde{\xi}(x) \ge \sqrt{\ln(1/\mu)}\right\} \le 2\sqrt{\mu}. \quad (3.15)$$

Taking together the bounds found in (3.14) and (3.15) we get from (3.1)

$$\mathbb{P}_0\{\mathcal{A}_1 \cup \mathcal{A}_2 \cup \mathcal{B}\} \to 0, \quad \text{as } \varepsilon \to 0. \quad (3.16)$$

II. Let us find the upper bound for the second error probability, i.e. for

$$\sup_{g\in\mathbb{H}(C\varphi_\varepsilon(\mathfrak{a}))} \mathbb{P}_g\{\bar{\mathcal{A}}_1 \cap \bar{\mathcal{A}}_2 \cap \bar{\mathcal{B}}\}.$$

For any $g \in \mathbb{H}(\mathfrak{a}, \mathfrak{L})$, $\tau \in \mathfrak{T}$ and $x \in \mathcal{D}$ let us denote by

$$\hat{B}_g(\tau) = \frac{1}{\lambda h[\lambda]}\int K\left(\frac{t_1 - x_1 - \lambda^{\varkappa} q\big((t_2-x_2)/\lambda\big)}{h[\lambda]}, \frac{t_2 - x_2}{\lambda}\right) f\big(G(t)\big)dt - f\big(G(x)\big);$$

$$\tilde{B}_g(x) = \frac{1}{\mu^2}\int K\left(\frac{t_1 - x_1}{\mu}, \frac{t_2 - x_2}{\mu}\right) f\big(G(t)\big)dt - f\big(G(x)\big).$$

It is clear that

$$\begin{aligned} \hat{B}_g(\tau) &= \mathbb{E}_g\left[\frac{\hat{g}\|K\|_2}{\sqrt{\lambda h[\lambda]}}\right] - f\big(G(x)\big), \\ \tilde{B}_g(x) &= \mathbb{E}_g\left[\frac{\tilde{g}_\varepsilon(x)\|K\|_2}{\mu}\right] - f\big(G(x)\big). \end{aligned} \quad (3.17)$$

Let $\mathcal{M} \triangleq \mathcal{M}(\beta, L_2)$ be the constant from Lemma D.1. Put $j = \overline{1, J+1}$,

$$\mathfrak{H}_0(x) = \left\{g \in \mathbb{H}(\mathfrak{a}, \mathfrak{L}) : \ \big|G_{1,0}(x)\big| \vee \big|G_{0,1}(x)\big| \le \frac{\varsigma}{\mathcal{M}^{1/\kappa}}\right\};$$

$$\mathfrak{H}_j(x) = \left\{g \in \mathbb{H}(\mathfrak{a}, \mathfrak{L}) : \ \left[\frac{\lambda_{j-1}}{\mathcal{M}}\right]^{1/\kappa} \le \big|G_{1,0}(x)\big| \vee \big|G_{0,1}(x)\big| \le \left[\frac{\lambda_j}{\mathcal{M}}\right]^{1/\kappa}\right\},$$

where, $\lambda_{J+1} \triangleq \mathcal{M}(L_2)^\kappa$, $G_{1,0}(x)$ and $G_{0,1}(x)$ are $\big[\partial G/\partial t_1\big](x)$ and $\big[\partial G/\partial t_2\big](x)$ respectively. First, for given $x \in \mathcal{D}$ let us find the upper bound for

$$\sup_{g\in\mathfrak{H}_0(x)} \Big|\tilde{B}_g(x)\Big|, \qquad \sup_{g\in\mathfrak{H}_j(x)} \Big|\hat{B}_g(\tau)\Big|, \ \tau = (q, \lambda_j, x), \ j = \overline{1, J+1}.$$

We denote by $S_1, S_2, \ldots,$ the functions uniformly bounded by 1 on $\mathcal{D}$ and by $C_1, C_2, \ldots,$ the absolute constants and without loss of generality we will assume that

$$\big|G_{1,0}(x)\big| = \big|G_{1,0}(x)\big| \vee \big|G_{0,1}(x)\big|. \quad (3.18)$$

II.1.a. *Upper bound for* $\sup_{g\in\mathfrak{H}_0(x)}\left|\tilde{B}_g(x)\right|$.

In view of Remark 1.1 we have $\forall t\in\mathcal{D}$

$$G(t)=G(x)+\mathcal{P}_x(t-x)+S_1(t)L_2||t-x||^{\beta},$$

where

$$\mathcal{P}_x(v)=\sum_{1\le|\vec{k}|\le\lfloor\beta\rfloor}\frac{\partial^{|\vec{k}|}G(x)}{\partial x_1^{k_1}\partial x_2^{k_2}}\prod_{j=1}^{2}\frac{(v_j)^{k_j}}{k_j!}$$

Since $f\in\mathbb{H}_1(\gamma,L_1)$ we have $\forall t\in\mathcal{D}$ (the summation with respect to empty set of indexes is supposed to be 0)

$$\begin{aligned}
f\big(G(t)\big)&=f\big(G(x)+\mathcal{P}_x(t-x)\big)+\sum_{l=1}^{\lfloor\gamma\rfloor}f^{(l)}\big(G(x)+\mathcal{P}_x(t-x)\big)\left[S_1(t)L_2||t-x||^{\beta}\right]^l\\
&\quad+S_2(t)L_1L_2^{\gamma}||t-x||^{\gamma\beta}\\
&=f\big(G(x)+\mathcal{P}_x(t-x)\big)\\
&\quad+\chi_1C_1S_3(t)L_1\left[L_2\vee 1\right]^{\gamma}||t-x||^{\beta}+S_2(t)L_1L_2^{\gamma}||t-x||^{\gamma\beta}\\
&=f\big(G(x)\big)+\sum_{l=1}^{\lfloor\gamma\rfloor}f^{(l)}\big(G(x)\big)\left[\mathcal{P}_x(t-x)\right]^l+L_1S_4(t)\left[\mathcal{P}_x(t-x)\right]^{\gamma}\\
&\quad+\chi_1C_1S_3(t)L_1\left[L_2\vee 1\right]^{\gamma}||t-x||^{\beta}+S_2(t)L_1L_2^{\gamma}||t-x||^{\gamma\beta},
\end{aligned}\tag{3.19}$$

where $\chi_1=0$ if $\gamma\le 1$ and $\chi_1=1$ if $\gamma>1$.

Taking into account that $\forall t\in[x_1-(1/2)\mu,x_1+(1/2)\mu]\times[x_2-(1/2)\mu,x_2+(1/2)\mu]$ and $\forall g\in\mathfrak{H}_0(x)$,

$$\left|\mathcal{P}_x(t-x)\right|\le C_2\varsigma\mu+\chi_2C_3L_2\mu^2,\quad \chi_2=\begin{cases}0, & \text{if } 1<\beta\le 2;\\ 1, & \text{if } \beta>2,\end{cases}$$

and using the assumptions **(K1)**, **(K2)** and **(K3)** with $N\ge\beta^2$ we obtain from (3.19):

$$\sup_{g\in\mathfrak{H}_0(x)}\left|\tilde{B}_g(x)\right|\le C_1(\mathfrak{a},\mathfrak{L})\left[(\varsigma\mu)^{\gamma}+\chi_2\mu^{2\gamma}+\chi_1\mu^{\beta}+\mu^{\gamma\beta}\right],\tag{3.20}$$

where $C_1(\mathfrak{a},\mathfrak{L})$ is the constant depending only on γ,β,L_1 and L_2.

Putting $\mathfrak{s}=2\gamma\wedge\beta\wedge\gamma\beta$ we note that (3.20) implies $\forall\mathfrak{a}\in\mathfrak{A}$

$$\sup_{g\in\mathfrak{H}_0(x)}\left|\tilde{B}_g(x)\right|\le 3C_1(\mathfrak{a},\mathfrak{L})\left[(\varsigma\mu)^{\gamma}+\mu^{\mathfrak{s}}\right]\le 6C_1(\mathfrak{a},\mathfrak{L})\varphi_{\varepsilon}(\mathfrak{a}).\tag{3.21}$$

II.1.b. *Upper bound for* $\sup_{g\in\mathfrak{H}_j(x)}\left|\hat{B}_g(\tau)\right|$, $j=\overline{1,J+1}$.

Let us denote by $q_x(\cdot)$ the implicit function being the solution of the equation $G(t)=G(x)$. In view of Lemma D.1 $\forall G\in\mathfrak{H}_j(x)$ the function $q_x(\cdot)$ is uniquely defined on the interval $I_j=\left[x_2-\lambda_j,x_2+\lambda_j\right], j=\overline{1,J}$ and $I_{J+1}=\left[x_2-1,x_2+1\right]$.

In view of Remark 1.1, taking into account that $G\big(q_x(t_2),t_2\big)=G(x),\forall t_2\in I_j$ we have $\forall t=(t_1,t_2)\in\mathcal{D}$ such that $t_2\in I_j$

$$G(t)=G(x)+\mathfrak{P}_x\big(t_1-q_x(t_2),t_2\big)+S_5(t)L_2\big|t_1-q_x(t_2)\big|^{\beta},$$

$$\mathfrak{P}_x(u,t_2)=\sum_{k=1}^{\lfloor\beta\rfloor}\left[\frac{\partial^k}{\partial x_1^k}G\big(q_x(t_2),t_2\big)\right]u^k\triangleq\sum_{k=1}^{\lfloor\beta\rfloor}\mathcal{Z}_{k,x}(t_2)u^k. \tag{3.22}$$

Therefore, repeating all lines in (3.19) we obtain $\forall t\in I_j\otimes\mathcal{I}_j$

$$\begin{aligned}f\big(G(t)\big)&=f\big(G(x)\big)+\sum_{l=1}^{\lfloor\gamma\rfloor}f^{(l)}\big(G(x)\big)\left[\mathfrak{P}_x\big(t_1-q_x(t_2),t_2\big)\right]^l\\&\quad+\chi_1C_4S_7(t)L_1\left[L_2\vee 1\right]^{\gamma}\big|t_1-q_x(t_2)\big|^{\beta}\\&\quad+L_1S_8(t)\left[\mathfrak{P}_x\big(t_1-q_x(t_2),t_2\big)\right]^{\gamma}+S_6(t)L_1L_2^{\gamma}\big|t_1-q_x(t_2)\big|^{\gamma\beta},\end{aligned} \tag{3.23}$$

Set

$$q_{j,x}(v)=\left[q_x\big(x_2+v\lambda_j\big)-x_1\right]\Big/\lambda_j^{\varkappa} \tag{3.24}$$

and let $\tau_{x,j}=\big(q_{j,x},\lambda_j,x\big)$. Note that

$$\begin{aligned}\hat{B}_g(\tau_{j,x})&=\frac{1}{\lambda_jh[\lambda_j]}\int K\left(\frac{t_1-q_x(t_2)}{h[\lambda_j]},\frac{t_2-x_2}{\lambda_j}\right)f\big(G(t)\big)dt-f\big(G(x)\big)\\&=\frac{1}{\lambda_jh[\lambda_j]}\int\mathcal{K}\left(\frac{t_1-q_x(t_2)}{h[\lambda_j]}\right)\mathcal{K}\left(\frac{t_2-x_2}{\lambda_j}\right)f\big(G(t)\big)dt-f\big(G(x)\big).\end{aligned} \tag{3.25}$$

Here we used, in particular, the definition of the kernel K. Let us also remark that the kernel $\mathcal{K}$ is vanished on $\mathbb{R}\setminus[-1/2,1/2]$ and, therefore, the integration w.r.t t_2 in (3.25) is done on the domain of definition of $q_x(\cdot)$ (see Lemma D.1 and the definition of λ_j).

Changing the variables in (3.25) $u=\big(t_1-q_x(t_2)\big)/h[\lambda_j]$, $v=\big(t_2-x_2\big)/\lambda_j$, noting that the functions $\mathcal{Z}_{k,x}(\,\cdot\,)$ in (3.22) are independent of u and using the assumptions **(K1)**, **(K2)** and **(K3)** with $N\geq\beta^2$, we obtain from (3.23) and (3.25) similarly to (3.20)

$$\sup_{g\in\mathfrak{H}_j(x)}\left|\hat{B}_g(\tau_{j,x})\right| \le C_2(\mathfrak{a},\mathfrak{L})\Big\{\Big(\big[\lambda_j\big]^{1/\kappa}h[\lambda_j]\Big)^{\gamma}+\chi_2\Big(h[\lambda_j]\Big)^{2\gamma}$$

$$+\chi_1\Big(h[\lambda_j]\Big)^{\beta}+\Big(h[\lambda_j]\Big)^{\gamma\beta}\Big\} \tag{3.26}$$

$$\le 3C_2(\mathfrak{a},\mathfrak{L})\left\{\Big(\big[\lambda_j\big]^{1/\kappa}h[\lambda_j]\Big)^{\gamma}+\Big(h[\lambda_j]\Big)^{\mathfrak{s}}\right\},$$

where $C_2(\mathfrak{a},\mathfrak{L})$ is the constant depending only on γ,β,L_1 and L_2.

To obtain the first inequality in (3.26) we used that, in view of Lemma D.1 (*2*) and Remark 1.2, there exists $C_5 = C_5(\beta, L_2)$ such that $\forall j=\overline{1,J+1}$ and $\forall g\in\mathfrak{H}_j$

$$\sup_{t_2\in I_j}\left|G_{0,1}\big(q_x(t_2),t_2\big)\right| \le C_5\lambda_j^{1/\kappa},$$

and,therefore, $\forall t\in\mathcal{D}$ such that $\left|t_1-q_x(t_2)\right|\le(1/2)h[\lambda_j]$ and $t_2\in\big[x_2-\lambda_j,x_2+\lambda_j\big]$

$$\left|\mathfrak{P}_x\big(t_1-q_x(t_2),t_2\big)\right| \le C_6\big(\lambda_j\big)^{1/\kappa}h[\lambda_j]+\chi_2C_7L_2\Big(h\big[\lambda_j\big]\Big)^2.$$

Now let us return to (3.26). First, we note that

$$\Big(\big[\lambda_j\big]^{1/\kappa}h[\lambda_j]\Big)^{\gamma} \le \big(\big[h[1]\big)^{\gamma}=\big(h^*\big)^{\gamma}\le\varphi_\varepsilon(\mathfrak{a}). \tag{3.27}$$

Next,

$$\Big(h[\lambda_j]\Big)^{\mathfrak{s}} \le \Big(h\big[\lambda_0\big]\Big)^{\mathfrak{s}} = \Big(h^*\big[\varsigma\big]^{\frac{\kappa(\varkappa-\beta)}{\beta+1}}\Big)^{\mathfrak{s}}$$

$$=\begin{cases}\big[\varphi_\varepsilon(\mathfrak{a})\big]^{z(\mathfrak{a})}\big[\ln(1/\varepsilon)\big]^{y}, & \text{if }\ \beta>2\gamma-1;\\ \big[\varphi_\varepsilon(\mathfrak{a})\big]^{z(\mathfrak{a})} & \text{if }\ \beta\le 2\gamma-1,\end{cases} \tag{3.28}$$

where

$$z(\mathfrak{a})=\begin{cases}\dfrac{\mathfrak{s}}{\gamma}\left[1-\dfrac{\kappa(\beta-\varkappa)(\beta-1)}{2\beta(\beta+1)}\right] & \text{if }\ \beta>2\gamma-1;\\ \dfrac{2\mathfrak{s}}{\beta+1}\left[1-\dfrac{\kappa(\beta-\varkappa)(\beta-\gamma)}{2\gamma\beta}\right] & \text{if }\ \beta\le 2\gamma-1.\end{cases}$$

and $y=y(\mathfrak{a})>0$ is the given number.

1. *Case* $\gamma\le 1, 1<\beta\le 2$.

$$z(\mathfrak{a})=\beta\left[1-\frac{\beta-1}{2\beta(\beta+1)}\right]=\frac{2\beta^2+\beta+1}{2(\beta+1)}>1.$$

2. *Case* $\gamma \le 1, \beta > 2$.

$$z(\mathfrak{a}) = 2\left[1 - \frac{(\beta-\alpha)(\beta-1)}{2\beta(\beta+1)}\right] \ge 2 - \frac{\beta-1}{\beta+1} > 1.$$

Note also that in these cases necessarily $\beta > 2\gamma - 1$.

3. *Case* $1 < \gamma < \beta \le 2,\ \beta > 2\gamma - 1$.

$$z(\mathfrak{a}) = \frac{\beta}{\gamma}\left[1 - \frac{\beta-1}{2\beta(\beta+1)}\right] \ge \frac{2\beta}{\beta+1}\left[1 - \frac{\beta-1}{2\beta(\beta+1)}\right] = \frac{2\beta^2+\beta+1}{(\beta+1)^2} > 1.$$

4. *Case* $1 < \gamma < \beta \le 2,\ \beta \le 2\gamma - 1$.

$$z(\mathfrak{a}) = \frac{2\beta}{\beta+1}\left[1 - \frac{\beta-\gamma}{2\gamma\beta}\right] \ge \frac{2\beta}{\beta+1}\left[1 - \frac{\beta-1}{2\beta(\beta+1)}\right] > 1.$$

5. *Case* $\gamma > 1, \beta > 2,\ \beta > 2\gamma - 1$.

$$z(\mathfrak{a}) = \left[2 \wedge \frac{\beta}{\gamma}\right]\left[1 - \frac{(\beta-\alpha)(\beta-1)}{2\beta(\beta+1)}\right] \ge \frac{2\beta}{\beta+1}\left[1 - \frac{\beta-1}{2(\beta+1)}\right]$$
$$= \frac{\beta^2+3\beta}{(\beta+1)^2} > 1.$$

6. *Case* $\gamma > 1, \beta > 2,\ \beta \le 2\gamma - 1$.

$$z(\mathfrak{a}) = \frac{2\beta}{\beta+1}\left[1 - \frac{(\beta-\alpha)(\beta-\gamma)}{2\gamma\beta}\right] \ge \frac{2\beta}{\beta+1}\left[\frac{3}{2} - \frac{\beta}{2\gamma}\right] \ge \frac{\beta^2+3\beta}{(\beta+1)^2} > 1.$$

Thus, we have from (3.26), (3.27) and (3.28) that

$$\sup_{j=\overline{1,J+1}} \sup_{g \in \mathfrak{H}_j(x)} \left|\hat{B}_g(\tau_{j,x})\right| \le 6C_2(\mathfrak{a}, \mathfrak{L})\varphi_\varepsilon(\mathfrak{a}). \tag{3.29}$$

II.2. Let $C > 0$ be the constant the choice of which will be done later. Since $g \in \mathbb{H}\big(C\varphi_\varepsilon(\mathfrak{a})\big)$ there exists $\breve{x} = \breve{x}(g) \in \mathcal{D}$ such that

$$\left|g(\breve{x})\right| \ge [C/2]\varphi_\varepsilon(\mathfrak{a}). \tag{3.30}$$

Let us introduce the following notations:

$$\breve{j} = \breve{j}(g) = \left\{j = \overline{0, J+1}:\ g \in \mathbb{H}\big(C\varphi_\varepsilon(\mathfrak{a})\big) \cap \mathfrak{H}_j(\breve{x})\right\};$$
$$\breve{q}(v) = q_{\breve{j},\breve{x}}(v) = \left[q_{\breve{x}}\left(\breve{x}_2 + v\lambda_{\breve{j}}\right) - \breve{x}_1\right] \Big/ \left(\lambda_{\breve{j}}\right)^{\varkappa}; \quad \breve{\tau} = \tau_{\breve{j},\breve{x}},$$

where $q_{j,x}$ is defined in (3.24). Note also that $\breve{j}$ is correctly defined because $\forall x \in \mathcal{D}$

$$\mathbb{H}(\mathfrak{a}, \mathfrak{L}) = \bigcup_{j=0}^{J+1} \mathfrak{H}_j(x), \quad \mathfrak{H}_j(x) \bigcap \mathfrak{H}_{j'}(x) = \emptyset,\ j \ne j',\ j, j' = \overline{0, J+1}.$$

In view of Lemma D.1 (*1*) and (*3*)

$$\breve{q} \in \mathbb{H}_1(\beta, \mathcal{S}), \text{ on } [-1,1]. \tag{3.31}$$

Put

$$\breve{\mathcal{A}}_i = \left\{ \left[\hat{g}_{\varepsilon,i}(\breve{\tau}) \left(h[\lambda_{\breve{j}}]/(\lambda_{\breve{j}})^{\varkappa} \right)^{1/2\beta} \right] < 3\varepsilon \mathcal{L}(\beta, L_2) \right\}, \ i = 1,2,$$

$$\breve{\mathcal{B}} = \left\{ \tilde{g}_\varepsilon(\breve{x}) < \left[3\sqrt{\ln(1/\mu)} + 2 \right] \varepsilon \right\}.$$

and note that due to (3.31)

$$\bar{\mathcal{A}}_i = \bigcap_{j=1}^{J+1} \bar{\mathcal{A}}_{\lambda_j, i} \subseteq \breve{\mathcal{A}}_i, \quad \bar{\mathcal{B}} \subseteq \breve{\mathcal{B}}.$$

Therefore, $\forall g \in \mathbb{H}\big(C\varphi_\varepsilon(\mathfrak{a})\big)$

$$\mathbb{P}_g\{\bar{\mathcal{A}}_1 \cap \bar{\mathcal{A}}_2 \cap \bar{\mathcal{B}}\} \le \mathbb{P}_g\{\breve{\mathcal{A}}_1 \cap \breve{\mathcal{A}}_2 \cap \breve{\mathcal{B}}\} \le \begin{cases} \mathbb{P}_g\{\breve{\mathcal{A}}_1 \cap \breve{\mathcal{A}}_2\} & \text{if } \breve{j} \ne 0, \\ \mathbb{P}_g\{\breve{\mathcal{B}}\} & \text{if } \breve{j} = 0. \end{cases} \tag{3.32}$$

Without loss of generality we will assume that (3.18) is true for $\breve{x}$ and, therefore it is sufficient to find the upper bounds for $\mathbb{P}_g\{\breve{\mathcal{A}}_1\}$ and $\mathbb{P}_g\{\breve{\mathcal{B}}\}$.

II.2.a. *Upper bound for* $\mathbb{P}_g\{\breve{\mathcal{A}}_1\}$.

First let us note that $\forall \lambda > 0$

$$\sqrt{\lambda h[\lambda]} \left(\frac{h[\lambda]}{\lambda^{\varkappa}} \right)^{\frac{1}{2\beta}} = \varepsilon \varphi^{-1}(\mathfrak{a}). \tag{3.33}$$

Next,

$$\hat{g}_{\varepsilon,1}(\breve{\tau}) = \frac{\sqrt{\lambda h[\lambda]}}{||K||_2} \left[f(G(\breve{x})) + \hat{B}_g(\breve{\tau}) \right] + \varepsilon \hat{\xi}(\breve{\tau}).$$

Thus, we have from (3.29), (3.11), (3.33) and the definition of $\breve{\tau}$

$$\begin{aligned}
\breve{\mathcal{A}}_1 &= \left\{ \left(h[\breve{\lambda}]/\breve{\lambda}^{\varkappa} \right)^{\frac{1}{2\beta}} \left| \hat{g}_{\varepsilon,1}(\breve{\tau}) \right| \le 3\varepsilon \mathcal{L}(\beta, L_2) \right\} \\
&\subseteq \left\{ \frac{\varepsilon \varphi^{-1}(\mathfrak{a})}{||K||_2} \left[|f(G(\breve{\tau}))| - |\hat{B}_g(\breve{\tau})| \right] - \varepsilon \left(h[\breve{\lambda}]/\breve{\lambda}^{\varkappa} \right)^{\frac{1}{2\beta}} |\hat{\xi}(\breve{\tau})| \le 3\varepsilon \mathcal{L}(\beta, L_2) \right\} \\
&\subseteq \left\{ |\hat{\xi}(\breve{\tau})| \ge \left(\breve{\lambda}^{\varkappa}/h[\breve{\lambda}] \right)^{\frac{1}{2\beta}} \left[(C/2 - 6C_2(\mathfrak{a}, \mathfrak{L}))/||K||_2 - 3\mathcal{L}(\beta, L_2) \right] \right\} \\
&\subseteq \left\{ |\hat{\xi}(\breve{\tau})| \ge \inf_{\lambda \in \Lambda} \left(\lambda^{\varkappa}/h[\lambda] \right)^{\frac{1}{2\beta}} \left[(C/2 - 6C_2(\mathfrak{a}, \mathfrak{L}))/||K||_2 - 3\mathcal{L}(\beta, L_2) \right] \right\} \\
&\subseteq \left\{ |\hat{\xi}(\breve{\tau})| \ge \varepsilon^{-a} \left[(C/2 - 6C_2(\mathfrak{a}, \mathfrak{L}))/||K||_2 - 3\mathcal{L}(\beta, L_2) \right] \right\},
\end{aligned}$$

where C is chosen such that $C > 12C_2(\mathfrak{a},\mathfrak{L}) + 3||K||_2\mathcal{L}(\beta, L_2)$ and a is defined in (3.11).

Taking into account that $\hat{\xi}(\check{\tau})$ is the standard normal random variable and denoting by Φ its distribution function we obtain $\forall g$ such that $\check{j} \neq 0$

$$\mathbb{P}_g\{\check{\mathcal{A}}_1\} \leq \Phi\Big(\varepsilon^{-a}\big[(C/2 - 6C_2(\mathfrak{a},\mathfrak{L}))/||K||_2 - 3\varepsilon\mathcal{L}(\beta, L_2)\big]\Big). \qquad (3.34)$$

II.2.b. *Upper bound for* $\mathbb{P}_g\{\check{\mathcal{B}}\}$.

First, let us note that

$$\mu = \varepsilon\sqrt{\ln(1/\varepsilon)}\varphi^{-1}(\mathfrak{a}), \quad \ln(1/\mu) \sim \mathcal{Z}(\mathfrak{a})\ln(1/\varepsilon), \qquad (3.35)$$

where

$$\mathcal{Z}(\mathfrak{a}) = \begin{cases} \frac{\beta+1}{2\gamma\beta+\beta+1} & \text{if } \beta > 2\gamma - 1, \\ \frac{1}{\beta+1} & \text{if } \beta \leq 2\gamma - 1. \end{cases}$$

Next,

$$\tilde{g}_\varepsilon(\check{x}) = \frac{\mu}{||K||_2}\Big[f(G(\check{x})) + \tilde{B}_g(\check{x})\Big] + \varepsilon\tilde{\xi}(\check{x}).$$

Similarly to previous calculation we get from (3.21) and (3.35)

$$\check{\mathcal{B}} = \Big\{\tilde{g}_\varepsilon(\check{x}) < \Big[3\sqrt{\ln(1/\mu)} + 2\Big]\varepsilon\Big\}$$

$$\subseteq \left\{\frac{\mu}{||K||_2}\Big[\big|f(G(\check{x}))\big| - \big|\tilde{B}_g(\check{x})\big|\Big] - \varepsilon|\tilde{\xi}(\check{x})| \leq \Big[3\sqrt{\ln(1/\mu)} + 2\Big]\varepsilon\right\}$$

$$\subseteq \left\{\frac{\varepsilon\sqrt{\ln(1/\varepsilon)}}{\varphi(\mathfrak{a})||K||_2}\Big[\big|f(G(\check{x}))\big| - \big|\tilde{B}_g(\check{x})\big|\Big] - \varepsilon|\tilde{\xi}(\check{x})| \leq \Big[3\sqrt{\ln(1/\mu)} + 2\Big]\varepsilon\right\}$$

$$\subseteq \left\{|\tilde{\xi}(\check{x})| \geq \sqrt{\ln(1/\varepsilon)}\Big[(C/2 - 6C_1(\mathfrak{a},\mathfrak{L}))/||K||_2 - 3\sqrt{\mathcal{Z}(\mathfrak{a})}\Big]\right\},$$

where C is chosen such that $C > 12C_1(\mathfrak{a},\mathfrak{L}) + 3||K||_2\sqrt{\mathcal{Z}(\mathfrak{a})}$.

Taking into account that $\tilde{\xi}(\check{x})$ is the standard normal random variable we obtain $\forall g$ such that $\check{j} = 0$

$$\mathbb{P}_g\{\check{\mathcal{B}}\} \leq \Phi\Big(\sqrt{\ln(1/\varepsilon)}\big[(C/2 - 6C_1(\mathfrak{a},\mathfrak{L}))/||K||_2 - 3\sqrt{\mathcal{Z}(\mathfrak{a})}\big]\Big). \qquad (3.36)$$

It remains to note that upper bounds in (3.34) and (3.36) are independent of g and therefore, we finally obtain from (3.32)

$$\limsup_{\varepsilon\to 0} \sup_{g\in\mathbb{H}(C\varphi_\varepsilon(\mathfrak{a}))} \mathbb{P}_g\{\bar{\mathcal{A}}_1 \cap \bar{\mathcal{A}}_2 \cap \bar{\mathcal{B}}\} = 0,$$

for all $C > \big[12C_2(\mathfrak{a},\mathfrak{L}) + 3||K||_2\mathcal{L}(\beta, L)\big] \vee \big[12C_1(\mathfrak{a},\mathfrak{L}) + 3||K||_2\sqrt{\mathcal{Z}(\mathfrak{a})}\big]$. □

3.2. Proof of Theorem 2. There are two different constructions for the lower bound. The first one is used in the case $\beta > 2\gamma - 1$ and it is based on a rather sophisticated random walk.

As to the second construction, used in the case $\beta \leq 2\gamma - 1$, it is absolutely standard. Putting $f(t) = t$ the problem is reduced to the testing under alternative described by the function G and, after that, one can use the result from [11].

Thus, only the case $\beta > 2\gamma - 1$ will be considered.

In the proof of the theorem $C_1, C_2, \dots$ denote absolute constants. Each time they appear, we explain the link between these constants and the constants $\mathfrak{a}$ and $\mathfrak{L}$.

<u>I *General remarks*</u>

As usual (see, for example, [12]), we will construct a lower bound for $\inf_{\tilde{\Delta}} \mathcal{R}(\tilde{\Delta}, C\varphi_\varepsilon)$. It involves a parametric subset of alternatives $\{f_\theta, \theta \in \Theta\} \subset \mathbb{H}(\mathfrak{a}, \mathfrak{L})$ indexed by Θ and endowed by a probability measure π. We have

$$\inf_{\tilde{\Delta}} \mathcal{R}(\tilde{\Delta}, C\varphi_\varepsilon) \geq 1 - \frac{1}{2} \left|\mathbb{P}_0 - \mathbb{P}_\pi\right|_2,$$

where $\mathbb{P}_\pi(A) = \int_\Theta \mathbb{P}_{f_\theta}(A)\, d\pi(\theta)$ and $|\mathbb{P}_0 - \mathbb{P}_\pi|_2^2 = \mathbb{E}_{\mathbb{P}_0}\left(\frac{d\mathbb{P}_\pi}{d\mathbb{P}_0} - 1\right)^2$.

The proof is done in the following way. First, the parametric subset in the alternative is defined. Next, we do some calculation on the L_2–distance. Finally, the probability π is chosen such that it leads to the expected rate of testing up to a logarithmic factor.

<u>**I.1** *Parametric subset of alternatives.*</u>

Let us choose a function $F : \mathbb{R} \longrightarrow \mathbb{R}^+$ such that

1. $F\left(\frac{1}{2}\right) > 0$.
2. $F(1/2 + u) = F(1/2 - u)$.
3. $F(u) = 0,\ u \notin [0,\ 1]$.
4. $F \in \mathbb{H}_1(\gamma, 1)$.

Let us also choose a function $\mathfrak{r} : \mathbb{R} \longrightarrow \mathbb{R}$ such that

(i). The function $\mathfrak{r}$ belongs to the Hölder space $\mathbb{H}_1(\beta, L_3)$ with L_3 a constant compatible with property (iii).

(ii). The function $\mathfrak{r}$ is non-decreasing.

(iii). $\mathfrak{r}(0) = 0, \mathfrak{r}(1) = 1$ and $\forall l < m = \lfloor \beta \rfloor : \mathfrak{r}^{(l)}(0) = \mathfrak{r}^{(l)}(1) = 0$.

Introduce the parameters λ, h and T such that

$$\frac{L_1}{L_3^\gamma} h^\gamma F(1/2) = C \frac{\varphi_\varepsilon(\mathfrak{a})}{\left(\ln(1/\varepsilon)^{2+\eta}\right)^{\frac{2\gamma\beta}{2\gamma\beta+\beta+1}}}, \quad h = \lambda^\beta, \quad T = [1/\lambda].$$

Let $\xi = (\xi_1, \dots, \xi_T) \in \Xi_T = \{0,\ 1\}^T$. Define

$$S_k(\xi) = \sum_{i=1}^k \xi_i, \quad S_0 = 0.$$

Set $\Delta_k = [-1 + (k-1)\lambda, \ -1 + k\lambda], k = \overline{1,T}$ and define $\forall y \in [-1, 0]$

$$\bar{\mathfrak{g}}_\xi(y) = \sum_{k=1}^{T} \left[S_{k-1} + \xi_k \mathfrak{r}\left(\frac{y-(k-1)\lambda}{\lambda}\right)\right] \mathbb{I}_{\Delta_k}(y).$$

Finally, we put $\mathfrak{g}_\xi(y) = \left[\bar{\mathfrak{g}}_\xi(y) + \bar{\mathfrak{g}}_\xi(-y)\right]/2, \ y \in [-1,1]$.
The parametric subset of alternatives is defined as follows.

$$r_\xi(x) = L_1 h^\gamma F\left(\frac{x_1 - \lambda^\beta \mathfrak{g}_\xi(x_2)}{h}\right), \quad \xi \in \Xi_T.$$

Clearly that $r_\xi = f(G)$, where

$$G(x_1, x_2) = \frac{L_2}{L_3}\left(x_1 - \lambda^\beta \mathfrak{g}_\xi(x_2)\right), \quad f(u) = \frac{L_1}{L_3^\gamma} h^\gamma F\left(\frac{L_3 u}{h} + \frac{1}{2}\right).$$

By construction $G \in \mathbb{H}_2(\beta, L_2)$, $f \in \mathbb{H}_1(\gamma, L_1)$. and, moreover,

$$\|r_\xi\|_\infty \geq C \frac{\varphi_\varepsilon(\mathfrak{a})}{\left(\ln(1/\varepsilon)^{2+\eta}\right)^{\frac{2\beta\gamma}{2\beta\gamma+\beta+1}}}.$$

I.2 *Calculation of the Bayesian likelihood ratio.*

Let π be a probability measure on Ξ_T which will be chosen later.

According to Girsanov formula, the Bayesian likelihood ratio under $\mathbb{P}_0$ is

$$l_\pi = \mathbb{E}_\pi\left(\exp\left(\frac{1}{\varepsilon}\int r_\xi(x)\ dW(x) - \frac{1}{2\varepsilon^2}\int r_\xi(x)^2\ dx\right)\right).$$

Put

$$\sigma^2 = \int_{\mathcal{D}_0 \cap (\mathbb{R}\times\Delta_k)} r_\xi(x)^2\ dx = C_1 h^{2\gamma+1}\lambda,$$

where the absolute constant C_1 depends only on L_1, L_2 and L_3.

Let $\tilde{\xi}$ be an independent version of ξ. Then,

$$\begin{aligned}
\mathbb{E}_0\left(l_\pi^2\right) &= \mathbb{E}_{\pi\times\pi}\mathbb{E}_0 \exp\left(\frac{1}{\varepsilon}\int\left[\mathfrak{r}_\xi(x) + \mathfrak{r}_{\tilde{\xi}}(x)\right] dW(x) - \frac{1}{2\varepsilon^2}\int\left[\mathfrak{r}_\xi^2(x) + \mathfrak{r}_{\tilde{\xi}}^2(x)\right] dx\right) \\
&= \mathbb{E}_{\pi\times\pi}\exp\left\{\varepsilon^{-2}\langle \mathfrak{r}_\xi, \mathfrak{r}_{\tilde{\xi}}\rangle\right\} \leq \mathbb{E}_{\pi\times\pi}\exp\left\{\left[\sigma^2/\varepsilon^2\right]\sum_{k=1}^{T}\mathbb{I}_{\left|S_k(\xi)-S_k(\tilde{\xi})\right|\leq 1}\right\} \\
&\leq \sum_{s=0}^{T} A_\varepsilon^s E_{\pi\times\pi}\left(\sum_{1\leq k_1<\ldots<k_s\leq T} I_{\left|S_{k_1-1}(\xi)-S_{k_1-1}(\tilde{\xi})\right|\leq 1}\cdots I_{\left|S_{k_s-1}(\xi)-S_{k_s-1}(\tilde{\xi})\right|\leq 1}\right),
\end{aligned}$$

where $A_\varepsilon = \exp\left(\varepsilon^2/\sigma^2\right) - 1$.

Note that up to this point, all formulas hold for any prior distribution π on Ξ_T.

I.3 *Prior distribution and final calculation.*

In order to get the rate of testing, we choose the prior distribution π as the distribution of the random walk defined in Lemma C.1. Putting

$$\mathcal{I}_{s,x} = \mathbb{I}_{S_{k_1-1}(\xi)=x_1} \cdots I_{S_{k_s-1}(\xi)=x_s}$$

we have

$$\mathbb{E}_{\pi\times\pi}\left(\sum_{1\le k_1<\ldots<k_s\le T} \mathbb{I}_{|S_{k_1-1}(\xi)-S_{k_1-1}(\tilde{\xi})|\le 1} \cdots \mathbb{I}_{|S_{k_s-1}(\xi)-S_{k_s-1}(\tilde{\xi})|\le 1}\right)$$

$$= \sum_{1\le k_1<\ldots<k_s\le T} \mathbb{E}_{\pi\times\pi}\left(\sum_{x_1,\ldots,x_s} \mathcal{I}_{s,x}\mathbb{I}_{|S_{k_1-1}(\tilde{\xi})-x_1|\le 1} \cdots \mathbb{I}_{|S_{k_s-1}(\tilde{\xi})-x_s|\le 1}\right)$$

$$= \sum_{1\le k_1<\ldots<k_s\le T} \mathbb{E}_{\pi\times\pi}\Bigg(\sum_{x_1,\ldots,x_s} \mathcal{I}_{s,x}\mathbb{I}_{|S_{k_1-1}(\tilde{\xi})-x_1|\le 1} \cdots \mathbb{I}_{|S_{k_{s-1}-1}(\tilde{\xi})-x_{s-1}|\le 1}$$

$$\mathbb{E}\left[\mathbb{I}_{|S_{k_s-1}(\tilde{\xi})-x_s|\le 1} | S_{k_1-1}\left(\tilde{\xi}\right), \ldots, S_{k_{s-1}-1}\left(\tilde{\xi}\right)\right]\Bigg)$$

$$\le \sum_{1\le k_1<\ldots<k_s\le T} \mathbb{E}_{\pi\times\pi}\left(\sum_{x_1,\ldots,x_s} \mathcal{I}_{s,x}\mathbb{I}_{|S_{k_1-1}(\tilde{\xi})-x_1|\le 1} \cdots \mathbb{I}_{|S_{k_{s-1}-1}(\tilde{\xi})-x_{s-1}|\le 1}\right)$$

$$\frac{3c\ln(k_s - k_{s-1}+1)^{1+\eta}}{k_s - k_{s-1}}$$

$$\le \sum_{1\le k_1<\ldots<k_s\le T} \frac{3c\ln(k_1)^{1+\eta}}{k_1-1}\frac{3c\ln(k_2-k_1+1)^{1+\eta}}{k_2-k_1}\cdots\frac{3c\ln(k_s-k_{s-1}+1)^{1+\eta}}{k_s-k_{s-1}}\mathbb{E}_{\pi\times\pi}\mathcal{I}_{s,x}$$

$$\le \sum_{1\le k_1<\ldots<k_s\le T} \frac{3c\ln(k_1)^{1+\eta}}{k_1-1}\frac{3c\ln(k_2-k_1+1)^{1+\eta}}{k_2-k_1}\cdots\frac{3c\ln(k_s-k_{s-1}+1)^{1+\eta}}{k_s-k_{s-1}}$$

$$\le \left(3c\ln(T)^{2+\eta}\right)^s \le \left(c_{RW}\ln(T)\right)^{s(2+\eta)}.$$

Thus, the following upper bound holds

$$\mathbb{E}_0\left(l_\pi^2\right) \le \sum_{s=0}^{T} A_\varepsilon^s \left(c_{RW}\ln T\right)^{(2+\eta)s}$$

$$\le \frac{1}{1-\left(c_{RW}\ln T\right)^{2+\eta}\left(\exp\left[\sigma^2/\varepsilon^2\right]-1\right)}. \tag{3.37}$$

The right-hand side of (3.37) is of order $1+\left(c_{RW}\ln T\right)^{(2+\eta)}\sigma^2/\varepsilon^2$. Recall that

$$h^\gamma = C\frac{L_3^\gamma}{L_1F(1/2)}\left(\varepsilon\Big/\left[\ln 1/\varepsilon\right]^{1+\eta/2}\right)^{\frac{2\beta}{2\gamma\beta+\beta+1}}.$$

Therefore,

$$\inf_{\tilde{\Delta}} \mathcal{R}(\tilde{\Delta}, C\varphi_\varepsilon) \geq 1 - \frac{1}{2}\sqrt{\frac{1}{1-(c_{RW}\ln T)^{(2+\eta)}\left(\exp\left[\sigma^2/\varepsilon^2\right]-1\right)} - 1}. \quad (3.38)$$

Thus,

$$\liminf_{\varepsilon\to 0} \inf_{\tilde{\Delta}} \mathcal{R}(\tilde{\Delta}, C\varphi_\varepsilon) \geq 1 - \frac{1}{2}\sqrt{\frac{C^{\frac{2\gamma\beta+\beta+1}{\gamma\beta}} C_3}{1 - C^{\frac{2\gamma\beta+\beta+1}{\gamma\beta}} C_3}}, \quad (3.39)$$

where the absolute constant C_3 depends only on $L_1, L_2, L_3, \beta, \gamma, F(1/2)$ and c_{RW}. It remains to note that the left-hand side of (3.39) goes to 1 if $C \to 0$. □

APPENDIX

A. Metric entropy. Let $(\mathfrak{T}, \rho)$ be metric space and let $\mathbb{T} \subset \mathfrak{T}$ be a compact set. For any $\delta > 0$ we denote by $\mathcal{E}^{\rho}_{\delta}(\mathbb{T})$ δ-entropy of $\mathbb{T}$ w.r.t ρ.

The result below can be found in [10].

LEMMA A.1. *If* $\rho = \|\cdot\|_\infty$ *(uniform norm on* $[-1,1]$*) then* $\forall \delta > 0$

$$\mathcal{E}^{\infty}_{\delta}\Big(\mathbb{H}_1(\nu, L)\Big) \leq C\left[\frac{L}{\delta}\right]^{\frac{1}{\nu}},$$

C is the constant independent on δ *and* L.

The upper bound given in the lemma is sharp. We do not present the lower bound because it is not used in the paper.

B. Gaussian random functions. In this section we present some results concerning the large deviation probability of the extrema of gaussian random function.

These results can be found, for example, in [1] and [17].

LEMMA B.1. *Let* $\eta(t), t \in \mathbb{T}$ *be a centered gaussian random function and* $\rho(s,t) = \sqrt{\mathbb{E}|\eta(t)-\eta(s)|^2}$ *its intrinsic semi-metrics. Then*

$$\mathbb{E}\sup_{t\in\mathbb{T}} \eta(t) \leq 4\sqrt{2}\int_0^{\sigma} \sqrt{\mathcal{E}^{\rho}_{u}(\mathbb{T})}du,$$

where $\sigma = \sqrt{sup_{t\in\mathbb{T}}\ Var\big(\eta(t)\big)}$.

Moreover, if Dudley integral is finite then $\eta(\cdot)$ *is a.s. bounded on* $\mathbb{T}$.

LEMMA B.2. *Let* $\eta(t), t \in \mathbb{T}$ *be a centered gaussian random function a.s. bounded on* $\mathbb{T}$. *Then* $\forall x > 0$

$$\mathbb{P}\Big\{\sup_{t\in\mathbb{T}} \eta(t) > \mathbb{E}\sup_{t\in\mathbb{T}} \eta(t) + x\Big\} \leq \exp\big\{-x^2/2\sigma^2\big\}.$$

C. Random walk. The following lemma deals with the existence of a random walk which is more unpredictable than the simple random walk. It is a direct consequence from [5] (Theorem 1.4).

Let $\zeta = \{\zeta_j, j \in \mathbb{N}\}$ denote a sequence of random variables taking values in $\{0,1\}$ and Let $\mathbb{S}_n(\zeta)$ denote the partial sum $\sum_{j=1}^n \zeta_j$.

LEMMA C.1. *For any $\eta > 0$, there exists a probability π on $[\{0\},\{1\}]^\infty$ and a constant $c > 0$ independent of n such that for any $k > 0$ and any $x \geq 0$:*

$$\sup_{n\geq 1} \pi\Big\{\mathbb{S}_{n+k}(\zeta) = x | \mathbb{S}_1(\zeta), \dots, \mathbb{S}_n(\zeta)\Big\} \leq \frac{c(\ln k)^{1+\eta}}{k}.$$

D. Implicit functions. In this section we establish some smoothness properties of implicit functions. In spite of the fact that the main argument here is Implicit Function Theorem [15], we were unable to find required results in the existing literature and we give below their direct proof.

Let $G \in \mathbb{H}_2(\beta, L)$ on $\mathcal{D}_0 \subseteq \mathbb{R}^2$, $x_0 \in \mathcal{D}_0$ be the fixed point, and let us consider the equation

$$G(t) = G(x_0) \tag{D.1}$$

Without loss of generality we will suppose that

$$\big|G_{1,0}(x_0,y_0)\big| = \big|G_{1,0}(x_0,y_0)\big| \vee \big|G_{0,1}(x_0,y_0)\big| \triangleq a \in (0,1].$$

LEMMA D.1. *There exist the positive constant $\mathcal{M} = \mathcal{M}(\beta, L)$ and $\mathcal{S} = \mathcal{S}(\beta, L)$ such that $\forall G \in \mathbb{H}_2(\beta, L)$ the implicit function $q(\cdot)$ satisfying (D.1)*

1. *exists and is uniquely defined on the set $\Big[y_0 - \mathcal{M}a^\kappa, y_0 + \mathcal{M}a^\kappa\Big]$;*
2. - *takes its values in $\Big[x_0 - \mathcal{M}(4a)^\kappa, x_0 + \mathcal{M}(4a)^\kappa\Big]$;*
   - *$|q'(y)| \leq K_1$, $\forall y \in \Big[y_0 - \mathcal{M}a^\kappa, y_0 + \mathcal{M}a^\kappa\Big]$, where the constant K_1 depend only on β;*
3. *belongs to $\mathbb{H}_1(\beta, \mathcal{S}a^{\alpha-\beta})$ on $\Big[y_0 - \mathcal{M}a^\kappa, y_0 + \mathcal{M}a^\kappa\Big]$,*

where, remind, $\kappa = \big[1/(\beta-1)\big] \vee 1$.

Proof. We will use the following notations: $K_1, K_2, \dots$ denote absolute constants depending only on β, $S_1(\cdot), S_2(\cdot), \dots$ denote bounded functions and F' denotes the derivative of the function F. Also without loss of generality we put $(x_0, y_0) = (0,0)$ and assume that $G_{1,0}(0,0) = a$.

Since $G_{1,0} \in \mathbb{H}_2(\beta - 1, L)$ the following inequality holds:

$$G_{1,0}(x,y) \geq a/2, \quad \forall |x|, |y| \leq (a/2Lm)^\kappa, \; m = \beta - \alpha. \tag{D.2}$$

This follows from $G_{1,0}(x,y) \geq G_{1,0}(0,0) - Lm\|x-y\|^{1/\kappa}$.

Put $\delta = \big(a/4\beta Lm\big)^{\kappa}$. We have

$$G(\delta,0) = G(0,0)+G_{1,0}(0,0)\delta+S_1(\delta)\delta^{\beta\wedge 2} \geq a\delta - L\delta^{\beta\wedge 2} \geq a^{\kappa+1}(1/L)^{\kappa}\rho(\beta),$$

where $\rho(\beta)$ is constant depending on β only. In the same way as above, we have

$$G(-\delta,0) \leq -a^{\kappa+1}(1/L)^{\kappa}\rho(\beta).$$

As the function $G \in \mathbb{H}_2(\beta, L)$, the following expansion at point $(\delta, 0)$ holds

$$G(\delta,y) = G(\delta,0) + G_{0,1}(\delta,0)y + S_2(y)|y|^{\beta\wedge 2}.$$

This implies

$$G(\delta,y) \geq \rho(\beta)a^{\kappa+1}(1/L)^{\kappa} - a|y| - Lm|y|^{\beta\wedge 2};$$
$$G(-\delta,y) \leq -\rho(\beta)a^{\kappa+1}(1/L)^{\kappa} + a|y| + Lm|y|^{\beta\wedge 2}.$$

Thus, $G(\delta,y) > 0, \quad G(-\delta,y) < 0, \quad \forall |y| \leq \big(a/16\beta mL\big)^{\kappa}$.

Let us fix $\bar{y} \in \big[-\big(a/16\beta mL\big)^{\kappa}, \big(a/16\beta mL\big)^{\kappa}\big]$ and consider the function $R(\cdot) = G(\cdot, \bar{y})$. We have

$$R(\delta)>0,\ R(-\delta)<0,\ R'(x)\geq a/2, \quad \forall x \in \big[-(a/4\beta mL)^{1/\beta-1}, (a/4\beta mL)^{\kappa}\big].$$

Then, there exists a unique point $\bar{x}$ such that $R(\bar{x}) = 0$. The implicit function $q(\cdot)$ is defined by

$$q(\bar{y}) = \bar{x}(\bar{y}).$$

Clearly it exists on the set $\big[-(a/16\beta mL)^{\kappa}, (a/16\beta mL)^{\kappa}\big]$ and takes its value in $\big[-(a/4\beta mL)^{\kappa}, (a/4\beta mL)^{\kappa}\big]$.

The implicit function theorem [15] asserts that the function $q(\cdot)$ is m times differentiable and we can calculate its derivatives using formula $G(q(y), y) \equiv 0$. In particular,

$$q'(y) = -\frac{G_{0,1}(q(y),y)}{G_{1,0}(q(y),y)}.$$

and in view of (D.2) $|q'(y)| \leq K_1, \quad \forall y \in \big[-(a/16\beta mL)^{\kappa}, (a/16\beta mL)^{\kappa}\big]$. where K_1 depends only on β.

Thus, the assertions (*1*) and (*2*) of the lemma are proved.

Let us now study the regularity of the implicit function $q(\cdot)$. Putting $G_{k,r} = \frac{\partial G^{k+r}}{\partial x_1^k \partial x_2^r}$, $k, r = \overline{1,m}$, and differentiating the identity $G(q(y), y) = 0$ we first prove that for $n = 1, \dots, m$:

$$q^{(n)}(y)G_{1,0}(q(y),y)+\mathcal{P}_n(q^{(1)}(y),\dots,q^{(n-1)}(y),y)+G_{0,n}(q(y),y)=0, \quad \text{(D.3)}$$

where $\mathcal{P}_n(z_1, \dots, z_{n-1}, y)$ is a polynomial such that each term is of the form

$$G_{k,r}(q(y), y) \prod_{i=1}^{n-1} z_i^{l_i}, \quad \sum_{i=1}^{n-1} i l_i \le n, \quad k + r \le n.$$

The relation is true for $n = 1$ and $n = 2$.

Assume that the relation is true up to n. Then the derivative of each term in the polynomial $\mathcal{P}_n$ is also a polynomial such that each term is of the form

$$G_{k,r}(q(y), y) \prod_{i=1}^{n} z_i^{l_i}, \quad \sum_{i=1}^{n} i l_i \le n+1, \quad k + r \le n+1.$$

Thus, we have

$$q^{(n+1)}(y) G_{1,0}(q(y), y) + \mathcal{P}_{n+1}(q^{(1)}(y), \dots, q^{(n+1)}(y), q^{(n)}(y), y) + G_{0,n+1}(q(y), y) + 0,$$

where $\mathcal{P}_{n+1}(z_1, \dots, z_{n-1}, z_n, y)$ is a polynomial such that each term is of the form

$$G_{k,r}(q(y), y) \prod_{i=1}^{n} z_i^{l_i}, \quad \sum_{i=1}^{n} i l_i \le n, \quad k + r \le n.$$

Therefore the relation holds for any $n = 1, \dots, m$.

In the same way as above we prove that $|q^{(n)}(y)| \le K (L/a)^{n-1}$. Indeed, this relation is true for $n = 1$ and $n = 2$.

Let us assume now that the relation is true up to n. Then, each term in the polynomial $\mathcal{P}_{n+1}$ is bounded

$$G_{k,r}(q(y), y) \prod_{i=1}^{n} q^{(i)}(y)^{l_i} \le K_4 L \prod_{i=1}^{n} (L/a)^{(i-1)l_i} \le K_4 L (L/a)^n.$$

It leads to $|q^{(n+1)}(y)| \le K_5 (L/a)^n$.

Let us now study the regularity of $q^{(m)}$. We use equality for $n = m$ at points y and z and proceed as follows.

$$\left| G_{k,r}(q(y), y) \prod_{i=1}^{m-1} q^{(i)}(v_i)^{l_i} - G_{k,r}(q(y), y) \prod_{i=1}^{m-1} q^{(i)}(w_i)^{l_i} \right|$$

$$\le L \prod_{\substack{i=1 \\ i \ne r}}^{m-1} (L/a)^{(i-1)l_i} (L/a)^r |y - z| K_6 (L/a)^{(r-1)(l_r - 1)} \le L (L/a)^{m-1} |y - z|;$$

$$\left| G_{k,r}(q(y), y) \prod_{i=1}^{m-1} q^{(i)}(z)^{l_i} - G_{k,r}(q(z), z) \prod_{i=1}^{m-1} q^{(i)}(z)^{l_i} \right| \le K_7 L |y - z| (L/a)^{m-1};$$

$$\left| G_{0,m}(q(y), y) - G_{0,m}(q(z), z) \right| \le K_8 L |y - z|^{\beta - m}.$$

Therefore summing all the differences and dividing by a we get

$$|q^{(m)}(y) - q^{(m)}(z)| \leq K_9 \, (L/a)^m \, |y - z|^{\beta - m},$$

where K_9 is a constant depending only on β.

These results entail that the implicit function is in $H_1(\beta, K_9 \, (L/a)^m)$. □

The authors are grateful to Fabienne Castell for suggesting to use the results from [5].

REFERENCES

[1] R.J. ADLER, *An Introduction to Continuity, Extrema and Related Topics for General Gaussian Processes*, IMS Lecture, Notes–Mongraph Series, 12 Institute of Mathematical Statistics, Hayward CA, 1990.

[2] L. BROWN AND M. LOW, *Asymptotic equivalence of nonparametric regression and white noise*, Ann. Statist., **24** (1996), pp. 2384–2398.

[3] H. CHEN, *Estimation of a projection–pursuit type regression model*, Ann. Statist., **19** (1991), pp. 142–157.

[4] G.K. GOLUBEV, *Asymptotically minimax estimation of a regression function in an additive model*, Problems Inform. Transmission, **28** (1992), pp. 101–112.

[5] O. HÄGGSTRÖM AND E. MOSSEL, *Nearest-neighbor walks with low predictability profile and percolation in $2 + \epsilon$ dimensions*, Ann. Probab., **26** (1998), pp. 1212–1231.

[6] P. HALL, *On projection–pursuit regression*, Ann. Statist., **17** (1989), pp. 573–588.

[7] M. HRISTACHE, A. JUDITSKY, AND V. SPOKOINY, *Direct estimation of the index coefficient in a single–index model*, Ann. Statist., **29** (2001), pp. 595–623.

[8] M. HRISTACHE, A. JUDITSKY, J. POLZEHL, AND V. SPOKOINY, *Structure adaptive approach for dimension reduction*, Ann. Statist., **29** (2001), pp. 1537–1566.

[9] P. HUBER, *Projection pursuit. With discussion.*, Ann. Statist., **13** (1985), pp. 435–525.

[10] I.A. IBRAGIMOV AND R.Z. HASMINSKII, *Statistical Estimation. Asymptotic Theory*, Springer-Verlag, 1981 (Originally published in Russian in 1979).

[11] YU.I. INGSTER *Asymptotically minimax testing of nonparametric hypotheses*, Probability theory and mathematical statistics, Vol. I (Vilnius, 1985), pp. 553–574, VNU Sci. Press, Utrecht, 1987.

[12] YU.I. INGSTER AND I.A. SUSLINA, *Nonparametric Goodness-of-Fit Testing Under Gaussian Models*, Lecture Notes in Statistics, Springer-Verlag, New York, 2003.

[13] A. IOUDITSKI, O.V. LEPSKI, AND A.B. TSYBAKOV, *Statistical estimation of composite functions*, Manuscript, 2006.

[14] A.N. KOLMOGOROV, *On representation of continuous functions of several variables as superposition of continuous functions of one variable and addition*, Dokl. Akad. Nauk SSSR, **114** (1957), pp. 369–373.

[15] S. LANG, *Analysis II*, Addison-Wesley, Reading, 1969.

[16] O.V. LEPSKI AND A.B. TSYBAKOV, *Asymptotically exact nonparametric hypothesis testing in sup-norm and at a fixed point*, Probab. Theory and Related Fields, **117** (2000), pp. 17–48.

[17] M. LIFSHITS, *Gaussian Random Functions*, Kluwer Academic Publishers, Dordrecht, 1995.

[18] M. REISS, *Asymptotic equivalence for nonparametric regression with multivariate and random design* (2006), Submitted.

[19] D.A. SPRECHER, *An improvement in the superposition theorem of Kolmogorov,*. J. Math. Anal. Appl., **38** (1972), pp. 208–213.

[20] C.J. Stone, *Optimal global rates of convergence for nonparametric regression*, Ann. Statist., **10** (1982), pp. 1040–1053.

[21] C.J. Stone, *Additive regression and other nonparametric models*, Ann. Statist., **13** (1985), pp. 689–705.

Part III: Stochastic Partial Differential Equations

ON PARABOLIC PDES AND SPDES IN SOBOLEV SPACES W^2_p WITHOUT AND WITH WEIGHTS

NICOLAI V. KRYLOV*

Abstract. We present a "streamlined" theory of solvability of parabolic PDEs and SPDEs in half spaces in Sobolev spaces with weights. The approach is based on interior estimates for equations in the whole space and is easier than and quite different from the standard one.

Key words. Stochastic partial differential equations, Sobolev spaces with weights, interior estimates.

AMS(MOS) subject classifications. Primary 60H15, 35K20.

1. Introduction. Let $\mathbb{R}^d$ be a d-dimensional Euclidean space consisting of points $x = (x^1, ..., x^d)$. We denote

$$D_i = \frac{\partial}{\partial x^i}, \quad D_{ij} = D_i D_j.$$

Introduce Du as the gradient of a function u with respect to x and $D^2 u$ as its Hessian matrix.

In domains $G \subset \mathbb{R}^d$ we are considering the following parabolic equation

$$\frac{\partial}{\partial t} u_t = L_t u_t + f_t, \quad t \in (0, T), \tag{1.1}$$

where $T \in (0, \infty]$ and

$$L_t u = (1/2) a^{ij}_t D_{ij} u + b^i_t D_i u + c_t u,$$

and the stochastic partial differential equation

$$du_t = (L_t u_t + f_t)\, dt + (\Lambda^k_t u_t + g^k_t)\, dw^k_t, \quad t \in (0, \tau), \tag{1.2}$$

where τ is a stopping time

$$\Lambda^k_t u = \sigma^{ik}_t D_i u + \nu^k_t u_t.$$

Here and everywhere below we are using the summation convention. The precise framework in which these equations are treated is given later in the article.

We present three sets of results:

Set 1. Interior W^2_p-estimates for solutions of (1.1) (Section 3) and (1.2) (Section 2) when $G = \mathbb{R}^d$.

---

*127 Vincent Hall, University of Minnesota, Minneapolis, MN, 55455. The work was partially supported by the NSF Grants DMS-0140405 and DMS-0653121.

Set 2. Boundary estimates for solutions of (1.1) (Section 5) and (1.2) (Section 4) in spaces like W^2_p with weights when $G = \mathbb{R}^d_+$, where

$$\mathbb{R}^d_+ = \{x = (x^1, x') \in \mathbb{R}^d : x^1 > 0, x' \in \mathbb{R}^{d-1}\}.$$

Set 3. Existence and uniqueness theorems for (1.1) (Section 7) and (1.2) (Section 8) in spaces like W^2_p with weights when $G = \mathbb{R}^d_+$.

In case of equation (1.1) we take $p \in (1, \infty)$ and in case of (1.2) we restrict p to $[2, \infty)$.

This is, of course, a quite narrow subclass of the class of problems treated in the modern theory of SPDEs and we do not try to overview this very widely developed theory referring the reader to the collection of articles [4], to [6], [16], [20], and many references therein.

Our main results collected in Set 3 are known from [7] and [9] (apart from Theorem 8.1) even for spaces of Bessel potentials H^γ_p, which coincide with W^γ_p if γ is a nonnegative integer. However, the proofs here are quite different and much more elementary albeit valid only for $\gamma = 2$ (see however Remark 2.1). Concerning earlier results on stochastic partial differential equations in *domains* the reader can consult [2], [1], [3], [5], [17], [18], and [19]. Notice that the approach in [2], [1], [3], and [5] is based on semigroup theory and requires, sometimes implicitly, certain compatibility of the data on the boundary. We work in Sobolev spaces with weights allowing the derivatives of solutions to blow up near the boundary, which enables us to avoid any compatibility conditions.

In a sense the result of Section 3 also belongs to "well-known results". However, we could not find it in the form we need and therefore give a complete proof. To the best of our knowledge, all other results of the paper are new.

We derive the results in Set 3 from those in Set 2, which in turn are derived from the results in Set 1. The approach in [7], [9], [17], [18], and [19] requires first developing the full theory of solvability in $\mathfrak{H}^2_{p,\theta}(\tau)$-spaces (see the definition of $\mathfrak{H}^2_{p,\theta}(\tau)$ in Section 4) for equations in $\mathbb{R}^d_+$ with coefficients independent of x. In case of (1.2) this leads to the fact that from the start one needs the following restriction on the parameter θ calibrating the rate with which the derivatives of solutions can blow up near $\partial\mathbb{R}^d_+$:

$$d - 1 + p\left[1 - \frac{1}{p(1-\tilde{\delta}) + \tilde{\delta}}\right] < \theta < d - 1 + p, \tag{1.3}$$

where $\tilde{\delta} \in (0, 1]$ is the constant of "parabolicity" of (1.2) in the direction of x^1 axis (see condition (8.6) below). By the way, observe that (1.3) is obviously satisfied if

$$d - 2 + p \le \theta < d - 1 + p.$$

Also if $\tilde{\delta} = 1$, then (1.3) becomes a condition necessary and sufficient for solvability theory of equations like (1.1) in domains: $d - 1 < \theta < d - 1 + p$.

Here from the very beginning we treat variable coefficients, allowing θ to be any number, and introduce condition (1.3) in the very end. Actually, apart from Section 8 and a few auxiliary results in Section 6 there is no restriction on θ whatsoever. We will see that (1.3) is needed for only one purpose: to estimate the zeroth order norm of solutions of equations with variable coefficients. Therefore, if for a *particular* equation and *particular* τ one can prove such an estimate for θ beyond the range in (1.3), then automatically, the solvability theory becomes available for *this* equation on $(0, \tau)$ (see Theorem 8.1). We will present some examples when it really happens in a subsequent article. There we will show an example in which for a θ (beyond the range in (1.3)) for the coefficients frozen at some points x_0 the equation is *not* solvable in $\mathfrak{H}^2_{p,\theta}(\tau)$, but the original equation with variable coefficients does admit a unique solution in $\mathfrak{H}^2_{p,\theta}(\tau)$.

One of motivations for writing this paper was to prepare necessary tools for extending to the case of variable coefficients the pointwise boundary Hölder regularity result from [13] proved there for equations with constant coefficients. For that purpose we prove Theorem 4.1 and say more about it in Remark 4.3.

One of advantages of our approach is seen from the proof of the uniqueness of solutions of SPDEs on random time intervals $[0, \tau]$. This issue was not properly addressed in quite a few articles. In [8] uniqueness for random time intervals was obtained by proving it first for deterministic times and then using a method of continuation of arbitrary solution on $[0, \tau]$ beyond τ. We do not use this roundabout here.

We have outlined above the contents of all sections in the article apart from Section 6 in which we collect some results needed to justify the way we prove the estimates of the zeroth norm of solutions.

2. Interior estimates for SPDEs in the whole space. Let $(\Omega, \mathcal{F}, P)$ be a complete probability space with a given filtration $(\mathcal{F}_t, t \geq 0)$ of σ-fields $\mathcal{F}_t \subset \mathcal{F}$ complete with respect to $\mathcal{F}, P$. Let $d_0 \geq 1$ be an integer, possibly taking the value ∞ and let w^k_t, defined for finite integers $1 \leq k \leq d_0$, be independent one-dimensional processes each of which is a Wiener process with respect to the filtration $\{\mathcal{F}_t, t \geq 0\}$. Let $\mathcal{P}$ be the predictable σ-field on $\Omega \times \mathbb{R}_+$, where $\mathbb{R}_+ = (0, \infty)$. Introduce ℓ_2 as the Hilbert space of sequences $y = (y^k, 1 \leq k \leq d_0, k < \infty)$ of real numbers with norm defined by

$$|y|^2_{\ell_2} = \sum_{k=1}^{d_0} |y^k|^2.$$

Let $d \geq 1$ be an integer and assume that for each $x \in \mathbb{R}^d$ and $i, j = 1, ..., d$ we are given $a^{ij}_t(x)$, $b^i_t(x)$, $c_t(x)$, which are real-valued processes defined for $t > 0$. We also assume that for each $x \in \mathbb{R}^d$ and $i = 1, ..., d$ we are given ℓ_2-valued processes $\sigma^i_t(x) = (\sigma^{ik}_t(x), k = 1, 2, ...)$ and $\nu^k_t(x) = (\nu^k_t(x), k = 1, 2, ...)$ defined for $t > 0$.

ASSUMPTION 2.1. *(i) The functions a, b, c, σ, and ν are $\mathcal{P} \times B(\mathbb{R}^d)$-measurable;*

(ii) The functions σ^{ik} and ν^k are continuously differentiable in x;

(iii) For some constants $\delta_0, \delta_1, \tilde{\delta} \in (0,1]$ for all values of the indices and arguments we have $a^{ij} = a^{ji}$ and for all unit $\lambda \in \mathbb{R}^d$,

$$\delta_0^{-1} \geq a_t^{ij}\lambda^i\lambda^j \geq \delta_0, \quad \alpha_t^{ij}\lambda^i\lambda^j \leq (1-\delta_1)a_t^{ij}\lambda^i\lambda^j, \tag{2.1}$$

where

$$\alpha_t^{ij} = (\sigma_t^i, \sigma_t^j)_{\ell_2}.$$

This assumption is supposed to be satisfied throughout the stochastic part of the article and in this section we are considering equation (1.2).

Equation (1.2) is understood in the sense introduced in [11] or [10] and the solutions are looked for in the space $\mathcal{H}_{p,0}^2(\tau)$ with $f \in \mathbb{L}_p(\tau)$ and $g = (g^k, k = 1, 2, ...) \in \mathbb{H}_p^1(\tau)$. We remind briefly the definition of these spaces. Everywhere in the stochastic part of the article unless specifically indicated otherwise we take

$$p \geq 2.$$

For a stopping time τ and integer $n = 0, 1, 2$ we take Sobolev spaces $W_p^n = W_p^n(\mathbb{R}^d)$ and we set

$$(0, \tau]\!] = \{(\omega, t) : 0 < t \leq \tau(\omega), t < \infty\},$$

$$\mathbb{H}_p^n(\tau) = L_p((0, \tau]\!], \mathcal{P}, W_p^n), \quad \mathbb{H}_p^n = \mathbb{H}_p^n(\infty), \quad \mathbb{L}_p ... = \mathbb{H}_p^0 ...$$

In an obvious way we introduce the norms in these spaces and extend these definitions for ℓ_2-valued functions such as $g = (g^k, k = 1, 2, ...)$.

If for any $(\omega, t) \in (0, \tau]\!]$ we are given a generalized function $u = u_t(\omega) = u_t$ on $\mathbb{R}^d$ such that $u \in \bigcap_{0<T<\infty} \mathbb{H}_p^2(\tau \wedge T)$, then we write $u \in \mathcal{H}_{p,0}^2(\tau)$ if and only if $D^2u \in \mathbb{L}_p(\tau)$ and there exist $f \in \mathbb{L}_p(\tau)$ and $g = (g^k, k = 1, 2, ...) \in \mathbb{H}_p^1(\tau)$ such that for any $\phi \in C_0^\infty(\mathbb{R}^d)$, the equation

$$(u_t, \phi) = \int_0^t (f_s, \phi)\, ds + \sum_{k=1}^{d_0} \int_0^t (g_s^k, \phi)\, dw_s^k \tag{2.2}$$

holds for all finite $t \leq \tau$ with probability 1. We set

$$\|u\|_{\mathcal{H}_{p,0}^2(\tau)} = \|D^2u\|_{\mathbb{L}_p(\tau)} + \|f\|_{\mathbb{L}_p(\tau)} + \|g\|_{\mathbb{H}_p^1(\tau)}.$$

If (2.2) holds we write

$$du_t = f_t\, dt + g_t^k\, dw_t^k \tag{2.3}$$

and this explains the sense in which (1.2) is understood. If G is a domain in $\mathbb{R}^d$ and (2.2) holds for any $\phi \in C_0^\infty(G)$, then we say that (2.3) holds in G.

To describe the first main result of this section we take an integer $d_1 \in [1, d]$ and for $x \in \mathbb{R}^d$ introduce

$$x' = (x^1, ..., x^{d_1}), \quad B'_R = \{x \in \mathbb{R}^d : |x'| < R\}.$$

THEOREM 2.1. *Take some $\varepsilon, R \in (0, \infty)$ and a stopping time τ and assume the following:*

(i) There exists a constant $K \in (0, \infty)$ such that for $x \in B'_R$ and $y \in \mathbb{R}^d$, such that $|x - y| \leq \varepsilon R$, and all values of the indices and other arguments

$$R|D\sigma_t^i(y)|_{\ell_2} + R|b_t^i(y)| + R|\nu_t(y)|_{\ell_2} + R^2|c_t(y)| + R^2|D\nu_t(y)|_{\ell_2} \leq K,$$
$$|\sigma_t^i(x) - \sigma_t^i(y)|_{\ell_2} + |a_t^{ij}(x) - a_t^{ij}(y)| \leq \beta_0,$$

where $\beta_0 = \beta_0(d, p, \delta_0, \delta_1) \in (0, 1]$ is a (small) constant from Lemma 2.1 below.

(ii) We are given a function $u \in \mathcal{H}_{p,0}^2(\tau)$ which is a solution of (1.2) with some $f \in \mathbb{L}_p(\tau)$ and $g = (g^k, k = 1, 2, ...) \in \mathbb{H}_p^1(\tau)$.

Finally, assume that $u_t(x) = 0$ if $x \notin B'_R$. Then there exist constants $N = N(d, p, \delta_0, \delta_1)$ and $N^* = N^*(d, p, \delta_0, \delta_1, \varepsilon, K)$ such that

$$\|D^2u\|_{\mathbb{L}_p(\tau)} + R^{-1}\|Du\|_{\mathbb{L}_p(\tau)}$$
$$\leq N(\|f\|_{\mathbb{L}_p(\tau)} + \|Dg\|_{\mathbb{L}_p(\tau)} + R^{-1}\|g\|_{\mathbb{L}_p(\tau)}) + N^*R^{-2}\|u\|_{\mathbb{L}_p(\tau)}. \tag{2.4}$$

To prove this theorem we need a lemma.

LEMMA 2.1. *Let τ be a stopping time and assume that we are given an $u \in \mathcal{H}_{p,0}^2(\tau)$ which satisfies (1.2) with some $f \in \mathbb{L}_p(\tau)$ and $g = (g^k, k = 1, 2, ...) \in \mathbb{H}_p^1(\tau)$. Assume that for a domain $G \subset \mathbb{R}^d$ (perhaps, $G = \mathbb{R}^d$) we have $u_t(x) = 0$ if $x \notin G$. Also assume that there are constants $\varepsilon, K \in (0, \infty)$ such that for all $x \in G$ and $y \in \mathbb{R}^d$, such that $|x - y| \leq \varepsilon$, and all values of other arguments we have*

$$|D\sigma_t^i(y)|_{\ell_2} + |b_t^i(y)| + |\nu_t(y)|_{\ell_2} + |c_t(y)| + |D\nu_t(y)|_{\ell_2} \leq K,$$
$$|\sigma_t^i(x) - \sigma_t^i(y)|_{\ell_2} + |a_t^{ij}(x) - a_t^{ij}(y)| \leq \beta_0,$$

where $\beta_0 = \beta_0(d, p, \delta_0, \delta_1) \in (0, 1]$ is a (small) constant an estimate from below for which can be obtained from the proof.

Then there exist constants

$$N = N(d, p, \delta_0, \delta_1), \quad N^* = N^*(d, p, \delta_0, \delta_1, \varepsilon, K)$$

such that

$$\|u\|_{\mathbb{H}_p^2(\tau)} \leq N(\|f\|_{\mathbb{L}_p(\tau)} + \|g\|_{\mathbb{H}_p^1(\tau)}) + N^*\|u\|_{\mathbb{L}_p(\tau)}. \tag{2.5}$$

Proof. Take a nonnegative $\zeta \in C_0^\infty(\mathbb{R}^d)$ with support in the ball of radius ε centered at the origin and such that

$$\int_{\mathbb{R}^d} \zeta^p(x)\,dx = 1.$$

Introduce $\zeta^y(x) = \zeta(x-y)$ and observe that with $N = N(d,p)$

$$|D^2u_t(x)|^p = \int_{\mathbb{R}^d} |\zeta^y(x)D^2u_t(x)|^p\,dy \le N\int_{\mathbb{R}^d} |D^2(\zeta^y(x)u(x))|^p\,dy \\ + N\int_{\mathbb{R}^d} |D\zeta^y(x)|^p|Du_t(x)|^p\,dy + N\int_{\mathbb{R}^d} |D^2\zeta^y(x)|^p|u_t(x)|^p\,dy. \tag{2.6}$$

For each $y \in \mathbb{R}^d$ the function $\zeta^y u_t$ satisfies

$$d(\zeta^y u_t) = (L_t^y(\zeta^y u_t) + \hat f_t^y)\,dt + (\Lambda_t^{yk}(\zeta^y u_t) + \hat g_t^{yk})\,dw_t^k,$$

where

$$L_t^y = (1/2)a_t^{ij}(y)D_{ij}, \quad \Lambda_t^{yk} = \sigma_t^{ik}(y)D_i,$$
$$\hat f_t^y = \zeta^y f_t + \zeta^y L_t u_t - L_t^y(\zeta^y u_t), \quad \hat g_t^{yk} = \zeta^y g_t^k + \zeta^y \Lambda_t^k u_t - \Lambda_t^{yk}(\zeta^y u_t).$$

By the results for equations with coefficients independent of x (see [10] or [11]), for each y

$$\|D^2(\zeta^y u)\|_{\mathbb{L}_p(\tau)}^p \le N(\|\hat f^y\|_{\mathbb{L}_p(\tau)}^p + \|D\hat g^y\|_{\mathbb{L}_p(\tau)}^p), \tag{2.7}$$

where here and below $N = N(d,p,\delta_0,\delta_1)$. Below we also denote by N^* generic constants depending only on $d, p, \delta_0, \delta_1, \varepsilon$, and K.

Observe that

$$\zeta^y(x)L_tu_t(x) - L_t^y(\zeta^y u_t)(x) = \zeta^y(x)(1/2)[a_t^{ij}(x) - a_t^{ij}(y)]D_{ij}u_t(x) \\ + \zeta^y(x)[L_tu_t(x) - (1/2)a_t^{ij}(x)D_{ij}u_t(x)] \\ + [\zeta^y(x)(1/2)a_t^{ij}(y)D_{ij}u_t(x) - L_t^y(\zeta^y u_t)(x)].$$

Here the right-hand side is zero unless $x \in G$ and $|x-y| \le \varepsilon$ and it follows from our assumptions that

$$|\zeta^y L_t u_t - L_t^y(\zeta^y u_t)|^p \le N\beta_0^p|\zeta^y|^p|D^2u_t|^p + N^*|\eta^y|^p(|Du_t|^p + |u_t|^p),$$

where

$$\eta^y = |\zeta^y|^p + |D\zeta^y| + |D^2\zeta^y|.$$

Therefore,

$$\|\hat f^y\|_{\mathbb{L}_p(\tau)}^p \le N\|\zeta^y f\|_{\mathbb{L}_p(\tau)}^p + N\beta_0^p\|\zeta^y D^2u_t\|_{\mathbb{L}_p(\tau)}^p \\ + N^*\|\eta^y(|Du_t| + |u_t|)\|_{\mathbb{L}_p(\tau)}^p. \tag{2.8}$$

Furthermore,

$$
\begin{aligned}
D(\zeta^y \Lambda_t u_t - \Lambda_t^y(\zeta^y u_t)) &= \zeta^y \Lambda_t D u_t - \Lambda_t^y(\zeta^y D u_t) \\
+ [D(\zeta^y \Lambda_t u_t) - \zeta^y \Lambda_t D u_t] &+ [\Lambda_t^y(\zeta^y D u_t) - D\Lambda_t^y(\zeta^y u_t)],
\end{aligned}
$$

and

$$
\begin{aligned}
|\zeta^y \Lambda_t D u_t - \Lambda_t^y(\zeta^y D u_t)|_{\ell_2}^p &\le N\beta_0^p |\zeta^y|^p |D^2 u_t|^p + N^* |\eta^y|^p |D u_t|^p, \\
|D(\zeta^y \Lambda_t u_t) - \zeta^y \Lambda_t D u_t|_{\ell_2}^p &\le N^* |\eta^y|^p (|D u_t|^p + |u_t|^p), \\
|\Lambda_t^y(\zeta^y D u_t) - D\Lambda_t^y(\zeta^y u_t)|_{\ell_2}^p &\le N^* |\eta^y|^p (|D u_t|^p + |u_t|^p).
\end{aligned}
$$

Hence,

$$
\begin{aligned}
\|D\hat{g}^y\|_{\mathbb{L}_p(\tau)}^p \le N(\|\zeta^y Dg\|_{\mathbb{L}_p(\tau)}^p &+ \|\eta^y g\|_{\mathbb{L}_p(\tau)}^p) + N\beta_0^p \|\zeta^y D^2 u_t\|_{\mathbb{L}_p(\tau)}^p \\
&+ N^* \|\eta^y (|Du_t| + |u_t|)\|_{\mathbb{L}_p(\tau)}^p.
\end{aligned}
$$

Coming back to (2.7) we conclude that for each $y \in \mathbb{R}^d$

$$
\begin{aligned}
\|D^2(\zeta^y u)\|_{\mathbb{L}_p(\tau)}^p \le N(\|\zeta^y f\|_{\mathbb{L}_p(\tau)}^p &+ \|\zeta^y Dg\|_{\mathbb{L}_p(\tau)}^p + \|\eta^y g\|_{\mathbb{L}_p(\tau)}^p) \\
&+ N\beta_0^p \|\zeta^y D^2 u_t\|_{\mathbb{L}_p(\tau)}^p + N^* \|\eta^y (|Du_t| + |u_t|)\|_{\mathbb{L}_p(\tau)}^p.
\end{aligned}
$$

We now integrate both part of (2.6) with respect to x, t, and ω. We also use simple observations like

$$
\begin{aligned}
\int_{\mathbb{R}^d} \|\zeta^y D^2 u_t\|_{\mathbb{L}_p(\tau)}^p \, dy &= E \int_0^\tau \int_{\mathbb{R}^d} \Big(\int_{\mathbb{R}^d} \zeta^p(x-y) |D^2 u_t(x)|^p \, dy \Big) \, dxdt \\
&= E \int_0^\tau \int_{\mathbb{R}^d} |D^2 u_t(x)|^p \, dxdt = \|D^2 u\|_{\mathbb{L}_p(\tau)}^p,
\end{aligned}
$$

$$
\int_{\mathbb{R}^d} \|\eta^y (|Du_t| + |u_t|)\|_{\mathbb{L}_p(\tau)}^p \, dy \le N^* (\|Du\|_{\mathbb{L}_p(\tau)}^p + \|u\|_{\mathbb{L}_p(\tau)}^p).
$$

Then we find

$$
\begin{aligned}
\|D^2 u\|_{\mathbb{L}_p(\tau)}^p \le N(\|f\|_{\mathbb{L}_p(\tau)}^p &+ \|g\|_{\mathbb{H}_p^1(\tau)}^p) \\
&+ N_1 \beta_0^p \|D^2 u\|_{\mathbb{L}_p(\tau)}^p + N^* \|u\|_{\mathbb{H}_p^1(\tau)}^p,
\end{aligned} \tag{2.9}
$$

where $N_1 = N_1(d, p, \delta_0, \delta_1)$. We can now specify the value of β_0 by taking it such that

$$
N_1 \beta_0^p \le 1/2.
$$

Then by collecting like terms in (2.9) we come to

$$
\|D^2 u\|_{\mathbb{L}_p(\tau)}^p \le N(\|f\|_{\mathbb{L}_p(\tau)}^p + \|g\|_{\mathbb{H}_p^1(\tau)}^p) + N_1^* \|u\|_{\mathbb{H}_p^1(\tau)}^p. \tag{2.10}
$$

Here by interpolation inequalities, for any $\gamma \in (0,1]$,

$$\|u\|^p_{\mathbb{H}^1_p(\tau)} \le \gamma\|D^2u\|^p_{\mathbb{L}_p(\tau)} + N(d,p)\gamma^{-1}\|u\|^p_{\mathbb{L}_p(\tau)},$$

so that by choosing γ to satisfy $N_1^*\gamma \le 1/2$ we get (2.5) from (2.10). The lemma is proved. □

Proof of Theorem 2.1. If $R = 1$ the result follows directly from Lemma 2.1. The case of general R we reduce to the particular one by using dilations. Introduce

$$\hat{\mathcal{F}}_t = \mathcal{F}_{R^2t}, \quad \hat{\tau} = R^{-2}\tau, \quad \hat{w}^k_t = R^{-1}w^k_{R^2t},$$
$$(\hat{a}_t, \hat{b}_t, \hat{c}_t, \hat{\sigma}_t, \hat{\nu}_t)(x) = (a_{R^2t}, Rb_{R^2t}, R^2c_{R^2t}, \sigma_{R^2t}, R\nu_{R^2t})(Rx),$$
$$\hat{u}_t(x) = u_{R^2t}(Rx), \quad \hat{f}_t(x) = R^2f_{R^2t}(Rx), \quad \hat{g}^k_t(x) = Rg^k_{R^2t}(Rx).$$

Also introduce the operators $\hat{L}_t$ and $\hat{\Lambda}^k_t$ constructing them from the above introduced coefficients. It is easily seen that $\hat{w}^k_t$ are independent $\hat{\mathcal{F}}_t$-Wiener processes, $\hat{\tau}$ is an $\hat{\mathcal{F}}_t$-stopping time, all the above processes with hats are predictable with respect to the filtration $\{\hat{\mathcal{F}}_t\}$, and $\hat{u} \in \hat{\mathcal{H}}^2_{p,0}(\hat{\tau})$, $\hat{f} \in \hat{\mathbb{L}}_p(\hat{\tau})$, $\hat{g} \in \hat{\mathbb{H}}^1_p(\hat{\tau})$, where the spaces with hats are defined on the basis of $\{\hat{\mathcal{F}}_t\}$.

Observe that for $t \le \hat{\tau}$

$$\begin{aligned}\hat{L}_t\hat{u}_t(x) &= (1/2)\hat{a}^{ij}(x)D_{ij}\hat{u}_t(x) + \hat{b}^i_t(x)D_i\hat{u}_t(x) + \hat{c}_t(x)\hat{u}_t(x)\\ &= R^2[(1/2)a^{ij}_{R^2t}D_{ij}u_{R^2t} + b^i_{R^2t}D_iu_{R^2t} + c_{R^2t}u_{R^2t}](Rx)\\ &= R^2L_{R^2t}u_{R^2t}(Rx),\end{aligned}$$

$$\int_0^t [\hat{L}_s\hat{u}_s(x) + \hat{f}_s(x)]\,ds = \int_0^{R^2t} [L_su_s(Rx) + f_s(Rx)]\,ds.$$

Of course, we understand this equality in the sense of distributions:

$$\int_0^t (\hat{L}_s\hat{u}_s + \hat{f}_s, \phi)\,ds = \int_0^{R^2t} ([L_su_s + f_s](R\cdot), \phi)\,ds$$

for any $\phi \in C_0^\infty(\mathbb{R}^d)$. One also knows that if $\hat{h}_t$ is an $\hat{\mathcal{F}}_t$-predictable process satisfying a natural integrability condition with respect to t, then

$$\int_0^t \hat{h}_s\,d\hat{w}^k_s = R^{-1}\int_0^{R^2t} \hat{h}_{R^{-2}s}\,dw^k_s \quad \text{(a.s.)}.$$

Therefore, (a.s.) for finite $t \le \hat{\tau}$

$$\begin{aligned}\int_0^t [\hat{\Lambda}^k_s\hat{u}_s + \hat{g}^k_s](x)\,d\hat{w}^k_s &= R\int_0^t [\Lambda^k_{R^2s}u_{R^2s} + g^k_{R^2s}](Rx)\,d\hat{w}^k_s\\ &= \int_0^{R^2t} [\Lambda^k_su_s + g^k_s](Rx)\,dw^k_s.\end{aligned}$$

It follows that (a.s.) for finite $t \leq \hat{\tau}$

$$\int_0^t [\hat{L}_s \hat{u}_s(x) + \hat{f}_s(x)]\, ds + \int_0^t [\hat{\Lambda}^k_s \hat{u}_s + \hat{g}^k_s](x)\, d\hat{w}^k_s = u_{R^2 t}(Rx) = \hat{u}_t(x),$$

so that $\hat{u}$ satisfies equation (1.2) with new operators and free terms. It is also easy to see that our objects with hats satisfy the assumptions of the theorem with $R = 1$. Therefore, by the result for $R = 1$

$$\|D^2 \hat{u}\|_{\mathbb{L}_p(\hat{\tau})} + \|D\hat{u}\|_{\mathbb{L}_p(\hat{\tau})} \leq N\big(\|\hat{f}\|_{\mathbb{L}_p(\hat{\tau})} + \|\hat{g}\|_{\mathbb{L}_p(\hat{\tau})} + \|D\hat{g}\|_{\mathbb{L}_p(\hat{\tau})}\big) + N^* \|\hat{u}\|_{\mathbb{L}_p(\hat{\tau})}.$$

Now it only remains to notice that changing variables shows that this inequality is precisely (2.4). The theorem is proved.

Here is an interior estimate.

THEOREM 2.2. *Take some $\varepsilon, R \in (0,\infty)$ and a stopping time τ and suppose that the assumptions (i) and (ii) of Theorem 2.1 are satisfied. Then, for any $r \in (0,R)$, we have*

$$\begin{aligned} \|I_{B'_r} D^2 u\|_{\mathbb{L}_p(\tau)} + R^{-1} \|I_{B'_r} Du\|_{\mathbb{L}_p(\tau)} \leq N\big(\|I_{B'_R} f\|_{\mathbb{L}_p(\tau)} + \|I_{B'_R} Dg\|_{\mathbb{L}_p(\tau)} \\ + (R-r)^{-1} \|I_{B'_R} g\|_{\mathbb{L}_p(\tau)}\big) + N^*(R-r)^{-2} \|I_{B'_R} u\|_{\mathbb{L}_p(\tau)}, \end{aligned} \tag{2.11}$$

where $N = N(\delta_0, \delta_1, d, p)$ and $N^* = N^*(K, \varepsilon, \delta_0, \delta_1, d, p)$.

Proof. We follow a usual procedure taken from the theory of PDEs. Let $\chi(s)$ be an infinitely differentiable function on $\mathbb{R}$ such that $\chi(s) = 1$ for $s \leq 1$ and $\chi(s) = 0$ for $s \geq 2$. For $m = 0, 1, 2, \ldots$ introduce, $(r_0 = r)$

$$r_m = r + (R-r) \sum_{j=1}^m 2^{-j}, \quad \xi_m(x) = \chi\big(2^{m+1}(R-r)^{-1}(|x'| - r_m) + 1\big).$$

As is easy to check, for

$$Q(m) = B'_{r_m},$$

it holds that

$$\zeta_m = 1 \quad \text{on} \quad Q(m), \quad \zeta_m = 0 \quad \text{outside} \quad Q(m+1).$$

Also (observe that $N 2^{m+1} = N_1 2^m$ with $N_1 = 2N$)

$$|D\zeta_m| \leq N 2^m (R-r)^{-1}, \quad |D^2 \zeta_m| \leq N 2^{2m} (R-r)^{-2}.$$

Next, the function $\zeta_m u_t$ is in $\mathcal{H}^2_{p,0}(\tau)$ and satisfies

$$d(\zeta_m u_t) = (L(\zeta_m u_t) + f_{mt})\, dt + (\Lambda^k(\zeta_m u_t) + g^k_{mt})\, dw^k_t,$$

where

$$f_{mt} = \zeta_m f_t + u(L_t - c_t)\zeta_m - a^{ij}_t D_i u_t D_j \zeta_m, \quad g^k_{mt} = \zeta_m g^k_t + u_t \sigma^{ik}_t D_i \zeta_m.$$

Since $\zeta_m u_t(x) = 0$ for $x \notin B'_R$, by Theorem 2.1

$$A_m := \|D^2(\zeta_m u)\|^p_{\mathbb{L}_p(\tau)} + R^{-p}\|D(\zeta_m u)\|^p_{\mathbb{L}_p(\tau)} \leq N(F' + R^{-p}G^0) \tag{2.12}$$
$$+ N^* R^{-2p}U^0 + NE\int_0^\tau (B_{mt} + C^1_{mt} + G^1_{mt} + R^{-p}C^0_{mt})\,dt,$$

where

$$F' = \|I_{B'_R} f\|^p_{\mathbb{L}_p(\tau)}, \quad G^0 = \|I_{B'_R} g\|^p_{\mathbb{L}_p(\tau)}, \quad U^0 = \|I_{B'_R} u\|^p_{\mathbb{L}_p(\tau)},$$
$$B_{mt} = \|u(L_t - c_t)\zeta_m - a^{ij}_t D_i u_t D_j \zeta_m\|^p_{L_p(\mathbb{R}^d)},$$
$$C^1_{mt} = \|D(u_t \sigma^i_t D_i \zeta_m)\|^p_{L_p(\mathbb{R}^d)}, \quad G^1_{mt} = \|D(\zeta_m g_t)\|^p_{L_p(\mathbb{R}^d)},$$
$$C^0_{mt} = \|u_t \sigma^i_t D_i \zeta_m\|^p_{L_p(\mathbb{R}^d)}.$$

Observe that by the above mentioned properties of ζ_m and the assumption on b, we have

$$|(L_t - c_t)\zeta_m| \leq N^* 2^{2m}(R-r)^{-2} + N^* R^{-1} 2^m (R-r)^{-1} \leq N^* 2^{2m}(R-r)^{-2}.$$

It follows that

$$B_{mt} \leq N^* 2^{2mp}(R-r)^{-2p} U^0_t + N^* 2^{mp}(R-r)^{-p}\|Du_t\|^p_{L_p(Q(m+1))},$$

where

$$U^0_t = \|u_t\|^p_{L_p(B'_R)}.$$

Furthermore, by interpolation inequalities for any $\gamma > 0$

$$\|Du_t\|^p_{L_p(Q(m+1))} \leq \|D(\zeta_{m+1}u_t)\|^p_{L_p(\mathbb{R}^d)}$$
$$\leq \gamma(R-r)^p 2^{-mp}\|D^2(\zeta_{m+1}u_t)\|^p_{L_p(Q(m+2))} + N\gamma^{-1}2^{mp}(R-r)^{-p}U^0_t,$$

so that for $\gamma \in (0,1)$ (with γ, perhaps, different from the one above)

$$B_{mt} \leq \gamma\|D^2(\zeta_{m+1}u_t)\|^p_{L_p(Q(m+2))} + N^*\gamma^{-1}2^{2mp}(R-r)^{-2p}U^0_t.$$

Similarly, for $\gamma \in (0,1)$

$$C^1_{mt} \leq \gamma\|D^2(\zeta_{m+1}u_t)\|^p_{L_p(Q(m+2))} + N^*\gamma^{-1}2^{2mp}(R-r)^{-2p}U^0_t$$

and almost obviously

$$C^0_{mt} \leq N2^{mp}(R-r)^{-p}U^0_t \leq R^p N\gamma^{-1}2^{2mp}(R-r)^{-2p}U^0_t$$
$$G^1_{mt} \leq N(\|Dg_t\|^p_{L_p(B'_R)} + 2^{mp}(R-r)^{-p}\|g_t\|^p_{L_p(B'_R)}).$$

Hence (2.12) yields

$$A_m \leq \gamma A_{m+1} + NF + N2^{mp}(R-r)^{-p}G^0 + N^*\gamma^{-1}2^{2mp}(R-r)^{-2p}U^0,$$

where

$$F = \|I_{B'_R} f\|^p_{\mathbb{L}_p(\tau)} + \|I_{B'_R} Dg\|^p_{\mathbb{L}_p(\tau)}.$$

Now we take $\gamma = 8^{-p}$ and get

$$\begin{aligned}\gamma^m A_m &\leq \gamma^{m+1} A_{m+1} + N\gamma^m F + N\gamma^m 2^{mp}(R-r)^{-p}G^0 \\ &\quad + N^*\gamma^m\gamma^{-1}2^{2mp}(R-r)^{-2p}U^0, \\ A_0 + \sum_{m=1}^{\infty}\gamma^m A_m &\leq \sum_{m=1}^{\infty}\gamma^m A_m + NF + N(R-r)^{-p}G^0 \\ &\quad + N^*(R-r)^{-2p}U^0.\end{aligned} \tag{2.13}$$

In order to cancel like terms in (2.13) we need the series to converge. Observe that

$$\|D^2(\zeta_m u_t)\|^p_{L_p} + R^{-p}\|D(\zeta_m u_t)\|^p_{L_p} \leq N(1 + 4^{mp}(R-r)^{-2p})\|u_t\|^p_{W^2_p},$$

and therefore

$$\sum_{m=1}^{\infty}\gamma^m A_m \leq N(1 + (R-r)^{-2p})\|u\|^p_{\mathbb{H}^2_p(\tau)}.$$

However, the right-hand side may be infinite. In order to circumvent this difficulty we take in the beginning of the proof $\tau \wedge T$ in place of τ, where $T \in (0, \infty)$. Then by the definition of $\mathcal{H}^2_p(\tau)$ we have

$$\|u\|^p_{\mathbb{H}^2_p(\tau\wedge T)} < \infty$$

and we get from (2.13) that A_0 for modified τ is less than the right-hand side of (2.11). Since its left-hand side, with $\tau \wedge T$ in place of τ, is obviously less than A_0, we obtain (2.11) with τ on the left replaced with $\tau \wedge T$. After this it only remains to let $T \to \infty$. The theorem is proved. □

REMARK 2.1. One can prove interior estimates for higher order derivatives of solutions if the coefficients are more regular. This is a routine matter and is achieved by considering equations for derivatives of solutions, which one obtains by differentiating the equation. On this way one would obtain more regular solutions of equations in half spaces. However, for brevity we do not pursue this issue.

3. Interior estimates for solutions of parabolic PDEs in the whole space. Assume that for $i, j = 1, ..., d$ we are given real-valued Borel measurable functions $a^{ij}_t(x)$, $b^i_t(x)$, $c_t(x)$ defined for $(t, x) \in \mathbb{R}_+ \times \mathbb{R}^d$, where $\mathbb{R}_+ = (0, \infty)$.

ASSUMPTION 3.1. *For all values of the indices and arguments $a^{ij} = a^{ji}$ and, for a constant $\delta_0 \in (0, 1]$ and all unit $\lambda \in \mathbb{R}^d$,*

$$\delta_0^{-1} \geq a^{ij}_t \lambda^i \lambda^j \geq \delta_0.$$

This assumption is supposed to be satisfied throughout the deterministic part of the paper. Here we take a constant $T \in (0, \infty]$ and consider equation (1.1)

The solutions of (1.1) are looked for in the space $\mathcal{H}^2_{p,0}(T)$ with $f \in \mathbb{L}_p(T)$. We remind briefly the definition of these spaces. Everywhere in the deterministic part of the article

$$p > 1.$$

For an integer $n = 0, 1, 2$ we take Sobolev spaces $W^n_p = W^n_p(\mathbb{R}^d)$ and we set

$$\mathbb{H}^n_p(T) = L_p((0,T), W^n_p), \quad \mathbb{H}^n_p = \mathbb{H}^n_p(\infty), \quad \mathbb{L}_p\ldots = \mathbb{H}^0_p\ldots$$

In an obvious way we introduce the norms in these spaces. Observe that we are using the same notation as in the stochastic case. Deterministic spaces are subspaces of stochastic ones and there is no real danger of confusion.

If for any finite $t \in [0, T]$ we are given a generalized function $u = u_t$ on $\mathbb{R}^d$ such that $u \in \bigcap_{0<S<\infty} \mathbb{H}^2_p(S \wedge T)$, then we write $u \in \mathcal{H}^2_{p,0}(T)$ if and only if $D^2 u \in \mathbb{L}_p(T)$ and there exist $f \in \mathbb{L}_p(T)$ such that for any $\phi \in C^\infty_0(\mathbb{R}^d)$, the equation

$$(u_t, \phi) = \int_0^t (f_s, \phi)\, ds$$

holds for all finite $t \le T$. In that case we write

$$\frac{\partial}{\partial t} u_t = f_t,$$

which explains the sense in which (1.1) is understood and set

$$\|u\|_{\mathcal{H}^2_{p,0}(T)} = \|D^2 u\|_{\mathbb{L}_p(T)} + \|f\|_{\mathbb{L}_p(T)}.$$

We take d_1, x', and B'_R from Section 2 and state the first main result of this section.

THEOREM 3.1. *Take some $\varepsilon, R \in (0,\infty)$ and assume the following:*

(i) There exists a constant $K \in (0,\infty)$ such that for $x \in B'_R$ and $y \in \mathbb{R}^d$, such that $|x - y| \le \varepsilon R$, and all values of the indices and other arguments

$$R|b^i_t(y)| + R^2|c_t(y)| \le K, \quad |a^{ij}_t(x) - a^{ij}_t(y)| \le \beta_0,$$

where $\beta_0 = \beta_0(d, p, \delta_0, \delta_1) > 0$ is a (small) constant from Lemma 3.1 below.

(ii) We are given a function $u \in \mathcal{H}^2_{p,0}(T)$ which is a solution of (1.1) *with an $f \in \mathbb{L}_p(T)$.*

Finally, assume that $u_t(x) = 0$ if $x \notin B'_R$. Then there exist constants $N = N(d, p, \delta_0)$ and $N^* = N^*(d, p, \delta_0, \varepsilon, K)$ such that

$$\|D^2 u\|_{\mathbb{L}_p(T)} + R^{-1}\|Du\|_{\mathbb{L}_p(T)} \le N\|f\|_{\mathbb{L}_p(T)} + N^* R^{-2}\|u\|_{\mathbb{L}_p(T)}.$$

Similarly to Theorem 2.1 this theorem is obtained by using the parabolic dilations from the following lemma in which one can take $G = B'_R$.

LEMMA 3.1. *Assume that we are given a* $u \in \mathcal{H}^2_{p,0}(T)$ *which satisfies* (1.1) *with an* $f \in \mathbb{L}_p(T)$. *Assume that for a domain* $G \subset \mathbb{R}^d$ *(perhaps,* $G = \mathbb{R}^d$*) we have* $u_t(x) = 0$ *if* $x \notin G$. *Also assume that there are constants* $\varepsilon, K \in (0, \infty)$ *such that for all* $x \in G$ *and* $y \in \mathbb{R}^d$, *with* $|x - y| \le \varepsilon$, *and all values of other arguments we have*

$$|b^i_t(y)| + |c_t(y)| \le K, \quad |a^{ij}_t(x) - a^{ij}_t(y)| \le \beta_0,$$

where $\beta_0 = \beta_0(d, p, \delta_0) > 0$ *is a (small) constant an estimate from below for which can be obtained from the proof.*

Then there exist constants

$$N = N(d, p, \delta_0), \quad N^* = N^*(d, p, \delta_0, \varepsilon, K)$$

such that

$$\|u\|_{\mathbb{H}^2_p(T)} \le N\|f\|_{\mathbb{L}_p(T)} + N^*\|u\|_{\mathbb{L}_p(T)}. \tag{3.1}$$

Proof. We take a nonnegative $\zeta \in C_0^\infty(\mathbb{R}^d)$ from the proof of Lemma 2.1 and use the facts that (2.6) holds and for each $y \in \mathbb{R}^d$ the function $\zeta^y u_t$ satisfies

$$\frac{\partial}{\partial t}(\zeta^y u_t) = L^y_t(\zeta^y u_t) + \hat{f}^y_t,$$

where

$$L^y_t = (1/2)a^{ij}_t(y)D_{ij}, \quad \hat{f}^y_t = \zeta^y f_t + \zeta^y L_t u_t - L^y_t(\zeta^y u_t).$$

By classical results for equations with coefficients independent of x, for each y

$$\|D^2(\zeta^y u)\|^p_{\mathbb{L}_p(T)} \le N\|\hat{f}^y\|^p_{\mathbb{L}_p(T)}, \tag{3.2}$$

where here and below $N = N(d, p, \delta_0)$. Below we also denote by N^* generic constants depending only on $d, p, \delta_0, \varepsilon$, and K.

By combining (2.8) with (3.2) we see that for each $y \in \mathbb{R}^d$

$$\|D^2(\zeta^y u)\|^p_{\mathbb{L}_p(T)} \le N(\|\zeta^y f\|^p_{\mathbb{L}_p(T)} + N\beta_0^p\|\zeta^y D^2 u_t\|^p_{\mathbb{L}_p(T)}$$
$$+N^*\|\eta^y(|Du_t| + |u_t|)\|^p_{\mathbb{L}_p(T)}.$$

We now integrate both part of (2.6) with respect to t, x. We also use simple observations like

$$\int_{\mathbb{R}^d} \|\zeta^y D^2 u_t\|^p_{\mathbb{L}_p(T)}\, dy = \int_0^T \int_{\mathbb{R}^d} \left(\int_{\mathbb{R}^d} \zeta^p(x - y)|D^2 u_t(x)|^p\, dy \right) dxdt$$
$$= \int_0^T \int_{\mathbb{R}^d} |D^2 u_t(x)|^p\, dxdt = \|D^2 u\|^p_{\mathbb{L}_p(T)},$$

$$\int_{\mathbb{R}^d} \|\eta^y(|Du_t| + |u_t|)\|^p_{\mathbb{L}_p(T)}\,dy \le N^*(\|Du\|^p_{\mathbb{L}_p(T)} + \|u\|^p_{\mathbb{L}_p(T)}).$$

Then we find

$$\|D^2u\|^p_{\mathbb{L}_p(T)} \le N\|f\|^p_{\mathbb{L}_p(T)} + N_1\beta_0^p\|D^2u\|^p_{\mathbb{L}_p(T)} + N^*\|u\|^p_{\mathbb{H}^1_p(T)}, \tag{3.3}$$

where $N_1 = N_1(d, p, \delta_0)$. We can now specify the value of β_0 by taking it such that

$$N_1\beta_0^p \le 1/2.$$

Then by collecting like terms in (3.3) we come to

$$\|D^2u\|^p_{\mathbb{L}_p(T)} \le N\|f\|^p_{\mathbb{L}_p(T)} + N_1^*\|u\|^p_{\mathbb{H}^1_p(T)}. \tag{3.4}$$

Here by interpolation inequalities, for any $\gamma \in (0, 1]$,

$$\|u\|^p_{\mathbb{H}^1_p(\tau)} \le \gamma\|D^2u\|^p_{\mathbb{L}_p(\tau)} + N(d,p)\gamma^{-1}\|u\|^p_{\mathbb{L}_p(\tau)},$$

so that by choosing γ such that $N_1^*\gamma \le 1/2$ we get (3.1) from (3.4). The lemma is proved.

Our next result is the following interior estimate.

THEOREM 3.2. *Take some $\varepsilon, R \in (0, \infty)$ and suppose that the assumptions (i) and (ii) of Theorem 3.1 are satisfied. Then, for any $r \in (0, R)$, we have*

$$\begin{aligned} &\|I_{B'_r}D^2u\|_{\mathbb{L}_p(T)} + R^{-1}\|I_{B'_r}Du\|_{\mathbb{L}_p(T)} \\ &\le N\|I_{B'_R}f\|_{\mathbb{L}_p(T)} + N^*(R-r)^{-2}\|I_{B'_R}u\|_{\mathbb{L}_p(T)}, \end{aligned} \tag{3.5}$$

where $N = N(\delta_0, d, p)$ and $N^* = N^*(K, \varepsilon, \delta_0, d, p)$.

Proof. We take r_m, ζ_m, and Q_m from the proof of Theorem 2.2 and notice that the function $\zeta_m u_t$ is in $\mathcal{H}^2_{p,0}(T)$ and satisfies

$$\frac{\partial}{\partial t}(\zeta_m u_t) = L(\zeta_m u_t) + f_{mt},$$

where

$$f_{mt} = \zeta_m f_t + u(L_t - c_t)\zeta_m - a^{ij}_t D_i u_t D_j \zeta_m.$$

By Theorem 3.1

$$\begin{aligned} A_m &:= \|D^2(\zeta_m u)\|^p_{\mathbb{L}_p(T)} + R^{-p}\|D(\zeta_m u)\|^p_{\mathbb{L}_p(T)} \\ &\le NF + N^*R^{-2p}U^0 + N\int_0^T B_{mt}\,dt, \end{aligned} \tag{3.6}$$

where

$$F = \|I_{B'_R} f\|^p_{\mathbb{L}_p(T)}, \quad U^0 = \|I_{B'_R} u\|^p_{\mathbb{L}_p(T)},$$
$$B_{mt} = \|u(L_t - c_t)\zeta_m - a^{ij}_t D_i u_t D_j \zeta_m\|^p_{L_p(\mathbb{R}^d)}.$$

Observe that by the properties of ζ_m and its derivatives and the assumption on b, we have

$$|(L_t - c_t)\zeta_m| \le N^* 2^{2m}(R-r)^{-2} + N^* R^{-1} 2^m (R-r)^{-1} \le N^* 2^{2m}(R-r)^{-2}.$$

It follows that

$$B_{mt} \le N^* 2^{2mp}(R-r)^{-2p} U^0_t + N^* 2^{mp}(R-r)^{-p}\|Du_t\|^p_{L_p(Q(m+1))},$$

where

$$U^0_t = \|u_t\|^p_{L_p(B'_R)}.$$

Furthermore, by interpolation inequalities for any $\gamma > 0$

$$\|Du_t\|^p_{L_p(Q(m+1))} \le \|D(\zeta_{m+1}u_t)\|^p_{L_p(\mathbb{R}^d)}$$
$$\le \gamma(R-r)^p 2^{-mp}\|D^2(\zeta_{m+1}u_t)\|^p_{L_p(Q(m+2))} + N\gamma^{-1}2^{mp}(R-r)^{-p}U^0_t,$$

so that for $\gamma \in (0,1)$ (with γ, perhaps, different from the one above)

$$B_{mt} \le \gamma\|D^2(\zeta_{m+1}u_t)\|^p_{L_p(Q(m+2))} + N^*\gamma^{-1}2^{2mp}(R-r)^{-2p}U^0_t.$$

Hence (3.6) yields

$$A_m \le \gamma A_{m+1} + NF + N^*\gamma^{-1}2^{2mp}(R-r)^{-2p}U^0.$$

Now we take $\gamma = 8^{-p}$ and get

$$\gamma^m A_m \le \gamma^{m+1}A_{m+1} + N\gamma^m F + N^*\gamma^m\gamma^{-1}2^{2mp}(R-r)^{-2p}U^0,$$

$$A_0 + \sum_{m=1}^{\infty}\gamma^m A_m \le \sum_{m=1}^{\infty}\gamma^m A_m + NF + N^*(R-r)^{-2p}U^0. \tag{3.7}$$

In order to cancel like terms in (3.7) we need the series to converge. Observe that

$$\|D^2(\zeta_m u_t)\|^p_{L_p} + R^{-p}\|D(\zeta_m u_t)\|^p_{L_p} \le N(1 + 4^{mp}(R-r)^{-2p})\|u_t\|^p_{W^2_p},$$

and therefore

$$\sum_{m=1}^{\infty}\gamma^m A_m \le N(1 + (R-r)^{-2p})\|u\|^p_{\mathbb{H}^2_p(T)}.$$

However, the right-hand side may be infinite if $T = \infty$. In order to circumvent this difficulty in that case we take in the beginning of the proof $S \in (0, \infty)$ in place of $T = \infty$. Then by the definition of $\mathcal{H}^2_p(\infty)$ we have

$$\|u\|^p_{\mathbb{H}^2_p(S)} < \infty$$

and we get from (3.7) that A_0 for modified T is less than the right-hand side of (3.5). Since its left-hand side, with S in place of T, is obviously less than A_0, we obtain (3.5) with T on the left replaced with S. After this it only remains to let $S \to \infty$. The theorem is proved.

REMARK 3.1. One can prove interior estimates for higher order derivatives of solutions if the coefficients are more regular. This is a routine matter and is achieved by considering equations for derivatives of solutions, which one obtains by differentiating the equation. On this way one would obtain more regular solutions of equations in half spaces. However, for brevity we do not treat this issue.

4. Local regularity near the boundary for SPDEs in half spaces. In the setting of Section 2 we will be considering (1.2) in $\mathbb{R}^d_+$. We need the stochastic Sobolev spaces $\mathbb{H}^n_{p,\theta}(\tau)$, $n = 0, 1, 2$, and $\mathfrak{H}^2_{p,\theta}(\tau)$ introduced in [15] in the following way. We take a number $\theta \in \mathbb{R}$ and introduce $L_{p,\theta}$ as the space of functions on $\mathbb{R}^d_+$ having finite norm given by

$$\|u\|^p_{L_{p,\theta}} = \int_{\mathbb{R}^d_+} |u(x)|^p (x^1)^{\theta - d}\, dx.$$

Then $H^1_{p,\theta}$ is the set of functions such that $u, MDu \in L_{p,\theta}$, where M is the operator of multiplying by x^1, and $H^2_{p,\theta}$ is the set of functions such that $u, MDu, M^2D^2u \in L_{p,\theta}$. The norms in these spaces are introduced in a natural way (cf. Remark 4.2 below) and, as before, we extend these definitions for ℓ_2-valued functions such as g. Naturally, we write $H^0_{p,\theta} = L_{p,\theta}$.

These definitions are used for all $p \in (1, \infty)$.

REMARK 4.1. In [12] the spaces $H^\gamma_{p,\theta}$ are introduce for all real γ and it is proved that $C^\infty_0(\mathbb{R}^d_+)$ is dense in $H^\gamma_{p,\theta}$. In our case that $\gamma = 0, 1, 2$ this is a rather simple fact.

To prove it, it suffices to show that the subset of $H^n_{p,\theta}$, $n = 0, 1, 2$, consisting of functions on $\mathbb{R}^d_+$ with compact support is dense in $H^n_{p,\theta}$. To do that, we take some nonnegative $\xi \in C^\infty_0(\mathbb{R})$ and $\eta \in C^\infty_0(\mathbb{R}^{d-1})$ such that $\xi(0) = 1$, $\xi(t) = 0$ for $|t| \geq 1$, and $\eta(0) = 1$ and for $m = 1, 2, ...$ define

$$\zeta_m(x) = \xi(m^{-1} \ln x^1)\eta(e^{-2m}x'), \quad x \in \mathbb{R}^d_+.$$

Notice that

$$|x^1 D_1 \zeta_m(x)| = |m^{-1}\xi'(m^{-1}\ln x^1)\eta(e^{-2m}x')|,$$
$$|x^1 D_i \zeta_m(x)| = x^1 e^{-2m} |\xi(m^{-1}\ln x^1)(D_i\eta)(e^{-2m}x')|, \quad i \geq 2,$$

and if $\xi(m^{-1}\ln x^1) \neq 0$, then $m^{-1}\ln x^1 \leq 1$ and $x^1 \leq e^m$. It follows that $|x^1 D\zeta_m(x)|$ tend to zero as $m \to \infty$ uniformly on $\mathbb{R}^d_+$.

The following formulas in which $i, j \geq 2$

$$\begin{aligned}
|(x^1)^2 D_{11}\zeta_m(x)| &= |m^{-2}\xi''(m^{-1}\ln x^1) - m^{-1}\xi'(m^{-1}\ln x^1)|\eta(e^{-2m}x'),\\
|(x^1)^2 D_{1i}\zeta_m(x)| &= x^1 e^{-2m} m^{-1}|\xi'(m^{-1}\ln x)(D_i\eta)(e^{-2m}x')|,\\
|(x^1)^2 D_{ij}\zeta_m(x)| &= (x^1)^2 e^{-4m}|\xi(m^{-1}\ln x^1)(D_{ij}\eta)(e^{-2m}x')|
\end{aligned}$$

show that $|(x^1)^2 D^2\zeta_m(x)|$ tend to zero as $m \to \infty$ uniformly on $\mathbb{R}^d_+$. Finally, observe that ζ_m are uniformly bounded and tend to 1 pointwise.

Now the dominated convergence theorem and the formulas

$$\begin{gathered}
|u - u\zeta_m| = |1 - \zeta_m|\,|u|,\\
|MDu - MD(\zeta_m u)| \leq |1 - \zeta_m|\,|Du| + |uMD\zeta_m|,\\
|M^2D^2u - M^2D^2(\zeta_m u)| \leq |1 - \zeta_m|\,|M^2D^2u|\\
+ 2|MD\zeta_m|\,|MDu| + |uM^2D^2\zeta_m|
\end{gathered}$$

easily show that, if $n \in \{0,1,2\}$ and $u \in H^n_{p,\theta}$, then $\zeta_m u \to u$ in $H^n_{p,\theta}$ as $m \to \infty$.

REMARK 4.2. A few times in the future we will be dealing with functions u such that $M^{-1}u \in H^2_{p,\theta}$. In connection with this it is useful to have in mind that

$$\begin{gathered}
MD_i(M^{-1}u) = D_i u - \delta^{1i}M^{-1}u,\\
M^2D_{ij}(M^{-1}u) = MD_{ij}u - \delta^{1i}D_j u - \delta^{1j}D_i u + 2\delta^{1i}\delta^{1j}M^{-1}u
\end{gathered}$$

implying that

$$\begin{gathered}
\|M^{-1}u\|_{H^1_{p,\theta}} = \|MD(M^{-1}u)\|_{L_{p,\theta}} + \|M^{-1}u\|_{L_{p,\theta}}\\
\leq N(\|Du\|_{L_{p,\theta}} + \|M^{-1}u\|_{L_{p,\theta}}), \quad \|Du\|_{L_{p,\theta}} \leq N\|M^{-1}u\|_{H^1_{p,\theta}},\\
\|M^{-1}u\|_{H^2_{p,\theta}} = \|M^2D^2(M^{-1}u)\|_{L_{p,\theta}} + \|M^{-1}u\|_{H^1_{p,\theta}}\\
\leq N(\|MD^2u\|_{L_{p,\theta}} + \|Du\|_{L_{p,\theta}} + \|M^{-1}u\|_{L_{p,\theta}}),\\
\|MD^2u\|_{L_{p,\theta}} + \|Du\|_{L_{p,\theta}} + \|M^{-1}u\|_{L_{p,\theta}} \leq N\|M^{-1}u\|_{H^2_{p,\theta}}.
\end{gathered}$$

It follows that the $H^2_{p,\theta}$ norm of $M^{-1}u$ is equivalent to

$$\|MD^2u\|_{L_{p,\theta}} + \|Du\|_{L_{p,\theta}} + \|M^{-1}u\|_{L_{p,\theta}}.$$

Next, for $n = 0,1,2$ and stopping times τ we define $\mathbb{H}^n_{p,\theta}(\tau)$ as the set of functions $f = f_t = f_t(\omega)$ on $(0,\tau]\!]$, $t < \infty$, with values in the set of distributions on $\mathbb{R}^d_+$ which are $\mathcal{P}$-measurable and have finite norm given by

$$\|f\|^p_{\mathbb{H}^n_{p,\theta}(\tau)} = E\int_0^\tau \|f_t\|^p_{H^n_{p,\theta}}\,dt.$$

Define $\mathbb{L}_{p,\theta}(\tau) = \mathbb{H}^0_{p,\theta}(\tau)$.

For a function $u = u_t = u_t(\omega)$ given on $(0,\tau]\!]$ for finite t with values in the set of distributions on $\mathbb{R}^d_+$ we write $u \in \mathfrak{H}^2_{p,\theta,0}(\tau)$ if and only if $M^{-1}u \in \mathbb{H}^2_{p,\theta}(\tau)$ and there exist a real valued $f \in \mathbb{L}_{p,\theta}(\tau)$ and an ℓ_2-valued $g = (g^k, k = 1,2,...) \in \mathbb{H}^1_{p,\theta}(\tau)$ such that for any $\phi \in C_0^\infty(\mathbb{R}^d_+)$ with probability 1 we have

$$(u_t, \phi) = \int_0^t (M^{-1}f_s, \phi)\, ds + \sum_{k=1}^\infty \int_0^t (g^k_s, \phi)\, dw^k_s$$

for all finite $t \in (0,\tau]$. We set

$$\|u\|_{\mathfrak{H}^2_{p,\theta}(\tau)} = \|M^{-1}u\|_{\mathbb{H}^2_{p,\theta}(\tau)} + \|f\|_{\mathbb{L}_{p,\theta}(\tau)} + \|g\|_{\mathbb{H}^1_{p,\theta}(\tau)}.$$

Recall that Assumptions 2.1 is supposed to be satisfied throughout the article. For $r > 0$ denote

$$Q_r = \{x \in \mathbb{R}^d : 0 < x^1 < r\}.$$

THEOREM 4.1. *Take an $R \in (0,\infty]$ and a stopping time τ and assume the following.*

(i) For some constants $\varepsilon \in (0,1]$ and $K \in (0,\infty)$ we have

$$\begin{aligned} &|x^1 b^i_t(x)| + |x^1 D\sigma^i_t(x)|_{\ell_2} + |x^1 \nu_t(x)|_{\ell_2} \\ &+ |(x^1)^2 D\nu_t(x)|_{\ell_2} + |(x^1)^2 c_t(x)| \le K, \\ &|a^{ij}_t(x) - a^{ij}_t(y)| + |\sigma^i_t(x) - \sigma^i_t(y)|_{\ell_2} \le \beta_0, \end{aligned} \tag{4.1}$$

whenever $x, y \in \mathbb{R}^d_+$, $x^1, y^1 \le R$, $|x - y| \le \varepsilon(x^1 \wedge y^1)$, $i,j = 1,...,d$, $t > 0$, where $\beta_0 = \beta_0(d,p,\delta_0,\delta_1) \in (0,1]$ is the constant from Theorem 2.1;

(ii) We have a function u such that $\phi u \in \mathcal{H}^2_{p,0}(\tau)$ for any $\phi \in C_0^\infty(Q_R)$ and u satisfies (1.2) in $\mathbb{R}^d_+$ with some $f \in \mathbb{L}_{p,\theta}(\tau)$, $g = (g^k, k = 1,2,...) \in \mathbb{H}^1_{p,\theta}(\tau)$.

Then for any $r \in (0, R/4)$

$$\begin{aligned} &\|I_{Q_r} M D^2 u\|_{\mathbb{L}_{p,\theta}(\tau)} + \|I_{Q_r} Du\|_{\mathbb{L}_{p,\theta}(\tau)} \le N \|I_{Q_R} M f\|_{\mathbb{L}_{p,\theta}(\tau)} \\ &+ N\|I_{Q_R} M Dg\|_{\mathbb{L}_{p,\theta}(\tau)} + N\|I_{Q_R} g\|_{\mathbb{L}_{p,\theta}(\tau)} + N^* \|I_{Q_R} M^{-1} u\|_{\mathbb{L}_{p,\theta}(\tau)}, \end{aligned} \tag{4.2}$$

where $N = N(d,p,\delta_0,\delta_1)$ and $N^* = N^*(d,p,\delta_0,\delta_1,\varepsilon,K)$.

Proof. We are going to apply Theorem 2.2 to shifted B'_R when $d_1 = 1$. For $n = -1,0,1,...$, set $r_n = 2^{-n/3} r$. Observe that if $n \ge 0$, then the half width of $Q_{r_{n-1}} \setminus Q_{r_{n+2}}$ equals $\rho_n := r_{n+2}/2$ and

$$r_{n-1} + \rho_n \le 2r_{-1} < 4r < R, \quad r_{n+2} - \rho_n = \rho_n.$$

It follows that for $x \in Q_{r_{n-1}} \setminus Q_{r_{n+2}}$ and y, such that $|x - y| \le \varepsilon\rho_n$, we have

$$\begin{gathered} R > r_{n-1} \ge x^1 \ge r_{n+2} \ge \rho_n, \quad y^1 \le x^1 + \varepsilon\rho_n < R, \quad y^1 \ge x^1 - \varepsilon\rho_n \ge \rho_n, \\ \rho_n \le x^1 \wedge y^1, \quad |x - y| \le \varepsilon(x^1 \wedge y^1), \end{gathered}$$

so that by our assumptions

$$\rho_n|b^i_t(y)| + \rho_n|D\sigma^i_t(y)|_{\ell_2} + \rho_n|\nu_t(y)|_{\ell_2} + \rho^2_n|c_t(y)| + \rho^2_n|D\nu_t(y)|_{\ell_2} \leq K,$$
$$|\sigma^i_t(x) - \sigma^i_t(y)|_{\ell_2} + |a^{ij}_t(x) - a^{ij}_t(y)| \leq \beta_0.$$

Furthermore, if $n \geq 0$, $\zeta \in C^\infty_0((0,R))$ and $\zeta(z) = 1$ for $r_{n+2} \leq z \leq r_{n-1}$, then ζu satisfies (1.2) in $\mathbb{R}^d$ with certain f and g which on $Q_{r_{n-1}} \setminus Q_{r_{n+2}}$ coincide with the original ones. Finally, if $n \geq 0$, then the distance between the boundaries of $Q_{r_n} \setminus Q_{r_{n+1}}$ and $Q_{r_{n-1}} \setminus Q_{r_{n+2}}$ is $(2^{1/3} - 1)r_{n+2}$.

It follows by Theorem 2.2 that for $n \geq 0$

$$\|I_{Q_{r_n}\setminus Q_{r_{n+1}}} D^2u\|^p_{\mathbb{L}_p(\tau)} + r^{-p}_{n-1}\|I_{Q_{r_n}\setminus Q_{r_{n+1}}} Du\|^p_{\mathbb{L}_p(\tau)}$$
$$\leq N\big(\|I_{Q_{r_{n-1}}\setminus Q_{r_{n+2}}} f\|^p_{\mathbb{L}_p(\tau)} + \|I_{Q_{r_{n-1}}\setminus Q_{r_{n+2}}} Dg\|^p_{\mathbb{L}_p(\tau)}$$
$$+ r^{-p}_{n+2}\|I_{Q_{r_{n-1}}\setminus Q_{r_{n+2}}} g\|^p_{\mathbb{L}_p(\tau)}\big) + N^* r^{-2p}_{n+2}\|I_{Q_{r_{n-1}}\setminus Q_{r_{n+2}}} u\|^p_{\mathbb{L}_p(\tau)}.$$

We multiply both parts by $r^{p+\theta-d}_{n+2}$ and use the facts that $r_{n-1} = 2r_{n+2}$ and on $Q_{r_{n-1}} \setminus Q_{r_{n+2}}$ the ratio x^1/r_{n+2} satisfies

$$1 \leq x^1/r_{n+2} \leq 2.$$

Then we obtain

$$\|I_{Q_{r_n}\setminus Q_{r_{n+1}}} MD^2u\|^p_{\mathbb{L}_{p,\theta}(\tau)} + \|I_{Q_{r_n}\setminus Q_{r_{n+1}}} Du\|^p_{\mathbb{L}_{p,\theta}(\tau)}$$
$$\leq N\big(\|I_{Q_{r_{n-1}}\setminus Q_{r_{n+2}}} Mf\|^p_{\mathbb{L}_{p,\theta}(\tau)} + \|I_{Q_{r_{n-1}}\setminus Q_{r_{n+2}}} MDg\|^p_{\mathbb{L}_{p,\theta}(\tau)}$$
$$+ \|I_{Q_{r_{n-1}}\setminus Q_{r_{n+2}}} g\|^p_{\mathbb{L}_{p,\theta}(\tau)}\big) + N^*\|I_{Q_{r_{n-1}}\setminus Q_{r_{n+2}}} M^{-1}u\|^p_{\mathbb{L}_{p,\theta}(\tau)}.$$

Upon summing up these inequalities over $n \geq 0$ we conclude

$$\|I_{Q_r} MD^2u\|^p_{\mathbb{L}_{p,\theta}(\tau)} + \|I_{Q_r} Du\|^p_{\mathbb{L}_{p,\theta}(\tau)} \leq N\big(\|I_{Q_{r_{-1}}} Mf\|^p_{\mathbb{L}_{p,\theta}(\tau)}$$
$$+ \|I_{Q_{r_{-1}}} MDg\|^p_{\mathbb{L}_{p,\theta}(\tau)} + \|I_{I_{Q_{r_{-1}}}} g\|^p_{\mathbb{L}_{p,\theta}(\tau)}\big) + N^*\|I_{I_{Q_{r_{-1}}}} M^{-1}u\|^p_{\mathbb{L}_{p,\theta}(\tau)},$$

which is somewhat sharper than (4.2). The theorem is proved. □

By letting $r \to \infty$ in (4.2) we get the following.

COROLLARY 4.1. *If the assumptions of Theorem 4.1 are satisfied with $R = \infty$, then*

$$\|MD^2u\|_{\mathbb{L}_{p,\theta}(\tau)} + \|Du\|_{\mathbb{L}_{p,\theta}(\tau)} \leq N\|Mf\|_{\mathbb{L}_{p,\theta}(\tau)}$$
$$+ N\|g\|_{\mathbb{H}^1_{p,\theta}(\tau)} + N^*\|M^{-1}u\|_{\mathbb{L}_{p,\theta}(\tau)}.$$

REMARK 4.3. This corollary for equations with constant coefficient is known from Lemma 3.6 of [14] for $d = 1$ and Lemma 3.8 of [15] for $d \geq 1$. For variable coefficients it is Lemma 4.1 of [8] if $d = 1$. However, there is a very important distinction between Corollary 4.1 and the above mentioned references, where from the start it is assumed that $u \in \mathfrak{H}^2_{p,\theta,0}(\tau)$.

As we have mentioned in the Introduction, we intend to use Theorem 4.1 to extend to the case of variable coefficients the pointwise boundary Hölder regularity result from [13] proved there for equations with constant coefficients. In a subsequent article we will show that solutions in $\mathfrak{H}^2_{p,\theta,0}(\tau)$-spaces with rather large θ admit estimates of $M^{-1}u$ in $\mathbb{L}_{p,\mu}(\tau)$ with much smaller μ if τ is chosen in a special way. This, Theorem 4.1, and embedding theorems will lead to proving the pointwise Hölder continuity of solutions near the boundary.

REMARK 4.4. We discuss assumption (4.1) in case that $R = \infty$. It implies that $|a^{ij}_t(x) - a^{ij}_t(y)| \leq \beta$ for a small $\beta > 0$ and all $x, y \in \mathbb{R}^d_+$ satisfying $|x - y| \leq x^1 \wedge y^1$. If $x^1 = y^1 = r$, then we need $|a^{ij}_t(r, x') - a^{ij}_t(r, y')| \leq \beta$ for $|x' - y'| \leq r$ and when r becomes larger we need $a^{ij}_t(r, x')$ to be close to constants on larger and larger balls in $\mathbb{R}^{d-1}$. Basically, we need $a^{ij}_t(r, x')$ to be independent of x' for large r. On the other hand, the behavior of $a^{ij}_t(x)$ near $\partial\mathbb{R}^d_+$ can be quite irregular.

For instance, take $d = 1$ and introduce the function

$$a(x) = 2 + \cos \ln x.$$

If $x \geq y > 0$ and $|x - y| \leq \varepsilon(x \wedge y)$, then $x \leq (1 + \varepsilon)y$, $\ln x \leq \ln y + \ln(1 + \varepsilon)$ and, for a $\xi \in (\ln y, \ln x)$, by the mean value theorem we have

$$|a(x) - a(y)| = |\sin \xi|(\ln x - \ln y) \leq \ln(1 + \varepsilon),$$

which can be made arbitrarily small by choosing an appropriate ε.

REMARK 4.5. It is worth emphasizing that in this section there is no restrictions on θ, say like (1.3).

5. Local regularity near the boundary for parabolic PDEs in half spaces. In the setting of Section 4 we will be considering (1.1) in $\mathbb{R}^d_+$. We need the weighted Sobolev spaces $\mathbb{H}^n_{p,\theta}(T)$, $n = 0, 1, 2$, and $\mathfrak{H}^2_{p,\theta}(T)$ introduced for $p > 1$ in the following way. We take a number $\theta \in \mathbb{R}$ and take $L_{p,\theta}$ and $H^n_{p,\theta}$ from Section 4.

Next, for $n = 0, 1, 2$ and $T \in (0, \infty]$ we define $\mathbb{H}^n_{p,\theta}(T)$ as the set of functions $f = f_t$ on $(0, T)$ with values in the set of distributions on $\mathbb{R}^d_+$ which are measurable and have finite norm given by

$$\|f\|^p_{\mathbb{H}^n_{p,\theta}(T)} = \int_0^T \|f_t\|^p_{H^n_{p,\theta}}\, dt.$$

Define $\mathbb{L}_{p,\theta}(T) = \mathbb{H}^0_{p,\theta}(T)$.

For a function $u = u_t$ given on $[0, T] \cap [0, \infty)$ with values in the set of distributions on $\mathbb{R}^d_+$ we write $u \in \mathfrak{H}^2_{p,\theta,0}(T)$ if and only if $M^{-1}u \in \mathbb{H}^2_{p,\theta}(T)$

and there exist a real valued $f \in \mathbb{L}_{p,\theta}(T)$ such that for any $\phi \in C_0^\infty(\mathbb{R}^d_+)$ we have

$$(u_t, \phi) = \int_0^t (M^{-1} f_s, \phi)\, ds$$

for all finite $t \in (0, T]$. In that case we write

$$\frac{\partial}{\partial t} u_t = M^{-1} f_t$$

and set

$$\|u\|_{\mathfrak{H}^2_{p,\theta}(T)} = \|M^{-1}u\|_{\mathbb{H}^2_{p,\theta}(T)} + \|f\|_{\mathbb{L}_{p,\theta}(T)}.$$

Recall that Assumptions 3.1 is supposed to be satisfied throughout the deterministic part of the article. For $r > 0$ denote

$$Q_r = \{x \in \mathbb{R}^d : 0 < x^1 < r\}.$$

THEOREM 5.1. *Take an $R \in (0, \infty]$ and assume the following.*
(i) For some constants $\varepsilon \in (0, 1]$ and $K \in (0, \infty)$ we have

$$|x^1 b^i_t(x)| + |(x^1)^2 c_t(x)| \leq K, \quad |a^{ij}_t(x) - a^{ij}_t(y)| \leq \beta_0,$$

whenever $x, y \in \mathbb{R}^d_+$, $x^1, y^1 \leq R$, $|x - y| \leq \varepsilon(x^1 \wedge y^1)$, $i, j = 1, ..., d$, $t > 0$, where $\beta_0 = \beta_0(d, p, \delta_0) \in (0, 1)$ is the constant from Theorem 3.1;

(ii) We have a function u such that $\phi u \in \mathcal{H}^2_{p,0}(T)$ for any $\phi \in C_0^\infty(Q_R)$ and u satisfies (1.1) in $\mathbb{R}^d_+$ with some $f \in \mathbb{L}_{p,\theta}(T)$.

Then for any $r \in (0, R/4)$

$$\begin{aligned} &\|I_{Q_r} M D^2 u\|_{\mathbb{L}_{p,\theta}(T)} + \|I_{Q_r} D u\|_{\mathbb{L}_{p,\theta}(T)} \\ &\leq N \|I_{Q_R} M f\|_{\mathbb{L}_{p,\theta}(T)} + N^* \|I_{Q_R} M^{-1} u\|_{\mathbb{L}_{p,\theta}(T)}, \end{aligned} \tag{5.1}$$

where $N = N(d, p, \delta_0)$ and $N^* = N^*(d, p, \delta_0, \varepsilon, K)$.

Proof. As in the proof of Theorem 4.1 for $n = -1, 0, 1, ...$ we set $r_n = 2^{-n/3} r$ and observe that if $n \geq 0$, then for $x \in Q_{r_{n-1}} \setminus Q_{r_{n+2}}$ and y, such that $|x - y| \leq \varepsilon \rho_n$, where ρ_n is the half width of $Q_{r_{n-1}} \setminus Q_{r_{n+2}}$, we have

$$\rho_n |b^i_t(y)| + \rho_n^2 |c_t(y)| \leq K, \quad |a^{ij}_t(x) - a^{ij}_t(y)| \leq \beta_0.$$

Furthermore, if $n \geq 0$, $\zeta \in C_0^\infty((0, R))$ and $\zeta(z) = 1$ for $r_{n+2} \leq z \leq r_{n-1}$, then ζu satisfies (1.1) in $\mathbb{R}^d$ with certain f, which on $Q_{r_{n-1}} \setminus Q_{r_{n+2}}$ coincides with the original one. Finally, if $n \geq 0$, then the distance between the boundaries of $Q_{r_n} \setminus Q_{r_{n+1}}$ and $Q_{r_{n-1}} \setminus Q_{r_{n+2}}$ is $(2^{1/3} - 1) r_{n+2}$.

It follows by Theorem 3.1 that for $n \geq 0$

$$\|I_{Q_{r_n}\setminus Q_{r_{n+1}}}D^2u\|^p_{\mathbb{L}_p(T)} + r^{-p}_{n-1}\|I_{Q_{r_n}\setminus Q_{r_{n+1}}}Du\|^p_{\mathbb{L}_p(T)}$$
$$\leq N\|I_{Q_{r_{n-1}}\setminus Q_{r_{n+2}}}f\|^p_{\mathbb{L}_p(T)} + N^* r^{-2p}_{n+2}\|I_{Q_{r_{n-1}}\setminus Q_{r_{n+2}}}u\|^p_{\mathbb{L}_p(T)}.$$

We multiply both parts by $r^{p+\theta-d}_{n+2}$ and use the facts that $r_{n-1} = 2r_{n+2}$ and on $Q_{r_{n-1}} \setminus Q_{r_{n+2}}$ the ratio x^1/r_{n+2} satisfies

$$1 \leq x^1/r_{n+2} \leq 2.$$

Then we obtain

$$\|I_{Q_{r_n}\setminus Q_{r_{n+1}}}MD^2u\|^p_{\mathbb{L}_{p,\theta}(T)} + \|I_{Q_{r_n}\setminus Q_{r_{n+1}}}Du\|^p_{\mathbb{L}_{p,\theta}(T)}$$
$$\leq N\|I_{Q_{r_{n-1}}\setminus Q_{r_{n+2}}}Mf\|^p_{\mathbb{L}_{p,\theta}(T)} + N^*\|I_{Q_{r_{n-1}}\setminus Q_{r_{n+2}}}M^{-1}u\|^p_{\mathbb{L}_{p,\theta}(T)}.$$

Upon summing up these inequalities over $n \geq 0$ we conclude

$$\|I_{Q_r}MD^2u\|^p_{\mathbb{L}_{p,\theta}(T)} + \|I_{Q_r}Du\|^p_{\mathbb{L}_{p,\theta}(T)}$$
$$\leq N\|I_{Q_{r_{-1}}}Mf\|^p_{\mathbb{L}_{p,\theta}(T)} + N^*\|I_{I_{Q_{r_{-1}}}}M^{-1}u\|^p_{\mathbb{L}_{p,\theta}(T)},$$

which is somewhat sharper than (5.1). The theorem is proved. □

By letting $r \to \infty$ in (5.1) we get the following.

COROLLARY 5.1. *If the assumptions of Theorem 5.1 are satisfied with* $R = \infty$, *then*

$$\|MD^2u\|_{\mathbb{L}_{p,\theta}(T)} + \|Du\|_{\mathbb{L}_{p,\theta}(T)} \leq N\|Mf\|_{\mathbb{L}_{p,\theta}(T)} + N^*\|M^{-1}u\|_{\mathbb{L}_{p,\theta}(T)}.$$

6. Auxiliary results. In this section

$$p > 1, \quad \gamma := \theta - d - p + 1.$$

Here we present some auxiliary facts needed for proving existence theorems in Sections 7 and 8.

Several times in the future we use the following Scheffé's lemma, which we prove for completeness of presentation.

LEMMA 6.1. *Let* (E, Σ, μ) *be a measure space,* $r \in [1, \infty)$, $u_n, u \in L_r(\mu)$, *and* $u_n \to u$ *in measure. Finally, let*

$$\|u_n\|_{L_r(\mu)} \to \|u\|_{L_r(\mu)}.$$

Then

$$\|u_n - u\|_{L_r(\mu)} \to 0. \tag{6.1}$$

Proof. We have

$$\big||u|^r - |u_n|^r\big| = (|u|^r - |u_n|^r)_+ - (|u|^r - |u_n|^r).$$

Upon integrating through this equation and observing that $(|u|^r - |u_n|^r)_+ \le |u|^r$ we conclude by the dominated convergence theorem that

$$\int_E \big||u|^r - |u_n|^r\big|\,\mu(dx) \to 0. \tag{6.2}$$

Next, if $|u_n - u| \ge 3|u|$, then $|u_n| + |u| \ge 3|u|$, $|u| \le (1/2)|u_n|$, $|u|^r \le (1/2)|u_n|^r$,

$$|u_n|^r - |u|^r \ge (1/2)|u_n|^r, \quad |u_n - u| \le |u_n| + |u| \le 2|u_n|,$$
$$|u_n - u|^r \le 2^r|u_n|^r \le 4^r(|u_n|^r - |u|^r),$$

which along with (6.2) imply that

$$\int_E |u_n - u|^r I_{|u_n - u| \ge 3|u|}\,\mu(dx) \to 0.$$

Furthermore,

$$\int_E |u_n - u|^r I_{|u_n - u| < 3|u|}\,\mu(dx) \to 0$$

by the dominated convergence theorem. By combining the last two relations we come to (6.1). The lemma is proved. □

COROLLARY 6.1. *Let (E, Σ, μ) be a measure space, $r, s \in (1, \infty)$, $r^{-1} + s^{-1} = 1$, $u_n, u \in L_r(\mu)$, $v_n, v \in L_s(\mu)$, $u_n \to u$ and $v_n \to v$ in measure. Finally, let*

$$\|u_n\|_{L_r(\mu)} \to \|u\|_{L_r(\mu)}, \quad \|v_n\|_{L_s(\mu)} \to \|u\|_{L_s(\mu)}.$$

Then

$$\int_E |u_n v_n - uv|\,\mu(dx) \to 0, \quad \int_E u_n v_n\,\mu(dx) \to \int_E uv\,\mu(dx).$$

Indeed, it suffices to use Hölder's inequality and the formula

$$u_n v_n - uv = (u_n - u)v + (v_n - v)u + (u_n - u)(v_n - v).$$

Next, we prove a version of Corollary 6.2 of [12].

LEMMA 6.2. *If we are given a $d \times d$ matrix $\bar{a}$, then, for any $u \in MH^2_{p,\theta}$, we have*

$$\int_{\mathbb{R}^d_+} (x^1)^{\gamma+1}|u|^{p-2}u\bar{a}^{ij}D_{ij}u\,dx = \gamma(\gamma+1)p^{-1}\bar{a}^{11}\int_{\mathbb{R}^d_+}(x^1)^{\gamma-1}|u|^p\,dx,$$
$$- (p-1)\int_{\mathbb{R}^d_+}(x^1)^{\gamma+1}|u|^{p-2}\bar{a}^{ij}(D_i u)D_j u\,dx, \tag{6.3}$$

and if $\bar{a}$ is nonnegative and symmetric and $\gamma < 0$, then

$$\int_{\mathbb{R}^d_+} (x^1)^{\gamma+1}|u|^{p-2}\bar{a}^{ij}(D_i u)D_j u\,dx \geq \gamma^2 p^{-2}\bar{a}^{11}\int_{\mathbb{R}^d_+}(x^1)^{\gamma-1}|u|^p\,dx, \quad (6.4)$$

where (if $p < 2$) by $0^{-1}0$ we mean 0.

Proof. If u has compact support, then, owing to the assumption that $\gamma < 0$, the result follows from Corollary 6.2 of [12], which is proved by using integration by parts and Hardy's inequality. Observe that the case $1 < p < 2$ is quite nontrivial.

In the general case we take ζ_m from Remark 4.1 apply (6.3) to $u_m = \zeta_m u$ in place of u and pass to the limit as $m \to \infty$. To justify this procedure we claim that

$$\int_{\mathbb{R}^d_+}(x^1)^{\gamma-1}|u|^p\,dx, \quad \int_{\mathbb{R}^d_+}(x^1)^{\gamma+1}|u|^{p-2}uD_{ij}u\,dx \quad (6.5)$$

are continuous functions on $ML_{p,\theta}$ and $MH^2_{p,\theta}$, respectively.

The first one is continuous because it is just the p-th power of the $L_{p,\theta}$-norm of $M^{-1}u$. To prove the continuity of the second one, take v_n such that $M^{-1}v_n \to M^{-1}u$ in $H^2_{p,\theta}$. Then

$$M^{-1}v_n \to M^{-1}u, \quad MD^2v_n \to MD^2u$$

in $L_{p,\theta}$ by Remark 4.2. By Lemma 6.1 and Corollary 6.1 we also have

$$|M^{-1}v_n|^{p-2}M^{-1}v_n \to |M^{-1}u|^{p-2}M^{-1}u \quad \text{in} \quad L_{p/(p-1),\theta},$$

$$\int_{\mathbb{R}^d_+}(x^1)^{\gamma+1}|v_n|^{p-2}v_nD_{ij}v_n\,dx$$
$$= \int_{\mathbb{R}^d_+}(x^1)^{\theta-d}(|M^{-1}v_n|^{p-2}M^{-1}v_n)(MD_{ij}v_n)\,dx$$
$$\to \int_{\mathbb{R}^d_+}(x^1)^{\theta-d}(|M^{-1}u|^{p-2}M^{-1}vu)(MD_{ij}u)\,dx$$

and this proves the continuity of the second function in (6.5).

If $p \geq 2$ then a similar argument shows that the last expression in (6.3) is continuous on $MH^2_{p,\theta}$ and this proves the lemma if $p \geq 2$. For arbitrary $p > 1$ we need to proceed differently.

Since (6.3) holds with u_m in place of u, we conclude that

$$(p-1)\bar{a}^{ij}\lim_{m\to\infty}\int_{\mathbb{R}^d_+}(x^1)^{\theta-d}I^{ij}_m\,dx = \gamma(\gamma+1)p^{-1}\bar{a}^{11}\int_{\mathbb{R}^d_+}(x^1)^{\gamma-1}|u|^p\,dx$$
$$-\int_{\mathbb{R}^d_+}(x^1)^{\gamma+1}|u|^{p-2}u\bar{a}^{ij}D_{ij}u\,dx, \quad (6.6)$$

where

$$
\begin{aligned}
I^{ij}_m &:= |M^{-1}u_m|^{p-2}(D_iu_m)D_ju_m \\
&= \zeta^p_m|M^{-1}u|^{p-2}(D_iu)D_ju \\
&\quad +|M^{-1}u|^{p-2}(M^{-1}u)(D_iu)\zeta_nMD_j\zeta_n \\
&\quad +|M^{-1}u|^{p-2}(M^{-1}u)(D_ju)\zeta_nMD_i\zeta_n \\
&\quad +|M^{-1}u|^{p}(MD_i\zeta_n)MD_j\zeta_n .
\end{aligned}
\tag{6.7}
$$

Here the factors of the functions

$$
\zeta_nMD_i\zeta_n, \quad \zeta_nMD_j\zeta_n, \quad M(D_i\zeta_n)MD_j\zeta_n
$$

are integrable against $(x^1)^{\theta-d}\,dx$, which follows from Hölder's inequality, and the functions themselves are uniformly bounded and tend to zero on $\mathbb{R}^d_+$. Therefore, (6.6) implies that

$$
\begin{aligned}
&(p-1)\lim_{m\to\infty}\int_{\mathbb{R}^d_+}(x^1)^{\theta-d}\zeta^p_m|M^{-1}u|^{p-2}\bar a^{ij}(D_iu)D_ju\,dx \\
&= \gamma(\gamma+1)p^{-1}\bar a^{11}\int_{\mathbb{R}^d_+}(x^1)^{\gamma-1}|u|^p\,dx-\int_{\mathbb{R}^d_+}(x^1)^{\gamma+1}|u|^{p-2}u\bar a^{ij}D_{ij}u\,dx.
\end{aligned}
\tag{6.8}
$$

By taking here $\bar a^{ij}=\delta^{ij}$ and using Fatou's lemma we see that

$$
\int_{\mathbb{R}^d_+}(x^1)^{\theta-d}|M^{-1}u|^{p-2}|Du|^2\,dx<\infty.
$$

After that by (6.7) and the dominated convergence theorem we obtain

$$
\lim_{m\to\infty}\int_{\mathbb{R}^d_+}(x^1)^{\theta-d}I^{ij}_m\,dx=\int_{\mathbb{R}^d_+}(x^1)^{\theta-d}|M^{-1}u|^{p-2}(D_iu)D_ju\,dx,
$$

which along with (6.6) lead to (6.3). The above argument also shows that one can pass to the limit in (6.4) written for u_m in place of u. This proves the lemma. □

COROLLARY 6.2. *Denote*

$$
P:=\int_{\mathbb{R}^d}(x^1)^{\gamma+1}|u|^{p-1}|D^2u|\,dx, \quad Q=\int_{\mathbb{R}^d_+}(x^1)^{\gamma+1}|u|^{p-2}|Du|^2\,dx.
$$

Then

$$
Q\le N(d,p)P+(p-1)^{-1}p^{-1}(\gamma^2+\gamma)\|M^{-1}u\|^p_{L_{p,\theta}}, \tag{6.9}
$$

$$
P\le\|M^{-1}u\|^{p-1}_{L_{p,\theta}}\|MD^2u\|_{L_{p,\theta}}. \tag{6.10}
$$

Indeed, estimate (6.10) follows from Hölder's inequality. By (6.3) for $\bar{a}^{ij} = \delta^{ij}$ we have

$$(p-1)Q = \gamma(\gamma+1)p^{-1}\|M^{-1}u\|^p_{L_{p,\theta}} - \int_{\mathbb{R}^d_+} (x^1)^{\gamma+1}|u|^{p-2}u\Delta u\,dx$$
$$\leq \gamma(\gamma+1)p^{-1}\|M^{-1}u\|^p_{L_{p,\theta}} + N(d)P$$

and (6.9) follows.

Next, we need an auxiliary function. Set

$$B'_R = \{x' \in \mathbb{R}^{d-1} : |x'| < R\}.$$

LEMMA 6.3. *For any $\delta > 0$ there exists an $R = R(\delta, d, p) > 1$ and a nonnegative function $\eta \in C_0^\infty(\mathbb{R}^d)$ vanishing outside $(1, R) \times B'_R$ such that $\ln R \leq N(d,p)\delta^{-1/2}$ and*

$$\int_{\mathbb{R}^d_+} \frac{1}{x^1}\eta^p(x)\,dx = 1, \quad \int_{\mathbb{R}^d_+} \eta^{p-2}(x)|D\eta(x)|^2 x^1\,dx \leq \delta. \tag{6.11}$$

Proof. Take a nonnegative $\xi \in C_0^\infty(\mathbb{R}^d)$ with unit norm in L_p, vanishing outside $(-1, 0) \times B'_1$, and then for some $r > 0$ set

$$\eta(x) = r^{-1/p}e^{-(d-1)r/p}\xi(r^{-1}\ln x^1, e^{-r}x')$$

Then

$$\int_{\mathbb{R}^d_+} \frac{1}{x^1}\eta^p\,dx = r^{-1}\int_{\mathbb{R}^d_+} \frac{1}{x^1}\xi^p(r^{-1}\ln x^1, x')\,dx$$
$$= r^{-1}\int_{\mathbb{R}^d} \xi^p(r^{-1}x^1, x')\,dx = 1,$$
$$\int_{\mathbb{R}^d_+} \eta^{p-2}|D_1\eta|^2 x^1\,dx = r^{-3}\int_{\mathbb{R}^d_+} (\xi^{p-2}|D_1\xi|^2)(r^{-1}\ln x^1, x')\frac{1}{x^1}\,dx$$
$$= r^{-2}\int_{\mathbb{R}^d} \xi^{p-2}|D_1\xi|^2\,dx,$$

where the last integral is finite even if $p < 2$ since $|D\xi|^2 \leq N(d)\xi \sup|D^2\xi|$ because ξ is nonnegative and smooth. Also, since $\xi(x) = 0$ for $x^1 > 0$ and by assumption $r > 0$, for $i \geq 2$ we have

$$\int_{\mathbb{R}^d_+} \eta^{p-2}|D_i\eta|^2 x^1\,dx = r^{-1}e^{-2r}\int_{\mathbb{R}^d_+} (\xi^{p-2}|D_i\xi|^2)(r^{-1}\ln x^1, x')x^1\,dx$$
$$= r^{-1}e^{-2r}\int_{\mathbb{R}^d} (\xi^{p-2}|D_i\xi|^2)(r^{-1}x^1, x')e^{2x^1}\,dx$$
$$\leq r^{-1}e^{-2r}\int_{\mathbb{R}^d} (\xi^{p-2}|D_i\xi|^2)(r^{-1}x^1, x')\,dx$$
$$= e^{-2r}\int_{\mathbb{R}^d} (\xi^{p-2}|D_i\xi|^2)(x^1, x')\,dx.$$

By observing that $e^{-2r} \le Nr^{-2}$, we have both relations in (6.11) satisfied for an $r = N(d,p)\delta^{-1/2}$.

Now we have to take care of the support of η. The one constructed above has support in

$$(e^{-r}, 1) \times B'_{e^r}.$$

It turns out that conditions (6.11) are dilation invariant. Therefore, the function $\eta(e^{-r}x)$ satisfies them as well. This function has support in

$$(1, e^r) \times B'_{e^{2r}}$$

and we see that we can take $R = e^{2r}$. The lemma is proved. □

Now we extend (6.4) for variable $\bar{a}$ by using a localization procedure.

LEMMA 6.4. *Assume that $\gamma < 0$ and let $\varepsilon, \beta, K \in (0,\infty)$ be some constants and let $\bar{a}(x)$ be a measurable function given on $\mathbb{R}^d_+$ with values in the set of symmetric nonnegative matrices and such that $|\bar{a}^{ij}| \le K$ and*

$$|\bar{a}^{ij}(x) - \bar{a}^{ij}(y)| \le \beta \tag{6.12}$$

whenever $x, y \in \mathbb{R}^d_+$ and $|x-y| \le \varepsilon(x^1 \wedge y^1)$. Then for any $u \in MH^1_{p,\theta}$ and $\delta > 0, \kappa \in (0,1]$ we have

$$\begin{aligned} I &:= \int_{\mathbb{R}^d_+} (x^1)^{\gamma+1} |u|^{p-2} \bar{a}^{ij} (D_i u) D_j u \, dx \\ &\ge (1-\kappa)\gamma^2 p^{-2} \int_{\mathbb{R}^d_+} (x^1)^{\gamma-1} \bar{a}^{11} |u|^p \, dx - N(\varepsilon^{-1}R + 1)\beta Q \qquad (6.13) \\ &\quad - N(\varepsilon^{-1}R\beta + \beta + \kappa^{-1}\delta) \|M^{-1}u\|^p_{L_{p,\theta}}, \end{aligned}$$

where $N = N(d,p,K,\theta)$, $\ln R = N(d,p)\delta^{-1/2}$ and Q is introduced in Corollary 6.2.

Proof. Take the function η from Lemma 6.3 and for any $y \in \mathbb{R}^d_+$ set

$$\eta^y(x) = \eta(y^1 x^1, y^1(y' - x'))(y^1)^{(d-2)/p},$$
$$\zeta^y(x) = \eta^p(y^1 x^1, y^1(y' - x'))(y^1)^{d-2} = [\eta^y(x)]^p.$$

Observe that, for any $x \in \mathbb{R}^d_+$ and $\alpha \in \mathbb{R}$

$$\begin{aligned} \int_{\mathbb{R}^d_+} (y^1)^{-\alpha} \zeta^y(x) \, dy &= \int_{\mathbb{R}^d_+} (y^1)^{-\alpha-1} \eta^p(y^1 x^1, y') \, dy \\ &= (x^1)^\alpha \int_{\mathbb{R}^d_+} (y^1)^{-\alpha-1} \eta^p(y^1, y') \, dy. \end{aligned} \tag{6.14}$$

For $\alpha = 0$ this yields

$$I = \int_{\mathbb{R}^d_+} I(y) \, dy,$$

where $I(y) = I_1(y) + I_2(y)$ and with $\bar{y} = ((y^1)^{-1}, y')$

$$I_1(y) = \bar{a}^{ij}(\bar{y}) \int_{\mathbb{R}^d_+} (x^1)^{\gamma+1} \zeta^y |u|^{p-2} (D_i u) D_j u \, dx,$$

$$I_2(y) = \int_{\mathbb{R}^d_+} (x^1)^{\gamma+1} \zeta^y |u|^{p-2} [\bar{a}^{ij}(x) - \bar{a}^{ij}(\bar{y})] (D_i u) D_j u \, dx.$$

Here and below on many occasions we drop the argument x but not y.

With R from Lemma 6.3, if $\zeta^y(x) \neq 0$, then $1 < y^1 x^1 < R$ and $y^1 |y' - x'| < R$ implying that

$$\bar{y}^1 < x^1 < R\bar{y}^1, \quad |\bar{y}' - x'| < R\bar{y}^1 = R(x^1 \wedge \bar{y}^1),$$
$$0 < x^1 - \bar{y}^1 < R\bar{y}^1, \quad |x^1 - \bar{y}^1| < R(x^1 \wedge \bar{y}^1), \quad |x - \bar{y}| < 2R(x^1 \wedge \bar{y}^1).$$

In case $2R \geq \varepsilon$ introduce $n := [4R\varepsilon^{-1}] + 1$ and find n points x_i, $i = 1, ..., n$, evenly spread on the straight segment connecting x and $\bar{y}$ and such that $x_1 = x$, $x_n = \bar{y}$. Then $n - 1 = [4R\varepsilon^{-1}] \geq 4R\varepsilon^{-1} - 1 \geq 2R\varepsilon^{-1}$ and

$$|x_i - x_{i+1}| = (n-1)^{-1} |x - \bar{y}| \leq (n-1)^{-1} 2R(x^1 \wedge \bar{y}^1)$$
$$\leq \varepsilon(x^1 \wedge \bar{y}^1) \leq \varepsilon(x^1_i \wedge x^1_{i+1}).$$

It follows that $|\bar{a}^{ij}(x) - \bar{a}^{ij}(\bar{y})| \leq (n-1)\beta$ and

$$|\bar{a}^{ij}(x) - \bar{a}^{ij}(\bar{y})| \leq 4(\varepsilon^{-1}R + 1)\beta, \tag{6.15}$$

if $2R \geq \varepsilon$. Estimate (6.15) is also true if $2R \leq \varepsilon$, because then $|x - \bar{y}| < 2R(x^1 \wedge \bar{y}^1) \leq \varepsilon(x^1 \wedge \bar{y}^1)$. Hence,

$$|I_2(y)| \leq N(\varepsilon^{-1}R + 1)\beta \int_{\mathbb{R}^d_+} (x^1)^{\gamma+1} \zeta^y |u|^{p-2} |Du|^2 \, dx$$

and by the construction of ζ and the definition of Q

$$\int_{\mathbb{R}^d_+} I_2(y) \, dy \leq N\varepsilon^{-1} R\beta Q.$$

Thus,

$$I \geq \int_{\mathbb{R}^d_+} I_1(y) \, dy - N(\varepsilon^{-1}R + 1)\beta Q. \tag{6.16}$$

To estimate $I_1(y)$ we introduce

$$v^y(x) = (\zeta^y(x))^{1/p} u(x) = \eta^y(x) u(x)$$

so that

$$\bar a^{ij}(\bar y)\zeta^y|u|^{p-2}(D_iu)D_ju = \bar a^{ij}(\bar y)|v^y|^{p-2}(D_iv^y - uD_i\eta^y)(D_jv^y - uD_j\eta^y)$$
$$\geq (1-\kappa)J_1^y - (\kappa^{-1}-1)J_2^y \geq (1-\kappa)J_1^y - \kappa^{-1}J_2^y,$$

where

$$J_1^y(x) = \bar a^{ij}(\bar y)|v^y|^{p-2}(D_iv^y)D_jv^y, \quad J_2^y(x) = \bar a^{ij}(\bar y)|v^y|^{p-2}u^2(D_i\eta^y)D_j\eta^y.$$

It follows that

$$\begin{aligned}\int_{\mathbb{R}^d_+} I_1(y)\,dy \geq & (1-\kappa)\int_{\mathbb{R}^d_+}\int_{\mathbb{R}^d_+}(x^1)^{\gamma+1}J_1^y(x)\,dxdy \\ & - \kappa^{-1}\int_{\mathbb{R}^d_+}\int_{\mathbb{R}^d_+}(x^1)^{\gamma+1}J_2^y(x)\,dxdy.\end{aligned} \tag{6.17}$$

Observe that

$$\int_{\mathbb{R}^d_+}(x^1)^{\gamma+1}J_2^y\,dx \leq N\int_{\mathbb{R}^d_+}(x^1)^{\gamma+1}(\eta^y)^{p-2}|D\eta^y|^2|u|^p\,dx,$$

$$\int_{\mathbb{R}^d_+}\int_{\mathbb{R}^d_+}(x^1)^{\gamma+1}J_2^y\,dxdy \leq N\int_{\mathbb{R}^d_+}(x^1)^{\gamma+1}|u|^p\Big(\int_{\mathbb{R}^d_+}(\eta^y)^{p-2}|D\eta^y|^2\,dy\Big)\,dx$$

and by the construction of η

$$\begin{aligned}\int_{\mathbb{R}^d_+}(\eta^y)^{p-2}|D\eta^y|^2\,dy &= \int_{\mathbb{R}^d_+}[\eta^{p-2}|D\eta|^2](y^1x^1, y^1(y'-x'))(y^1)^d\,dy \\ &= \int_{\mathbb{R}^d_+}[\eta^{p-2}|D\eta|^2](y^1x^1, y')y^1\,dy \\ &= (x^1)^{-2}\int_{\mathbb{R}^d_+}\eta^{p-2}|D\eta|^2y^1\,dy \leq \delta(x^1)^{-2},\end{aligned}$$

$$\int_{\mathbb{R}^d_+}\int_{\mathbb{R}^d_+}(x^1)^{\gamma+1}J_2^y\,dxdy \leq N\delta\int_{\mathbb{R}^d_+}(x^1)^{\gamma-1}|u|^p\,dx = N\delta\|M^{-1}u\|^p_{L_{p,\theta}}.$$

We now conclude from (6.17) that

$$\int_{\mathbb{R}^d_+} I_1(y)\,dy \geq (1-\kappa)\int_{\mathbb{R}^d_+}\int_{\mathbb{R}^d_+}(x^1)^{\gamma+1}J_1^y(x)\,dxdy - \kappa^{-1}N\delta\|M^{-1}u\|^p_{L_{p,\theta}}.$$

Next, apply Lemma 6.2 for each y to get

$$\int_{\mathbb{R}^d_+}(x^1)^{\gamma+1}J_1^y\,dx \geq \gamma^2p^{-2}\bar a^{11}(\bar y)\int_{\mathbb{R}^d_+}(x^1)^{\gamma-1}|v^y|^p\,dx$$

$$= \gamma^2p^{-2}\int_{\mathbb{R}^d_+}(x^1)^{\gamma-1}\bar a^{11}(\bar y)\zeta^y|u|^p\,dx = \gamma^2p^{-2}\int_{\mathbb{R}^d_+}(x^1)^{\gamma-1}\bar a^{11}\zeta^y|u|^p\,dx$$
$$+ \gamma^2p^{-2}\int_{\mathbb{R}^d_+}(x^1)^{\gamma-1}[\bar a^{11}(\bar y) - \bar a^{11}(x)]\zeta^y|u|^p\,dx.$$

We use again (6.15) on the support of ζ^y and then integrate with respect to y. Then we conclude

$$\int_{\mathbb{R}^d_+}\int_{\mathbb{R}^d_+}(x^1)^{\gamma+1}J^y_1\,dxdy \geq \gamma^2 p^{-2}\int_{\mathbb{R}^d_+}(x^1)^{\gamma-1}\bar{a}^{11}\zeta^y|u|^p\,dx$$
$$- N(\varepsilon^{-1}R+1)\beta\|M^{-1}u\|^p_{L_{p,\theta}},$$
$$\int_{\mathbb{R}^d_+} I_1(y)\,dy \geq (1-\kappa)\gamma^2 p^{-2}\int_{\mathbb{R}^d_+}(x^1)^{\gamma-1}\bar{a}^{11}|u|^p\,dx$$
$$- N[(1-\kappa)(\varepsilon^{-1}R+1)\beta + \kappa^{-1}\delta]\|M^{-1}u\|^p_{L_{p,\theta}},$$

which in combination with (6.16) leads to (6.13). The lemma is proved. □

We now generalize (6.3) for variable $\bar{a}$.

LEMMA 6.5. *Let $\varepsilon, \beta \in (0,\infty)$ be some constants. Assume that we are given bounded measurable functions $\bar{a}^{ij}(x)$ defined on $\mathbb{R}^d_+$ for $i,j = 1,...,d$ and such that* (6.12) *holds for all $i,j = 1,...,d$ and $x,y \in \mathbb{R}^d_+$ such that $|x-y| \leq \varepsilon(x^1 \wedge y^1)$. Then for any $u \in MH^2_{p,\theta}$ we have*

$$\begin{aligned} I :=& \int_{\mathbb{R}^d_+}(x^1)^{\gamma+1}|u|^{p-2}u\bar{a}^{ij}D_{ij}u\,dx \\ &\leq N\beta(\|MD^2u\|_{L_{p,\theta}}\|M^{-1}u\|^{p-1}_{L_{p,\theta}} + \varepsilon^{-2}\chi\|M^{-1}u\|^p_{L_{p,\theta}}) \\ &- (p-1)\int_{\mathbb{R}^d_+}(x^1)^{\gamma+1}|u|^{p-2}\bar{a}^{ij}(D_iu)D_ju\,dx \\ &+ p^{-1}\gamma(\gamma+1)\int_{\mathbb{R}^d_+}(x^1)^{\gamma-1}\bar{a}^{11}|u|^p\,dx, \end{aligned} \tag{6.18}$$

where $\chi = (\gamma^2+1)(1+\varepsilon^2)$ and $N = N(d,p)$.

Proof. As in the proof of Lemma 6.2 we may assume that the support of u is a compact subset of $\mathbb{R}^d_+$. Without losing generality we may also assume that $\bar{a}^{ij} = \bar{a}^{ji}$ for all i,j.

Next, take a nonnegative $\zeta \in C^\infty_0(\mathbb{R}^d_+)$ with unit integral and such that $\zeta(x) = 0$ if $x^1 \notin (1, 1+\varepsilon/2)$ or $|x'| \geq \varepsilon/2$. On can construct such functions in a unified way, first choosing such a function, say ζ_1 corresponding to $\varepsilon = 1$ and then for any $\varepsilon > 0$ setting

$$\zeta(x) = \varepsilon^{-d}\zeta_1(1+\varepsilon^{-1}(x^1-1), \varepsilon^{-1}x').$$

Below we are also using the function

$$\eta(x) = (x^1)^{\gamma+2}\zeta(x).$$

One shows easily that for the indicator function ψ of the set $(1,3/2)\times B'_{1/2}$ we have

$$\begin{aligned} |D^2\eta(x)| &\leq N\varepsilon^{-2-d}(\gamma^2+1)(x^1)^\gamma[(x^1)^2+\varepsilon^2]\psi(1+\varepsilon^{-1}(x^1-1), \varepsilon^{-1}x') \\ &\leq N(x^1)^\gamma\varepsilon^{-2-d}\chi\psi(1+\varepsilon^{-1}(x^1-1), \varepsilon^{-1}x'), \end{aligned}$$

where $N = N(d)$. In particular,

$$\int_{\mathbb{R}^d_+} (x^1)^{-\gamma} |D^2 \eta| \, dx \le N \varepsilon^{-2} \chi \int_{\mathbb{R}^d_+} \psi \, dx. \tag{6.19}$$

This estimate will be used later and now for $y \in \mathbb{R}^d$ define

$$\zeta^y(x) = (x^1)^{\gamma+2} \zeta(y^1 x^1, y^1(y' - x'))(y^1)^{d-1}.$$

Observe that similarly to (6.14) we have

$$\int_{\mathbb{R}^d_+} \zeta^y(x) \, dy = (x^1)^{\gamma+1} \int_{\mathbb{R}^d_+} \zeta(y) \, dy = (x^1)^{\gamma+1}. \tag{6.20}$$

It follows that

$$I = \int_{\mathbb{R}^d_+} I(y) \, dy,$$

where $I(y) = I_1(y) + I_2(y)$,

$$I_1(y) = \int_{\mathbb{R}^d_+} \zeta^y |u|^{p-2} u \bar{a}^{ij}(\bar{y}) D_{ij} u \, dx, \quad \bar{y} = ((y^1)^{-1}, y'),$$

$$I_2(y) = \int_{\mathbb{R}^d_+} \zeta^y |u|^{p-2} u [\bar{a}^{ij}(x) - \bar{a}^{ij}(\bar{y})] D_{ij} u \, dx.$$

By the choice of ζ we have that if $\zeta^y(x) \ne 0$, then $1 < y^1 x^1 < 1 + \varepsilon/2$ and $y^1 |y' - x'| < \varepsilon/2$ implying that

$$\bar{y}^1 < x^1 < (1 + \varepsilon/2) \bar{y}^1, \quad |\bar{y}' - x'| < \bar{y}^1 \varepsilon/2 = (\varepsilon/2)(x^1 \wedge \bar{y}^1),$$
$$0 < x^1 - y^1 < \bar{y}^1 \varepsilon/2, \quad |x^1 - y^1| < (\varepsilon/2)(x^1 \wedge \bar{y}^1),$$
$$|x - \bar{y}| < \varepsilon(x^1 \wedge \bar{y}^1), \quad |\bar{a}^{ij}(x) - \bar{a}^{ij}(\bar{y})| \le \beta.$$

Hence,

$$|I_2(y)| \le N\beta \int_{\mathbb{R}^d_+} \zeta^y |u|^{p-1} |D^2 u| \, dx,$$

and by (6.20) and Corollary 6.2

$$\left| \int_{\mathbb{R}^d_+} I_2(y) \, dy \right| \le N\beta \|MD^2 u\|_{L_{p,\theta}} \|M^{-1} u\|^{p-1}_{L_{p,\theta}}.$$

We conclude

$$I \le \int_{\mathbb{R}^d_+} I_1(y) \, dy + N\beta \|MD^2 u\|_{L_{p,\theta}} \|M^{-1} u\|^{p-1}_{L_{p,\theta}}. \tag{6.21}$$

To estimate $I_1(y)$ we integrate by parts which is a very simple matter if $p \geq 2$ and is also justified for $p \in (1,2)$ in Lemma 2.17 of [12]. Then we find that $I_1(y) = J_1(y) + J_2(y)$, where

$$J_1(y) = -(p-1)\int_{\mathbb{R}^d_+} \zeta^y |u|^{p-2}\bar{a}^{ij}(\bar{y})(D_i u)D_j u\,dx,$$

$$\begin{aligned} J_2(y) &= -\int_{\mathbb{R}^d_+} (D_j\zeta^y)|u|^{p-2}u\bar{a}^{ij}(\bar{y})D_i u\,dx \\ &= -p^{-1}\int_{\mathbb{R}^d_+} (D_j\zeta^y)\bar{a}^{ij}(\bar{y})D_i|u|^p\,dx \\ &= p^{-1}\int_{\mathbb{R}^d_+} |u|^p\bar{a}^{ij}(\bar{y})D_{ij}\zeta^y(x)\,dx. \end{aligned}$$

As above

$$|J_1(y) + (p-1)\int_{\mathbb{R}^d_+} \zeta^y|u|^{p-2}\bar{a}^{ij}(D_i u)D_j u\,dx| \leq N\beta\int_{\mathbb{R}^d_+} \zeta^y|u|^{p-2}|Du|^2\,dx,$$

$$\left|\int_{\mathbb{R}^d_+} J_1(y)\,dy + (p-1)\int_{\mathbb{R}^d_+} (x^1)^{\gamma+1}|u|^{p-2}\bar{a}^{ij}(D_i u)D_j u\,dx\right| \leq N\beta Q,$$

where Q is taken from Corollary 6.2. It follows that

$$\begin{aligned} &\left|\int_{\mathbb{R}^d_+} J_1(y)\,dy + (p-1)\int_{\mathbb{R}^d_+} (x^1)^{\gamma+1}|u|^{p-2}\bar{a}^{ij}(D_i u)D_j u\,dx\right| \\ &\leq N\beta\|MD^2u\|_{L_{p,\theta}}\|M^{-1}u\|^{p-1}_{L_{p,\theta}} + N\beta(\gamma^2+|\gamma|)\|M^{-1}u\|^p_{L_{p,\theta}}. \end{aligned}$$

Therefore, (6.21) yields

$$\begin{aligned} I \leq &N\beta(\|MD^2u\|_{L_{p,\theta}}\|M^{-1}u\|^{p-1}_{L_{p,\theta}} + (\gamma^2+1)\|M^{-1}u\|^p_{L_{p,\theta}}) \\ &-(p-1)\int_{\mathbb{R}^d_+} (x^1)^{\gamma+1}|u|^{p-2}\bar{a}^{ij}(D_i u)D_j u\,dx + \int_{\mathbb{R}^d_+} J_2(y)\,dy. \end{aligned} \tag{6.22}$$

Similarly, to the above estimates

$$\left|J_2(y) - p^{-1}\int_{\mathbb{R}^d_+} |u(x)|^p\bar{a}^{ij}D_{ij}\zeta^y\,dx\right| \leq N\beta\int_{\mathbb{R}^d_+} |u|^p|D^2\zeta^y|\,dx. \tag{6.23}$$

To estimate $|D^2\zeta^y|$ introduce the function

$$\xi(z,y) = (z^1)^{\gamma+2}\zeta(z^1, y^1y' - z')(y^1)^{d-\gamma-3}.$$

Obviously, $\zeta^y(x) = \xi(y^1x, y)$. Therefore,

$$\begin{aligned} |D^2\zeta^y(x)| &= (y^1)^2|(D^2_z\xi)(y^1x,y)|, \\ |(D^2_z\xi)(z,y)| &= |(D^2\eta)(z^1, y^1y' - z')|(y^1)^{d-\gamma-3}, \\ |D^2\zeta^y(x)| &= (y^1)^{d-1-\gamma}|(D^2\eta)(x^1y^1, y^1(y'-x'))|. \end{aligned}$$

In particular, owing to (6.19)

$$\int_{\mathbb{R}^d_+} |D^2\zeta^y|\,dy = (x^1)^{\gamma-1}\int_{\mathbb{R}^d_+} (y^1)^{-\gamma}|D^2\eta(y)|\,dy \le N(x^1)^{\gamma-1}\varepsilon^{-2}\chi.$$

It follows that

$$\int_{\mathbb{R}^d_+}\Big(\int_{\mathbb{R}^d_+} |u|^p|D^2\zeta^y|\,dx\Big)\,dy \le N\varepsilon^{-2}\chi\|M^{-1}u\|^p_{L_{p,\theta}},$$

$$\int_{\mathbb{R}^d_+} J_2(y)\,dy \le N\beta\varepsilon^{-2}\chi\|M^{-1}u\|^p_{L_{p,\theta}} + p^{-1}\int_{\mathbb{R}^d_+} |u|^p\bar a^{ij}\Big(\int_{\mathbb{R}^d_+} D_{ij}\zeta^y\,dy\Big)\,dx, \tag{6.24}$$

where we have used Fubini's theorem which is possible since the right-hand side of (6.24) is finite.

Finally, owing to (6.20)

$$\int_{\mathbb{R}^d_+} D_{ij}\zeta^y(x)\,dy = D_{ij}\int_{\mathbb{R}^d_+}\zeta^y(x)\,dy = D_{ij}(x^1)^{\gamma+1} = \gamma(\gamma+1)(x^1)^{\gamma-1}\delta_{1i}\delta_{1j},$$

so that

$$\int_{\mathbb{R}^d_+} J_2(y)\,dy \le N\beta\varepsilon^{-2}\chi\|M^{-1}u\|^p_{L_{p,\theta}} + p^{-1}\gamma(\gamma+1)\int_{\mathbb{R}^d_+}(x^1)^{\gamma-1}\bar a^{11}|u|^p\,dx.$$

By combining this with (6.22) we come to (6.19) and the lemma is proved. □

7. Existence and uniqueness for parabolic PDEs in half spaces. We take

$$p\in(1,\infty),\quad T\in(0,\infty),\quad d-1<\theta<d-1+p \tag{7.1}$$

and prove the following version of Theorem 2.14 of [9].

THEOREM 7.1. *There exist constants $\beta\in(0,1]$ and $N<\infty$, depending only on δ_0, d, p, and θ, such that if*

$$|a^{ij}_t(x)-a^{ij}_t(y)| + |x^1 b^i_t(x)| + |(x^1)^2 c_t(x)| \le \beta, \tag{7.2}$$

whenever $x,y\in\mathbb{R}^d_+$, $|x-y|\le x^1\wedge y^1$, $i,j=1,...,d$, $t\in(0,T)$, then for any $f\in M^{-1}\mathbb{L}_{p,\theta}(T)$ there exists a unique $u\in\mathfrak{H}^2_{p,\theta,0}(T)$ satisfying (1.1) in $\mathbb{R}^d_+$. Furthermore, for any $u\in\mathfrak{H}^2_{p,\theta,0}(T)$ we have

$$\|M^{-1}u\|_{\mathbb{H}^2_{p,\theta}(T)} \le N\|M(Lu-\partial u/\partial t)\|_{\mathbb{L}_{p,\theta}(T)}. \tag{7.3}$$

REMARK 7.1. Part of condition (7.2) could have been replaced with the requirement that for an $\varepsilon\in(0,1]$

$$|a^{ij}_t(x)-a^{ij}_t(y)|\le\varepsilon\beta$$

for all x, y satisfying $|x-y| \le \varepsilon(x^1 \wedge y^1)$ and all i, j, t. However, this would not make the theorem more general as follows from the argument we used to obtain (6.15).

REMARK 7.2. Theorem 7.1 is used in [9] to prove the solvability of parabolic equations in Sobolev spaces with weights in bounded C^1-domains. The proof for spaces like $\mathfrak{H}^n_{p,\theta}(T)$ with $n=2$ is quite standard and is based on partitions of unity and flattening the boundary. Then equations are solved in small neighborhoods of the boundary, where conditions like (7.2) allow one to treat equations with quite irregular a^{ij} and blowing up b and c. The reader is sent to [9] for details.

To prove the Theorem 7.1 we need a lemma, in which G is a domain in $\mathbb{R}^d$ and by $L_{p,loc}(G)$ we mean the set of functions u such that $u\phi \in L_p(G)$ for any $\phi \in C_0^\infty(G)$.

LEMMA 7.1. *Let u_t and f_t be measurable functions given for $t \le T$ with values in $L_{p,loc}(G)$. Assume that for a continuous function $\psi = \psi(x)$ given on G we have $\psi > 0$,*

$$\int_0^T (\|\psi u_t\|^p_{L_p(G)} + \|\psi^{-1} f_t\|^p_{L_p(G)})\, dt < \infty,$$

and for $t \le T$ and each $\phi \in C_0^\infty(G)$ we have

$$(\phi, u_t) = \int_0^t (\phi, f_s)\, ds. \tag{7.4}$$

Then

$$\infty > \int_0^T \int_G \psi^{p-2} |u_t|^{p-2} u_t f_t \, dx dt = p^{-1} \int_G \psi^{p-2} |u_T|^p \, dx. \tag{7.5}$$

Proof. The inequality in (7.5) follows by Hölder's inequality:

$$\int_0^T \int_G |\psi u_t|^{p-1} |\psi^{-1} f_t| \, dx dt$$
$$\le \Big(\int_0^T \|\psi u_t\|^p_{L_p(G)}\, dt \Big)^{1-1/p} \Big(\int_0^T \|\psi^{-1} f_t\|^p_{L_p(G)}\, dt \Big)^{1/p}.$$

Next, we reduce the general situation to the one in which u_t has compact support in G. Let $\zeta \in C_0^\infty(G)$ and $\zeta \ge 0$. Then

$$\partial(u_t \zeta)/\partial t = f_t \zeta\ .$$

If (7.5) is true for $u_t\zeta$, then

$$\int_0^T \int_G \zeta^p \psi^{p-2} |u_t|^{p-2} u_t f_t \, dx dt = p^{-1} \int_G \psi^{p-2} |\zeta u_T|^p \, dx. \tag{7.6}$$

The above estimates show that if we take $\zeta = \zeta_n \uparrow 1$, then (7.5) would follow from (7.6).

Therefore, in the rest of the proof we assume that there is a compact set $\Gamma \subset G$ such that $u_t(x) = f_t(x) = 0$ if $x \notin \Gamma$. In that case the values of ψ outside Γ become irrelevant and we may assume that $G = \mathbb{R}^d$, ψ is bounded away from zero and infinity, u_t, f_t are measurable and L_p-valued, and

$$\int_0^T (\|u_t\|_{L_p}^p + \|f_t\|_{L_p}^p)\,dt < \infty.$$

We now take a nonnegative $\zeta \in C_0^\infty(\mathbb{R}^d)$ which integrates to one and for an $x \in \mathbb{R}^d$ substitute $\varepsilon^{-d}\zeta(\varepsilon^{-1}(x-\cdot))$ in place of ϕ in (7.4). Then by using the notation

$$v^{(\varepsilon)}(x) = \int_{\mathbb{R}^d} v(x-\varepsilon y)\zeta(y)\,dy$$

we find that for any $x \in \mathbb{R}^d$ and $t \le T$

$$u_t^{(\varepsilon)}(x) = \int_0^t f_s^{(\varepsilon)}(x)\,ds.$$

Then we have

$$p^{-1}|u_T^{(\varepsilon)}(x)|^p = \int_0^T |u_s^{(\varepsilon)}(x)|^{p-2}u_s^{(\varepsilon)}(x)f_s^{(\varepsilon)}(x)\,ds.$$

Now we integrate through this relation over $x \in \mathbb{R}^d$ after multiplying it by $\psi^{p-2}(x)$. It turns out that we can interchange all integrals since, say $u_t^{(\varepsilon)}(x)$ is infinitely differentiable in x, measurable with respect to t, by Minkowskii's inequality

$$\|u_t^{(\varepsilon)}\|_{L_p} \le \|u_t\|_{L_p}, \tag{7.7}$$

and the estimates as in the beginning of the proof are valid. Then we conclude that

$$\int_0^T \int_G \psi^{p-2}|u_s^{(\varepsilon)}|^{p-2}u_s^{(\varepsilon)}f_s^{(\varepsilon)}\,dxds = p^{-1}\int_G \psi^{p-2}|u_T^{(\varepsilon)}|^p\,dx \tag{7.8}$$

We also observe that (for any $t \le T$) we have $u_t^{(\varepsilon)} \to u_t$ in L_p. Similar relation holds for f_t. In particular, $u_t^{(\varepsilon)} \to u_t$ in measure

$$\mu(dtdx) = \psi^{p-2}\,dxdt$$

on $[0,T] \times \mathbb{R}^d$. Furthermore, (7.7) and the dominated convergence theorem imply that

$$\|u^{(\varepsilon)}\|_{L_p(\mu)} \to \|u\|_{L_p(\mu)}$$

Similar relation holds for f. By Lemma 6.1 and Corollary 6.1

$$
\begin{aligned}
|u^{(\varepsilon)}|^p \to |u|^p \quad &\text{in} \quad L_1(\mu), \\
|u^{(\varepsilon)}|^{p-1} \to |u|^{p-1} \quad &\text{in} \quad L_{p/(p-1)}(\mu), \\
|u^{(\varepsilon)}|^{p-2}u^{(\varepsilon)} \to |u|^{p-2}u \quad &\text{in} \quad L_{p/(p-1)}(\mu), \\
|u^{(\varepsilon)}|^{p-2}u^{(\varepsilon)}f^{(\varepsilon)} \to |u|^{p-2}uf \quad &\text{in} \quad L_1(\mu).
\end{aligned}
$$

Thus, passing to the limit as $\varepsilon \to 0$ in (7.8) leads to (7.6). The lemma is proved. □

Proof of Theorem 7.1. Rather elementary arguments (cf. the proof of Lemma 5.7 of [12]) show that if $f \in C_0^\infty(\mathbb{R}_+ \times \mathbb{R}^d_+)$, then the unique classical bounded solution of $\partial u/\partial t = \Delta u + f$ in $\mathbb{R}_+ \times \mathbb{R}^d_+$ with zero initial and boundary condition belongs to $\mathfrak{H}^2_{p,\theta,0}(T)$. This gives us a starting point in the method of continuity and, due to the fact that $\mathfrak{H}^2_{p,\theta,0}(T)$ is a Banach space, shows that to prove the theorem we need only prove the apriori estimate

$$
\|u\|_{\mathfrak{H}^2_{p,\theta}(T)} \leq N\|M(Lu - \partial u/\partial t)\|_{\mathbb{L}_{p,\theta}(T)}
$$

for any $u \in \mathfrak{H}^2_{p,\theta,0}(T)$, where and below we denote by N generic constants depending only on $\delta_0, d, p, \theta, K$. We will keep track of β up until we will see how to choose it. By the definition of the norm in $\mathfrak{H}^2_{p,\theta,0}(T)$ we have

$$
\|u\|_{\mathfrak{H}^2_{p,\theta}(T)} = \|M^{-1}u\|_{\mathbb{H}^2_{p,\theta}(T)} + \|M\partial u/\partial t\|_{\mathbb{L}_{p,\theta}(T)}
$$

and

$$
\|M\partial u/\partial t\|_{\mathbb{L}_{p,\theta}(T)} \leq \|M(Lu - \partial u/\partial t)\|_{\mathbb{L}_{p,\theta}(T)} + \|MLu\|_{\mathbb{L}_{p,\theta}(T)},
$$

where by our assumptions on b and c

$$
\|MLu\|_{\mathbb{L}_{p,\theta}(T)} \leq NJ^{1/p}
$$

with

$$
J^{1/p} := \|MD^2u\|_{\mathbb{L}_{p,\theta}(T)} + \|Du\|_{\mathbb{L}_{p,\theta}(T)} + \|M^{-1}u\|_{\mathbb{L}_{p,\theta}(T)}.
$$

Finally, by recalling that by Remark 4.2 the left-hand side of (7.3) is equivalent to $J^{1/p}$, we see that we need only estimate $J^{1/p}$ in terms of the right-hand side of (7.3).

By Corollary 5.1 we have

$$
\|MD^2u\|_{\mathbb{L}_{p,\theta}(T)} + \|Du\|_{\mathbb{L}_{p,\theta}(T)} \leq N\|Mf\|_{\mathbb{L}_{p,\theta}(T)} + N\|M^{-1}u\|_{\mathbb{L}_{p,\theta}(T)}, \quad (7.9)
$$

where $f = (\partial/\partial t - L)u$.

Next, we use Lemma 7.1 with $G = \mathbb{R}^d_+$, $\psi = (x^1)^{-1}$, and $M^{(\theta-d)/p}u_t$ and $M^{(\theta-d)/p}f_t$ in place of u_t and f_t, respectively. We also observe that

$$\|\psi M^{(\theta-d)/p}u_t\|_{L_p(G)} = \|M^{-1}u_t\|_{L_{p,\theta}}.$$

Also

$$\|\psi^{-1}M^{(\theta-d)/p}(L_tu_t + f_t)\|_{L_p(G)} = \|M(L_tu_t + f_t)\|_{L_{p,\theta}}$$
$$\le N\|MD^2u_t\|_{L_{p,\theta}} + \|M|b_t|Du_t\|_{L_{p,\theta}} + \|M^2c_t(M^{-1}u_t)\|_{L_{p,\theta}} + \|Mf_t\|_{L_{p,\theta}},$$

so that by our assumptions and Remark 4.2

$$\|\psi^{-1}M^{(\theta-d)/p}(L_tu_t + f_t)\|_{L_p(G)} \le N\|M^{-1}u_t\|_{H^2_{p,\theta}} + \|Mf_t\|_{L_{p,\theta}}.$$

It follows that Lemma 7.1 is applicable and by noting that

$$\psi^{p-2}|M^{(\theta-d)/p}u_t|^{p-2}(M^{(\theta-d)/p}u_t)(M^{(\theta-d)/p}(L_tu_t + f_t))$$
$$= \psi^{p-2}M^{\theta-d}|u_t|^{p-2}(L_tu_t + f_t) = M^{\gamma+1}|u_t|^{p-2}(L_tu_t + f_t),$$

where $\gamma = \theta - d - p + 1$, we obtain

$$\int_0^T \int_{\mathbb{R}^d_+} (x^1)^{\gamma+1}|u_t|^{p-2}u_t(L_tu_t + f_t)\,dxdt \ge 0. \tag{7.10}$$

Observe that by (7.2)

$$2|u_t|^{p-2}u_t(L_tu_t + f_t) \le |u_t|^{p-2}u_ta^{ij}_tD_{ij}u_t + N\beta M^{-1}|u_t|^{p-1}|Du_t|$$
$$+N\beta M^{-2}|u_t|^p + N|u_t|^{p-1}|f_t|,$$

and by Young's inequality

$$M^{-1}|u_t|^{p-1}|Du_t| = (M^{-2(p-1)/p}|u_t|^{p-1})(M^{(p-2)/p}|Du_t|)$$
$$\le M^{-2}|u_t|^p + M^{p-2}|Du_t|^p,$$
$$|u_t|^{p-1}|f_t| \le \beta M^{-2}|u_t|^p + \beta^{1-p}M^{2(p-1)}|f_t|^p.$$

Furthermore,

$$\int_0^T \int_{\mathbb{R}^d_+} (x^1)^{\gamma+1}M^{-2}|u_t|^p\,dxdt = \|M^{-1}u\|^p_{\mathbb{L}_{p,\theta}(T)} =: I \le J,$$
$$\int_0^T \int_{\mathbb{R}^d_+} (x^1)^{\gamma+1}M^{p-2}|Du_t|^p\,dxdt = \|Du\|^p_{\mathbb{L}_{p,\theta}(T)}, \tag{7.11}$$
$$\int_0^T \int_{\mathbb{R}^d_+} (x^1)^{\gamma+1}M^{2(p-1)}|f_t|^p\,dxdt = \|Mf\|^p_{\mathbb{L}_{p,\theta}(T)}.$$

Therefore, coming back to (7.10) we get

$$N\beta J + N\beta^{1-p}\|Mf\|^p_{\mathbb{L}_{p,\theta}(T)} + \int_0^T \int_{\mathbb{R}^d_+} (x^1)^{\gamma+1}|u_t|^{p-2}u_t a^{ij}_t D_{ij}u_t \, dxdt \geq 0.$$

By combining this with Lemma 6.5 we obtain

$$N\beta J + N\beta^{1-p}\|Mf\|^p_{\mathbb{L}_{p,\theta}(T)} + p^{-1}\gamma(\gamma+1)\int_0^T \int_{\mathbb{R}^d_+} (x^1)^{\gamma-1}a^{11}_t|u_t|^p \, dxdt$$
$$- (p-1)\int_0^T \int_{\mathbb{R}^d_+} (x^1)^{\gamma+1}|u_t|^{p-2}a^{ij}_t(D_iu_t)D_ju_t \, dxdt \geq 0.$$

Lemma 6.4 with $\delta \in (0,1]$ and Corollary 6.2 allow us to estimate the last term and we conclude that for any $\kappa \in (0,1]$

$$\begin{aligned} N(\kappa^{-1}\delta + e^{N\delta^{-1/2}}\beta)J + N\beta^{-p}\|Mf\|^p_{\mathbb{L}_{p,\theta}(T)} \\ +\nu\int_0^T \int_{\mathbb{R}^d_+} (x^1)^{\gamma-1}a^{11}_t|u_t|^p \, dxdt \geq 0, \end{aligned} \tag{7.12}$$

where

$$\begin{aligned} \nu &= p^{-1}\gamma(\gamma+1) - (p-1)(1-\kappa)\gamma^2p^{-2} \\ &= \kappa p^{-2}(p-1)\gamma^2 + p^{-2}(\theta-d-p+1)\big(\theta-(d-1)\big). \end{aligned}$$

If we consider ν as a function of κ, then by virtue of (7.1) we have $\nu < 0$ if $\kappa = 0$. It follows that there exists a $\kappa = \kappa(\theta,p,d) > 0$ such that $\nu = -1/N$, where $N = N(\theta,p,d) \in (0,\infty)$. Then (7.12) and the assumption that $a^{11} \geq \delta_0$ imply that

$$\|M^{-1}u\|_{\mathbb{L}_{p,\theta}(T)} \leq N(\delta + e^{N\delta^{-1/2}}\beta)J + N\beta^{-p}\|Mf\|^p_{\mathbb{L}_{p,\theta}(T)}.$$

By using (7.9) we conclude

$$I \leq N_1(\delta + e^{N_1\delta^{-1/2}}\beta)I + N(\delta + e^{N_1\delta^{-1/2}}\beta + \beta^{-p})\|Mf\|^p_{\mathbb{L}_{p,\theta}(T)}. \tag{7.13}$$

Now we can specify the choice of $\delta = \delta(\theta,d,p,\delta_0,K), \beta = \beta(\theta,d,p,\delta_0,K) \in (0,1)$. We take them so that

$$N_1\delta \leq 1/4, \quad N_1e^{N_1\delta^{-1/2}}\beta \leq 1/4$$

and then (7.13) yields

$$I \leq N\|Mf\|^p_{\mathbb{L}_{p,\theta}(T)},$$

which along with (7.9) lead to (7.3) and the theorem is proved. □

8. Existence and uniqueness for SPDEs in half spaces. In this section τ is a fixed stopping time, $p \geq 2$. We start with a "conditional" result we were talking about in the Introduction. Here we only have a "natural" assumption on θ. Of course, we are working in the setting of Section 2.

THEOREM 8.1. *Let*

$$d-1 < \theta < d-1+p.$$

Assume that for a constant $K \in (0,\infty)$ we have

$$|x^1 \nu_t(x)|_{\ell_2} + |x^1 D\sigma^i_t(x)|_{\ell_2} + |(x^1)^2 D\nu_t(x)|_{\ell_2} \leq K.$$

Let β be the smallest of the constants called β_0 in Theorem 4.1 and β in Theorem 7.1 and assume that

$$|a^{ij}_t(x) - a^{ij}_t(y)| + |\sigma^i_t(x) - \sigma^i_t(y)|_{\ell_2} + |x^1 b^i_t(x)| + |(x^1)^2 c_t(x)| \leq \beta,$$

whenever $x, y \in \mathbb{R}^d_+$, $|x-y| \leq x^1 \wedge y^1$, $i,j = 1,...,d$, $t \in (0,\tau)$. Fix an $f \in \mathbb{L}_{p,\theta}(\tau)$.

Assume that there is a constant $N_0 < \infty$ such that for any

$$\lambda \in [0,1], \quad g \in \mathbb{H}^1_{p,\theta}(\tau), \quad u \in \mathfrak{H}^2_{p,\theta,0}(\tau) \tag{8.1}$$

we have the apriori estimate

$$\|M^{-1}u\|_{\mathbb{L}_{p,\theta}(\tau)} \leq N_0\big(\|f\|_{\mathbb{L}_{p,\theta}(\tau)} + \|g\|_{\mathbb{H}^1_{p,\theta}(\tau)}\big) \tag{8.2}$$

provided that

$$du_t = (L_t u_t + f_t)\,dt + \lambda(\Lambda^k_t u_t + g^k_t)\,dw^k_t, \quad t \in (0,\tau) \tag{8.3}$$

in $\mathbb{R}^d_+$ (estimate (8.2) is not supposed to hold if there is no solution $u \in \mathfrak{H}^2_{p,\theta,0}(\tau)$ of (8.3)).

Then for any $g \in \mathbb{H}^1_{p,\theta}(\tau)$ there exists a unique $u \in \mathfrak{H}^2_{p,\theta,0}(\tau)$ satisfying (8.3) with $\lambda = 1$. Furthermore, for this solution

$$\|M^{-1}u\|_{\mathbb{H}^2_{p,\theta}(\tau)} \leq N\|Mf\|_{\mathbb{L}_{p,\theta}(\tau)} + N\|g\|_{\mathbb{H}^1_{p,\theta}(\tau)}, \tag{8.4}$$

where N depends only on $d, p, \theta, \delta_0, \delta_1, K$, and N_0.

Proof. We know that $\mathfrak{H}^2_{p,\theta,0}(\tau)$ is a Banach space. From Theorem 7.1 we also know that for $\lambda = 0$ equation (8.3) is uniquely solvable in $\mathfrak{H}^2_{p,\theta,0}(\tau)$. Therefore, by the method of continuity to prove the unique solvability of (8.3) for $\lambda = 1$ we only need to show that there is a constant N such that for any objects in (8.1) we have

$$\|u\|_{\mathfrak{H}^2_{p,\theta}(\tau)} \leq N\big(\|Mf\|_{\mathbb{L}_{p,\theta}(\tau)} + \|g\|_{\mathbb{H}^1_{p,\theta}(\tau)}\big), \tag{8.5}$$

provided that (8.3) holds in $\mathbb{R}^d_+$.

By the definition of the norm in $\mathfrak{H}^2_{p,\theta,0}(T)$ we have

$$\|u\|_{\mathfrak{H}^2_{p,\theta}(\tau)} = \|M^{-1}u\|_{\mathbb{H}^2_{p,\theta}(\tau)} + \|M(Lu+f)\|_{\mathbb{L}_{p,\theta}(\tau)} + \lambda\|\Lambda u + g\|_{\mathbb{H}^1_{p,\theta}(\tau)},$$

where as in the proof of Theorem 7.1

$$\|M(Lu+f)\|_{\mathbb{L}_{p,\theta}(\tau)} \le \|Mf\|_{\mathbb{L}_{p,\theta}(\tau)} + N\|M^{-1}u\|_{\mathbb{H}^2_{p,\theta}(\tau)}.$$

Also by our assumptions on σ and ν

$$\|\Lambda u + g\|_{\mathbb{H}^1_{p,\theta}(\tau)} \le \|g\|_{\mathbb{H}^1_{p,\theta}(\tau)} + N\big(\|M^{-1}u\|_{\mathbb{L}_{p,\theta}(\tau)} + \|Du\|_{\mathbb{L}_{p,\theta}(\tau)} + \|MD^2u\|_{\mathbb{L}_{p,\theta}(\tau)}\big).$$

By combining this with Remark 4.2 we see that to prove (8.5) it suffices to prove (8.4) or else that

$$\|M^{-1}u\|_{\mathbb{L}_{p,\theta}(\tau)} + \|Du\|_{\mathbb{L}_{p,\theta}(\tau)} + \|MD^2u\|_{\mathbb{L}_{p,\theta}(\tau)} \le N\|Mf\|_{\mathbb{L}_{p,\theta}(\tau)} + N\|g\|_{\mathbb{H}^1_{p,\theta}(\tau)}.$$

However, the $\mathbb{L}_{p,\theta}(\tau)$-norms of MD^2u and Du are estimated through the $\mathbb{L}_{p,\theta}(\tau)$-norm of $M^{-1}u$ in Corollary 4.1 and the latter admits an estimate by assumption (8.2). The theorem is proved. □

Here is a version of one of the results of [7] in which we assume that for a constant $\tilde{\delta} \in (0,1]$ and all t and unit $\lambda \in \mathbb{R}^d$ we have

$$(a^{ij}_t - \alpha^{ij}_t)\lambda^i\lambda^j \ge \tilde{\delta}\frac{1}{a^{11}_t}\Big(\sum_j a^{1j}_t\lambda^j\Big)^2, \tag{8.6}$$

REMARK 8.1. In light of (2.1) one can always take $\tilde{\delta} = \delta_1$ in (8.6) because for the positive definite matrix (a^{ij}_t) and $\eta = (1,0,...,0)$ it holds that

$$\Big(\sum_j a^{1j}_t\lambda^j\Big)^2 = \Big(\sum_j a^{ij}_t\eta^i\lambda^j\Big)^2 \le (a^{ij}_t\eta^i\eta^j)a^{ij}_t\lambda^i\lambda^j = a^{11}_t a^{ij}_t\lambda^i\lambda^j.$$

On the other hand sometimes $\tilde{\delta} = 1$ and δ_1 may be less than 1. This happens, for instance, if $\alpha^{1j}_t \equiv 0$ for all j and $a^{1j}_t \equiv 0$ for $j \ne 1$.

THEOREM 8.2. *Let (1.3) be satisfied and assume that for a constant $K \in (0,\infty)$ we have*

$$|x^1 D\sigma^i_t(x)|_{\ell_2} + |(x^1)^2 D\nu_t(x)|_{\ell_2} \le K.$$

We assert that there exist constants $\beta > 0$ and $N < \infty$ depending only on $\delta_0, \delta_1, \tilde{\delta}, d, p, \theta$, and K, such that if

$$|a^{ij}_t(x) - a^{ij}_t(y)| + |\sigma^i_t(x) - \sigma^i_t(y)|_{\ell_2} + |x^1 b^i_t(x)| + |x^1\nu_t(x)|_{\ell_2} + |(x^1)^2 c_t(x)| \le \beta, \tag{8.7}$$

whenever $x, y \in \mathbb{R}^d_+$, $|x-y| \leq x^1 \wedge y^1$, $i,j = 1, ..., d$, $t \in (0, \tau)$, *then for any* $f \in M^{-1}\mathbb{L}_{p,\theta}(\tau)$, $g = (g^k, k = 1, 2, ...) \in \mathbb{H}^1_{p,\theta}(\tau)$ *there exists a unique* $u \in \mathfrak{H}^2_{p,\theta,0}(\tau)$ *satisfying* (1.2) *in* $\mathbb{R}^d_+$. *Furthermore,*

$$\|M^{-1}u\|_{\mathbb{H}^2_{p,\theta}(\tau)} \leq N\|Mf\|_{\mathbb{L}_{p,\theta}(\tau)} + N\|g\|_{\mathbb{H}^1_{p,\theta}(\tau)}.$$

REMARK 8.2. For $d = d_0 = 1$ an example of σ suitable for Theorem 8.2 can be obtained in the following way. Take a smooth function $f \geq 1$ on $(0, \infty)$ such that $f(x) = -\ln x$ for small x and $f(x) = 1$ for x large. Then $|xf'(x)|$ is bounded, say by a constant N_0. Next, for $\delta \in (0,1)$ introduce $\sigma(x) = \cos f^\delta(x)$. Then

$$|\sigma'(x)| = \delta f^{\delta-1}(x)|f'(x)|\,|\sin f^\delta(x)| \leq N_0\delta x^{-1},$$

and if $0 < y < x$ and $|x-y| \leq x \wedge y$, then $|x-y| \leq y$ and, for a point ξ between x and y, so that $y \leq \xi$, we have

$$|\sigma(x) - \sigma(y)| = \delta f^{\delta-1}(\xi)|f'(\xi)|\,|\sin f^\delta(\xi)|\,|x-y| \leq N_0\delta\xi^{-1}y \leq N_0\delta,$$

which can be made arbitrarily small if δ is chosen appropriately.

REMARK 8.3. One could have introduced the parameter ε as in Remark 7.1 and as there this would not make the theorem any more general.

REMARK 8.4. Theorem 8.2 is used in [7] to prove the solvability of SPDEs in Sobolev spaces with weights in bounded C^1-domains. The proof for spaces like $\mathfrak{H}^n_{p,\theta}(\tau)$ with $n = 2$ is quite standard and is based on partitions of unity and flattening the boundary. Then equations are solved in small neighborhoods of the boundary, where conditions like (8.7) allow one to treat equations with quite irregular a^{ij}, σ^{ik} and blowing up b, c, and ν (cf. Remark 4.4). The reader is sent to [7] for details.

To prove Theorem 8.2 we need the following counterpart of Lemma 7.1 in which we use the same notation.

LEMMA 8.1. *Let* $u_t = u_t(\omega)$, f_t, *and* $g_t = (g^1_t, g^2_t, ...)$ *be predictable processes given for* $t < \tau$ *with values in* $L_{p,loc}(G)$. *Assume that for a continuous function* $\psi = \psi(x)$ *given on* G *we have* $\psi > 0$,

$$E\int_0^\tau (\|\psi u_t\|^p_{L_p(G)} + \|\psi^{-1}f_t\|^p_{L_p(G)} + \|g_t\|^p_{L_p(G)})\,dt < \infty,$$

and

$$du_t = f_t\,dt + g^k_t\,dw^k_t$$

in G *for* $t < \tau$ *in the sense that for each* $\phi \in C^\infty_0(G)$ *(a.s.) for all* $t \in [0, \tau)$ *at once we have*

$$(\phi, u_t) = \int_0^t (\phi, f_s)\,ds + \int_0^t (\phi, g^k_s)\,dw^k_s.$$

Then

$$E\int_0^{\tau}\int_G \psi^{p-2}[2|u_t|^{p-2}u_t f_t+(p-1)|u_t|^{p-2}|g_t|_{\ell_2}^2]\,dxdt\ge 0. \tag{8.8}$$

Proof. We mimic the proof of Lemma 7.1. First, we observe that the left-hand side of (8.8) makes sense, since by Hölder's inequality ($p\ge 2$)

$$E\int_0^{\tau}\int_G |\psi u_t|^{p-2}|g_t|_{\ell_2}^2\,dt$$
$$\le \Big(E\int_0^{\tau}\|\psi u_t\|_{L_p(G)}^p\,dt\Big)^{1-2/p}\Big(E\int_0^{\tau}\|g_t\|_{L_p(G)}^p\,dt\Big)^{2/p}$$

and the term containing f_t is taken care of as in the proof of Lemma 7.1. Next, as in that proof we reduce the general situation to the one in which u_t has compact support in G.

More precisely, in the rest of the proof we assume that there is a compact set $\Gamma\subset G$ such that $u_t(x)=f_t(x)=g_t^k(x)=0$ if $x\notin\Gamma$. In that case the values of ψ outside Γ become irrelevant and we may assume that $G=\mathbb{R}^d$, ψ is bounded away from zero and infinity, u_t, f_t, g_t are predictable and L_p-valued, and

$$E\int_0^{\tau}(\|u_t\|_{L_p}^p+\|f_t\|_{L_p}^p+\|g_t\|_{L_p}^p)\,dt<\infty.$$

In addition one can replace here p with 2 if one replaces τ with $\tau\wedge T$, where T is a constant from $(0,\infty)$.

Then by using the notation from the proof of Lemma 7.1 we find that for any $x\in\mathbb{R}^d$ with probability one for all $t<\tau$

$$u_t^{(\varepsilon)}(x)=\int_0^t f_s^{(\varepsilon)}(x)\,ds+\int_0^t g_s^{(\varepsilon)k}(x)\,dw_s^k.$$

By Itô's formula for any constant $T\in(0,\infty)$ (here we use again that $p\ge 2$)

$$(2/p)|u_{\tau\wedge T}^{(\varepsilon)}(x)|^p=\int_0^{\tau\wedge T}[2|u_s^{(\varepsilon)}(x)|^{p-2}u_s^{(\varepsilon)}(x)f_s^{(\varepsilon)}(x)\,ds$$
$$+(p-1)|u_s^{(\varepsilon)}(x)|^{p-2}|g_s^{(\varepsilon)}|_{\ell_2}^2]\,ds+m_{t\wedge T}(x),$$

where $m_t(x)$ is a local martingale starting at zero. Observe that the left-hand side is nonnegative and

$$E\int_0^{\tau}[|u_s^{(\varepsilon)}(x)|^{p-1}|f_t^{(\varepsilon)}(x)|+|u_s^{(\varepsilon)}(x)|^{p-2}|g_s^{(\varepsilon)}(x)|_{\ell_2}^2]\,ds<\infty$$

because of the same reasons as in the beginning of the proof and the estimate

$$|v^{(\varepsilon)}(x)|\le N(\varepsilon,p)\|v\|_{L_p},$$

that follows from Hölder's inequality.

Then we find that

$$E\int_0^{\tau\wedge T}[2|u_s^{(\varepsilon)}(x)|^{p-2}u_s^{(\varepsilon)}(x)f_s^{(\varepsilon)}(x)\,ds+(p-1)|u_s^{(\varepsilon)}(x)|^{p-2}|g_s^{(\varepsilon)}|_{\ell_2}^2]\,ds\geq 0.$$

Now we integrate through this relation over $x\in\mathbb{R}^d$ after multiplying it by $\psi^{p-2}(x)$. It turns out that we can interchange all integrals since, say $u_t^{(\varepsilon)}(x)$ is infinitely differentiable in x, measurable with respect to other variables, by Minkowskii's inequality

$$\|u_t^{(\varepsilon)}\|_{L_p}\leq\|u_t\|_{L_p},\tag{8.9}$$

and the estimates as in the beginning of the proof are valid.

After integrating with respect to x and setting $T\to\infty$ we obtain

$$E\int_0^{\tau}\int_{\mathbb{R}^d}\psi^{p-2}[2|u_t^{(\varepsilon)}|^{p-2}u_t^{(\varepsilon)}f_t^{(\varepsilon)}+(p-1)|u_t^{(\varepsilon)}|^{p-2}|g_t^{(\varepsilon)}|_{\ell_2}^2]\,dxdt\geq 0.\tag{8.10}$$

Now notice that (for any ω and t such that $t<\tau$) we have $u_t^{(\varepsilon)}\to u_t$ in L_p. Similar relations hold for f_t and g_t. In particular, $u_t^{(\varepsilon)}\to u_t$ in measure

$$\mu(d\omega dtdx)=\psi^{p-2}\,P(d\omega)dtdx$$

on $(0,\tau]\!]\times\mathbb{R}^d$. Furthermore, (8.9) and the dominated convergence theorem imply that

$$\|u^{(\varepsilon)}\|_{L_p(\mu)}\to\|u\|_{L_p(\mu)}$$

Similar relations hold for f and g. By Lemma 6.1 and Corollary 6.1

$$|u^{(\varepsilon)}|^{p-2}\to|u|^{p-2}\quad\text{in}\quad L_{p/(p-2)}(\mu),$$
$$|u^{(\varepsilon)}|^{p-2}|g^{(\varepsilon)}|_{\ell_2}\to|u|^{p-2}|g|_{\ell_2}\quad\text{in}\quad L_1(\mu).$$

The term with $f^{(\varepsilon)}$ in (8.10) is taken care of in the same way as in the proof of Lemma 7.1.

Thus, passing to the limit as $\varepsilon\to 0$ in (8.10) leads to (8.8). The lemma is proved. □

Proof of Theorem 8.2. By Theorem 8.1 we only need to find $\beta=\beta(\delta_0,\delta_1,\tilde{\delta},d,p,\theta,K)\in(0,1]$ such that if the conditions of the theorem are satisfied with this β, then (8.2) holds with N_0 depending only on $\delta_0,\delta_1,\tilde{\delta},d,p,\theta$, and K, whenever (8.1) and (8.3) are satisfied.

Therefore, we concentrate on estimating

$$I^{1/p}:=\|M^{-1}u\|_{\mathbb{L}_{p,\theta}(\tau)}.$$

In Lemma 8.1 we take $G=\mathbb{R}^d_+$, $\psi(x)=(x^1)^{-1}$, and replace u_t,f_t,g_t with $M^{(\theta-d)/p}u_t$, $M^{(\theta-d)/p}(L_tu_t+f_t)$, $M^{(\theta-d)/p}(\Lambda_tu_t+g_t)$, respectively. We also observe that as in the proof of Theorem 7.1

$$\|\psi^{-1}M^{(\theta-d)/p}(L_tu_t+f_t)\|_{L_p(G)}\leq N\|M^{-1}u_t\|_{H^2_{p,\theta}}+\|Mf_t\|_{L_{p,\theta}},$$

where and below we denote by N generic constants depending only on $\delta_0,\delta_1,d,p,\theta,K$. We keep track of β up until we will see how to choose it.

Finally,

$$\begin{aligned}\|M^{(\theta-d)/p}(\Lambda_tu_t+g_t)\|_{L_p(G)}&=\|\Lambda_tu_t+g_t\|_{L_{p,\theta}}\\&\leq N\|Du_t\|_{L_{p,\theta}}+\|M\nu_t(M^{-1}u_t)\|_{L_{p,\theta}}+\|g_t\|_{L_{p,\theta}}\\&\leq N\|M^{-1}u_t\|_{H^2_{p,\theta}}+\|g_t\|_{L_{p,\theta}}.\end{aligned}$$

It follows that Lemma 8.1 is applicable and noting that

$$\psi^{p-2}|M^{(\theta-d)/p}u_t|^{p-2}(M^{(\theta-d)/p}u_t)(M^{(\theta-d)/p}(L_tu_t+f_t))$$
$$=\psi^{p-2}M^{\theta-d}|u_t|^{p-2}(L_tu_t+f_t)=M^{\gamma+1}|u_t|^{p-2}(L_tu_t+f_t),$$
$$\psi^{p-2}|M^{(\theta-d)/p}u_t|^{p-2}|M^{(\theta-d)/p}(\Lambda_tu_t+g_t)|^2_{\ell_2}$$
$$=\psi^{p-2}M^{\theta-d}|u_t|^{p-2}|\Lambda_tu_t+g_t|^2_{\ell_2}=M^{\gamma+1}|u_t|^{p-2}|\Lambda_tu_t+g_t|^2_{\ell_2},$$

where $\gamma=\theta-d-p+1$, we obtain

$$\begin{aligned}E\int_0^\tau\int_{\mathbb{R}^d_+}&(x^1)^{\gamma+1}\big[2|u_t|^{p-2}u_t(L_tu_t+f_t)\\&+(p-1)|u_t|^{p-2}|\Lambda_tu_t+g_t|^2_{\ell_2}\big]\,dxdt\geq0.\end{aligned}\tag{8.11}$$

Next, set

$$J^{1/p}:=\|MD^2u\|_{\mathbb{L}_{p,\theta}(\tau)}+\|Du\|_{\mathbb{L}_{p,\theta}(\tau)}+\|M^{-1}u\|_{\mathbb{L}_{p,\theta}(\tau)}$$

and note for the future that by Corollary 4.1 we have

$$J^{1/p}\leq N\|Mf\|_{\mathbb{L}_{p,\theta}(\tau)}+N\|g\|_{\mathbb{H}^1_{p,\theta}(\tau)}+NI^{1/p}.\tag{8.12}$$

We deal with the term containing Lu_t+f_t in (8.11) as in the proof of Theorem 7.1 and conclude that

$$\begin{aligned}N\beta J+N\beta^{1-p}\|Mf\|^p_{\mathbb{L}_{p,\theta}(\tau)}+E\int_0^\tau\int_{\mathbb{R}^d_+}&(x^1)^{\gamma+1}\big[|u_t|^{p-2}u_ta^{ij}_tD_{ij}u_t\\&+(p-1)|u_t|^{p-2}|\Lambda_tu_t+g_t|^2_{\ell_2}\big]\,dxdt\geq0.\end{aligned}$$

By combining this with Lemma 6.5 we obtain

$$\begin{aligned}-(p-1)E\int_0^\tau\int_{\mathbb{R}^d_+}&(x^1)^{\gamma+1}|u_t|^{p-2}\big[a^{ij}_t(D_iu_t)D_ju_t-|\Lambda_tu_t+g_t|^2_{\ell_2}\big]\,dxdt\\&+p^{-1}\gamma(\gamma+1)E\int_0^\tau\int_{\mathbb{R}^d_+}(x^1)^{\gamma-1}a^{11}_t|u_t|^p\,dxdt\\&+N\beta J+N\beta^{1-p}\|Mf\|^p_{\mathbb{L}_{p,\theta}(\tau)}\geq0.\end{aligned}\tag{8.13}$$

Here

$$|\Lambda_t u_t + g_t|_{\ell_2}^2 = \alpha_t^{ij}(D_i u_t)D_j u_t + |\nu_t|_{\ell_2}^2 |u_t|^2 + |g_t|_{\ell_2}^2 + 2(\sigma_t^i, \nu_t)_{\ell_2} u_t D_i u_t$$
$$+2(\sigma_t^i, g_t)_{\ell_2} D_i u_t + 2(\nu_t, g_t)_{\ell_2} u_t.$$

By our assumptions

$$(a_t^{ij} - \alpha_t^{ij})(D_i u_t)D_j u_t \geq \tilde{\delta}(a_t^{11})^{-1}(a_t^{1j} D_j u_t)^2,$$
$$|\nu_t|_{\ell_2}^2 |u_t|^p \leq \beta M^{-2}|u_t|^p,$$
$$|u_t|^{p-2}(\sigma_t^i, \nu_t)_{\ell_2} u_t D_i u_t \leq N\beta M^{-1}|u_t|^{p-1}|Du_t|$$
$$\leq N\beta(M^{-2}|u_t|^p + M^{p-2}|Du_t|^p),$$
$$|u_t|^{p-2}(\sigma_t^i, g_t)_{\ell_2} D_i u_t \leq N|u_t|^{p-2}|Du_t|\,|g_t|_{\ell_2} \leq N\beta^{1-p}M^{p-2}|g_t|_{\ell_2}^p$$
$$+ N\beta M^{(2-p)/(p-1)}|u_t|^{p(p-2)/(p-1)}|Du_t|^{p/(p-1)} \leq N\beta^{1-p}M^{p-2}|g_t|_{\ell_2}^p$$
$$+ N\beta(M^{-2}|u_t|^p + M^{p-2}|Du_t|^p),$$
$$|u_t|^{p-2}(\nu_t, g_t)_{\ell_2} u_t \leq N\beta M^{-1}|u_t|^{p-1}|g_t|_{\ell_2}$$
$$\leq N\beta M^{-2}|u_t|^p + N\beta M^{p-2}|g_t|_{\ell_2}^p.$$

By using the computations in (7.11), adding that

$$E\int_0^\tau \int_{\mathbb{R}_+^d} (x^1)^{\gamma+1} M^{p-2}|g_t|_{\ell_2}^p \, dxdt = \|g\|_{\mathbb{L}_{p,\theta}(\tau)}^p,$$

and concentrating on $\beta \leq 1$, we infer from (8.13) that

$$N\beta J + N\beta^{-p}(\|Mf\|_{\mathbb{L}_{p,\theta}(\tau)}^p + \|g\|_{\mathbb{L}_{p,\theta}(\tau)}^p)$$
$$+ p^{-1}\gamma(\gamma+1)E\int_0^\tau \int_{\mathbb{R}_+^d} (x^1)^{\gamma-1} a_t^{11}|u_t|^p \, dxdt$$
$$- (p-1)\tilde{\delta}E\int_0^\tau \int_{\mathbb{R}_+^d} (x^1)^{\gamma+1}|u_t|^{p-2}(a_t^{11})^{-1}(a_t^{1j} D_j u_t)^2 \, dxdt \geq 0.$$

Lemma 6.4 with $\delta \in (0,1]$ and Corollary 6.2 allow us to estimate the last term and we conclude that for any $\kappa \in (0,1]$

$$N(\kappa^{-1}\delta + e^{N\delta^{-1/2}}\beta)J + N\beta^{-p}(\|Mf\|_{\mathbb{L}_{p,\theta}(\tau)}^p + \|g\|_{\mathbb{L}_{p,\theta}(\tau)}^p)$$
$$+\nu E\int_0^\tau \int_{\mathbb{R}_+^d} (x^1)^{\gamma-1} a_t^{11}|u_t|^p \, dxdt \geq 0, \tag{8.14}$$

where

$$\nu = p^{-1}\gamma(\gamma+1) - (p-1)\tilde{\delta}(1-\kappa)\gamma^2 p^{-2} = \kappa p^{-2}(p-1)\gamma^2\tilde{\delta}$$
$$+ p^{-2}[p(1-\tilde{\delta}) + \tilde{\delta}](\theta - d - p + 1)\Big(\theta - (d-1) - p[1 - \frac{1}{p(1-\tilde{\delta}) + \tilde{\delta}}]\Big).$$

If we consider ν as a function of κ, then by virtue of (1.3) we have $\nu < 0$ if $\kappa = 0$. It follows that there exists a $\kappa = \kappa(\theta, \tilde{\delta}, p, d) > 0$ such that $\nu = -1/N$, where $N = N(\theta, \tilde{\delta}, p, d) \in (0, \infty)$. Then (8.14) and the assumption that $a^{11} \geq \delta_0$ imply that

$$\|M^{-1}u\|_{\mathbb{L}_{p,\theta}(\tau)} \leq N(\delta + e^{N\delta^{-1/2}}\beta)J + N\beta^{-p}(\|Mf\|^p_{\mathbb{L}_{p,\theta}(\tau)} + \|g\|^p_{\mathbb{L}_{p,\theta}(\tau)}).$$

By using (8.12) we conclude

$$\begin{aligned} I &\leq N_1(\delta + e^{N_1\delta^{-1/2}}\beta)I \\ &+ N(\delta + e^{N_1\delta^{-1/2}}\beta + \beta^{-p})(\|Mf\|^p_{\mathbb{L}_{p,\theta}(\tau)} + \|g\|^p_{\mathbb{H}^1_{p,\theta}(\tau)}). \end{aligned} \tag{8.15}$$

Now we can specify the choice of $\delta, \beta \in (0,1)$ depending only on $\theta, d, p, \delta_0, \delta_1, \tilde{\delta}$, and K. We take them so that

$$N_1\delta \leq 1/4, \quad N_1 e^{N_1\delta^{-1/2}}\beta \leq 1/4$$

and then (8.15) yields

$$I \leq N(\|Mf\|^p_{\mathbb{L}_{p,\theta}(T)} + \|g\|^p_{\mathbb{H}^1_{p,\theta}(\tau)}),$$

which is (8.2) and the theorem is proved. □

REFERENCES

[1] Z. Brzeźniak, *Stochastic partial differential equations in M-type 2 Banach spaces*, Potential Anal. (1995), **4**(1): 1–45.

[2] Z. Brzeźniak, *On stochastic convolution in Banach spaces and applications*, Stochastics Stochastics Rep. (1997), **61**(3–4): 245–295.

[3] G. Da Prato and A. Lunardi, *Maximal regularity for stochastic convolutions in L^p spaces*, Atti Accad. Naz. Lincei Cl. Sci. Fis. Mat. Natur. Rend. Lincei (9) Mat. Appl. (1998), **9**(1): 25–29.

[4] G. Da Prato and L. Tubaro eds, "Stochastic partial differential equations and applications—VII". Papers from the 7th Meeting held in Levico Terme, January 2004, Lecture Notes in Pure and Applied Mathematics, Vol. 245 (2006), Chapman & Hall/CRC, Boca Raton, FL.

[5] F. Flandoli, *Dirichlet boundary value problem for stochastic parabolic equations: Compatibility relations and regularity of solutions*, Stochastics and Stochastics Reports (1990), **29**(3): 331–357.

[6] M. Hairer and J.C. Mattingly, *Ergodicity of the 2D Navier-Stokes equations with degenerate stochastic forcing*, Ann. of Math. (2), (2006), **164**(3): 993–1032.

[7] Kyeong-Hun Kim, *On stochastic partial differential equations with variable coefficients in C^1 domains*, Stochastic Process. Appl. (2004), **112**(2): 261–283.

[8] Kyeong-Hun Kim and N.V. Krylov, *On SPDEs with variable coefficients in one space dimension*, Potential Analysis (2004), **21**(3): 209–239.

[9] Kyeong-Hun Kim and N.V. Krylov, *On the Sobolev space theory of parabolic and elliptic equations in C^1 domains*, SIAM J. Math. Anal. (2004), **36**(2): 618–642.

[10] N.V. Krylov, *On L_p-theory of stochastic partial differential equations in the whole space*, SIAM J. Math. Anal. (1996), **27**(2): 313–340.

[11] N.V. Krylov, *An analytic approach to SPDEs*, pp. 185–242 in Stochastic Partial Differential Equations: Six Perspectives, Mathematical Surveys and Monographs, Vol. 64, AMS, Providence, RI, 1999.

[12] N.V. Krylov, *Weighted Sobolev spaces and Laplace's equation and the heat equations in a half space*, Comm. in PDE (1999), **24**(9–10): 1611–1653.

[13] N.V. Krylov, *Maximum principle for SPDEs and its applications*, in "Stochastic Differential Equations: Theory and Applications, A Volume in Honor of Professor Boris L. Rozovskii", P.H. Baxendale, S.V. Lototsky, eds., Interdisciplinary Mathematical Sciences, Vol. **2**, World Scientific, 2007. http://arxiv.org/math.PR/0604125.

[14] N.V. Krylov and S.V. Lototsky, *A Sobolev space theory of SPDEs with constant coefficients on a half line*, SIAM J. Math. Anal. (1999), **30**(2): 298–325.

[15] N.V. Krylov and S.V. Lototsky, *A Sobolev space theory of SPDEs with constant coefficients in a half space*, SIAM J. Math. Anal. (1999), **31**(1): 19–33.

[16] S.B. Kuksin, *Remarks on the balance relations for the two-dimensional Navier-Stokes equation with random forcing*, J. Stat. Phys. (2006), **122**(1): 101–114.

[17] S.V. Lototsky, *Dirichlet problem for stochastic parabolic equations in smooth domains*, Stochastics and Stochastics Reports (1999), **68**(1–2): 145–175.

[18] S.V. Lototsky, *Sobolev spaces with weights in domains and boundary value problems for degenerate elliptic equations*, Methods and Applications of Analysis (2000), **1**(1): 195–204.

[19] S.V. Lototsky, *Linear stochastic parabolic equations, degenerating on the boundary of a domain*, Electron. J. Probab. (2001), **6**(24): 14.

[20] S. Lototsky and B. Rozovskii, *Wiener chaos solutions of linear stochastic evolution equations*, Ann. Probab. (2006), **34**(2): 638–662.

STOCHASTIC PARABOLIC EQUATIONS OF FULL SECOND ORDER*

SERGEY V. LOTOTSKY† AND BORIS L. ROZOVSKII‡

Abstract. A procedure is described for defining a generalized solution for stochastic differential equations using the Cameron-Martin version of the Wiener Chaos expansion. Existence and uniqueness of this Wiener Chaos solution is established for parabolic stochastic PDEs such that both the drift and the diffusion operators are of the second order.

1. Introduction. Consider a stochastic evolution equation

$$du(t) = (\mathcal{A}u(t) + f(t))dt + (\mathcal{M}u(t) + g(t))dW(t), \tag{1.1}$$

where $\mathcal{A}$ and $\mathcal{M}$ are differential operators, and W is a Wiener process on a probability space $(\Omega, \mathcal{F}, \mathbb{P})$. Traditionally, this equation was studied under the following assumptions:

- **A1.** The operator $\mathcal{A}$ is elliptic, the order of the operator $\mathcal{M}$ is less than the order of $\mathcal{A}$, and $\mathcal{A} - \frac{1}{2}\mathcal{M}\mathcal{M}^{\star}$ is elliptic (possibly degenerate) operator,

In fact, it is well known that unless assumption **A1** holds, Equation (1.1) has no solutions in $L_2(\Omega; X)$ for any reasonable choice of the state space X.

It was shown recently (see [4, 5, 6] and the references therein) that if only the operator $\mathcal{A}$ is elliptic and the order of $\mathcal{M}$ is *smaller* than the order of A, then there exists a unique generalized solution of Equation (1.1). This solution is often referred to as Wiener Chaos solution. It is given by the Wiener chaos expansion $u(t) = \sum_{|\alpha|<\infty} u_\alpha(t)\xi_\alpha$, where $\{\xi_\alpha\}_{|\alpha|<\infty}$ is the Cameron-Martin orthonormal basis in the space $L_2(\Omega; \mathcal{F}^W; X)$ of square integrable random elements in X measurable with respect to to the sigma-algebra $\mathcal{F}^W$ generated by the Wiener process. The Cameron-Martin basis $\{\xi_\alpha\}$ is indexed by multiindices $\alpha = (\alpha_1, \alpha_2, ...)$. It was shown that for certain positive weights $Q = \{q(\alpha)\}_{|\alpha|<\infty}$, the weighted norm

$$\|u\|_{Q,X}^2 := \sum_{|\alpha|<\infty} q^2(\alpha)\, \|u_\alpha\|_{L_2((0,T);X)}^2 < \infty,$$

*Sergey V. Lototsky acknowledges support from NSF CAREER award DMS-0237724. Boris L. Rozovskii acknowledges support from NSF Grant DMS 0604863, ARO Grant W911NF-07-1-0044, and ONR Grant N00014-07-1-0044.

†Department of Mathematics, USC, Los Angeles, CA 90089 (lototsky@math.usc.edu), http://www-rcf.usc.edu/~lototsky.

‡Division of Applied Mathematics, Brown University, Providence, RI 02912 (rozovsky@dam.brown.edu).

where X is the appropriate Hilbert space characterizing the "regularity" of the solution. Note that without assumption **A1**

$$E\|u\|^2_{L_2((0,T);X)} = \sum_{|\alpha|<\infty} \|u_\alpha(t)\|^2_{L_2((0,T);X)} = \infty.$$

In this paper, we consider the Cauchy problem for the following stochastic partial differential equation:

$$\begin{aligned} du =&(a_{ij}D_iD_ju + b_iD_iu + cu + f)dt \\ &+ (\rho_{ij}D_iD_ju + \sigma_iD_iu + \nu u + g)dW, \quad t \in [0,T], \ x \in \mathbb{R}. \end{aligned} \tag{1.2}$$

In contrast to the previous work, this is a parabolic SPDE of the *full* second order, in that the drift and diffusion operators have the same order 2. We construct a scale of weighted Wiener chaos spaces (related but not identical to Kondratiev's spaces) and prove the existence and uniqueness of the solution in the spaces from this scale.

2. Constructing a solution: an example. Let $\mathbb{F} = (\Omega, \mathcal{F}, \{\mathcal{F}_t\}_{0\le t\le T}, \mathbb{P})$ be a stochastic basis with the usual assumptions and $W = W(t)$, $0 \le t \le T$, a standard Wiener process on $\mathbb{F}$. For a Hilbert space X, denote by $L_2(\mathbb{W};X)$ the collection of X-valued random elements that are square integrable ($\mathbb{E}\|\cdot\|^2_X < \infty$) and are measurable with respect to the sigma-algebra generated by $W(t), t \in [0,T]$.

Consider the Itô equation

$$\begin{aligned} u(t,x) =&e^{-x^2/2} + \int_0^t u_{xx}(s,x)ds + \int_0^t u_{xx}(s,x)dW(s), \\ &t \in [0,T], \ x \in \mathbb{R}. \end{aligned} \tag{2.1}$$

If there is a solution, its Fourier transform in space, $\widehat{u}(t,y) = (1/\sqrt{2\pi})\int_{\mathbb{R}} e^{-ixy}u(t,x)dx$ satisfies

$$\begin{aligned} \widehat{u}(t,y) =&e^{-y^2/2} - y^2\int_0^t \widehat{u}(s,y)ds - y^2\int_0^t \widehat{u}(s,y)dW(t), \\ &t \in [0,T], \ y \in \mathbb{R}. \end{aligned} \tag{2.2}$$

For each fixed y, (2.2) defines a geometric Brownian motion:

$$\widehat{u}(t,y) = e^{-(1+t)y^2-(y^4/2)t-y^2W(t)}. \tag{2.3}$$

Let $H^\gamma(\mathbb{R})$ be the Sobolev space

$$\left\{f : \int_{\mathbb{R}}(1+|y|^2)^\gamma|\widehat{f}(y)|^2dy < \infty\right\}. \tag{2.4}$$

Since

$$\mathbb{E}|\widehat{u}(t,y)|^2 = e^{-2(1+t)y^2+y^4t}, \tag{2.5}$$

the solution of (2.1) cannot be an element of $L_2(\mathbb{W}; L_2((0,T); H^\gamma(\mathbb{R})))$ for *any* $\gamma \in \mathbb{R}$, even though the initial condition is non-random and is an element of $H^\gamma(\mathbb{R})$ for *every* $\gamma \in \mathbb{R}$.

Let us try another approach. Once again, assuming that the solution exists, we apply the Itô formula to the product $u(t,x)\mathcal{E}_h(t)$, where

$$\mathcal{E}_h(t) = \exp\left(\int_0^t h(s)dW(s) - \frac{1}{2}\int_0^t h^2(s)ds\right) \tag{2.6}$$

and $h = h(t)$ is a smooth deterministic function. Since

$$\mathcal{E}_h(t) = 1 + \int_0^t \mathcal{E}_h(s)h(s)dW(s), \tag{2.7}$$

we conclude that the function

$$u_h(t,x) = \mathbb{E}\big(u(t,x)\mathcal{E}_h(t)\big), \tag{2.8}$$

if defined, must satisfy the heat equation

$$u_h(t,x) = e^{-x^2/2} + \int_0^t (1+h(s))\frac{\partial^2 u_h(s,x)}{\partial x^2}ds. \tag{2.9}$$

If $\sup_t |h(t)| < 1$, then this equation has a unique solution in every $H^\gamma(\mathbb{R})$ and

$$u_h(t,x) = \mathbb{E}\exp(-(X(t,x))^2/2), \tag{2.10}$$

where

$$X(t,x) = x + \int_0^t \sqrt{2(1+h(s))}\,dW(s). \tag{2.11}$$

In other words, while existence of a solution of Equation (2.1) is still unclear, we now have a family of functions $u_h(t,x)$ defined by (2.10). All we need now is a systematic procedure of relating the family of deterministic functions $u_h = u_h(t,x)$ to a random process $u = u(t,x)$; then this process is natural to call a solution of (2.1).

Here is a possible way of constructing a stochastic process from u_h. Let $\mathfrak{m} = \{m_k,\ k \geq 1\}$ be the Fourier cosine basis in $L_2((0,T))$:

$$m_1(s) = \frac{1}{\sqrt{T}};\ m_k(t) = \sqrt{\frac{2}{T}}\cos\left(\frac{\pi(k-1)t}{T}\right),\ k>1;\ 0 \leq t \leq T. \tag{2.12}$$

Then

$$h(t) = \sum_{k\geq 1} h_k m_k(t). \tag{2.13}$$

For every fixed $t \in [0,T]$ and $\gamma \in \mathbb{R}$, we can now interpret the function $u_h(t,\cdot)$ as a mapping from the set of sequences $h = (h_1, h_2, \ldots)$ to the space $H^\gamma(\mathbb{R}^d)$, and, as equalities (2.10) and (2.11) suggest, this mapping is analytic in the region $\{h : \sum_{k\geq 1} h_k^2 < \varepsilon\}$ for sufficiently small ε. We will now compute the derivatives of this mapping.

Let $\mathcal{J}$ be the collection of multi-indices $\alpha = \{\alpha_k,\ k \geq 1\}$. Each $\alpha \in \mathcal{J}$ has non-negative integer elements α_k and

$$|\alpha| = \sum_k \alpha_k < \infty. \tag{2.14}$$

We also use the notation

$$\alpha! = \prod_k \alpha_k! \tag{2.15}$$

and consider special multi-indices, $\alpha = (0)$ with $|\alpha| = 0$ and $\alpha = \varepsilon_i$, with $|\alpha| = 1$, $\alpha_i = 1$.

For each $\alpha \in \mathcal{J}$ define

$$u_\alpha(t,x) = \frac{1}{\sqrt{\alpha!}} \frac{\partial^{|\alpha|} u_h(t,x)}{\partial h_1^{\alpha_1} \partial h_2^{\alpha_2} \cdots}\bigg|_{h=0}. \tag{2.16}$$

Then

$$u_h(t,x) = \sum_{\alpha \in \mathcal{J}} u_\alpha(t,x) \frac{h^\alpha}{\sqrt{\alpha!}}, \tag{2.17}$$

where

$$h^\alpha = \prod_{k\geq 1} h_k^{\alpha_k}. \tag{2.18}$$

On the other hand, by direct computation,

$$\mathcal{E}_h(t) = \mathbb{E}(\mathcal{E}_h(T)|\mathcal{F}_t^W) = \sum_{\alpha \in \mathcal{J}} \frac{h^\alpha}{\sqrt{\alpha!}} \xi_\alpha(t), \tag{2.19}$$

where

$$\xi_\alpha(t) = \mathbb{E}(\xi_\alpha|\mathcal{F}_t^W),\ \xi_\alpha = \frac{1}{\sqrt{\alpha!}} \prod_{k\geq 1} H_{\alpha_k}\left(\int_0^T m_k(t)dW(t)\right), \tag{2.20}$$

and

$$H_n(x) = (-1)^n e^{x^2/2} \frac{d^n}{dx^n} e^{-x^2/2} \tag{2.21}$$

is n-th Hermite polynomial. It is a standard fact [1] that the collection $\{\xi_\alpha,\ \alpha \in \mathcal{J}\}$ is an orthonormal basis in $L_2(\mathbb{W};\mathbb{R})$.

The functions $u_\alpha(t,x)$, $\alpha \in \mathcal{J}$, uniquely determine $u_h(t,x)$ according to (2.17). On the other hand, if

$$\sum_{\alpha\in\mathcal{J}} \|u_\alpha(t)\|^2_{H^\gamma(\mathbb{R})} < \infty, \tag{2.22}$$

then the $H^\gamma(\mathbb{R})$-valued random process

$$u(t,x) = \sum_{\alpha\in\mathcal{J}} u_\alpha(t,x)\xi_\alpha \tag{2.23}$$

satisfies $\mathbb{E}(u(t,x)\mathcal{E}_h(t)) = u_h(t,x)$; if, in addition, u is $\mathcal{F}_t^W$-adapted, then also

$$u(t,x) = \sum_{\alpha\in\mathcal{J}} u_\alpha(t,x)\xi_\alpha(t). \tag{2.24}$$

If condition (2.22) fails, then (2.23) is a formal series, which we *define* to be the stochastic process corresponding to the family u_h.

As (2.5) suggests, if u_h is the solution of (2.9), then (2.22) fails for every γ. Let us now see how fast the series diverges. Equality (2.9) implies

$$\begin{aligned} u_{(0)}(t,x) &= e^{-x^2/2} + \int_0^t \frac{\partial^2 u_{(0)}(s,x)}{\partial x^2} ds, \ |\alpha| = 0;\\ u_{\epsilon_i}(t) &= \int_0^t \frac{\partial^2 u_{\epsilon_i}(s,x)}{\partial x^2} ds + \int_0^t \frac{\partial^2 u_{(0)}(s,x)}{\partial x^2} m_i(s) ds, \ |\alpha| = 1; \\ u_\alpha(t) &= \int_0^t \frac{\partial^2 u_\alpha(s,x)}{\partial x^2} ds + \sum_{k=1}^\infty \sqrt{\alpha_k} \int_0^t \frac{\partial^2 u_{\alpha-\epsilon_k}(s,x)}{\partial x^2} m_k(s) ds, \ |\alpha| > 1. \end{aligned} \tag{2.25}$$

Equations of the type (2.25) have been studied [4, Section 6 and References]. In particular, it is known that

$$\sum_{|\alpha|=n} \|u_\alpha(t)\|^2_{H^\gamma(\mathbb{R})} = \frac{t^n}{n!} \|D_x^{2n}\Phi_t u_0\|^2_{H^\gamma(\mathbb{R})}, \tag{2.26}$$

where $D_x = \partial/\partial x$, Φ_t is the heat semigroup, and $u_0(x) = e^{-x^2/2}$. To simplify further computation, let us assume that $\gamma = 0$. Then, switching to the Fourier transform,

$$\|D_x^{2n}\Phi_t u_0\|^2_{L_2(\mathbb{R}^d)} = \int_{\mathbb{R}} |y|^{4n} e^{-y^2(t+1)} dy = \frac{\Gamma\left(2n+\frac{1}{2}\right)}{(1+t)^{2n}}. \tag{2.27}$$

Using Stirling's formula for the Gamma function Γ,

$$\sum_{|\alpha|=n} \|u_\alpha(t)\|^2_{L_2(\mathbb{R})} = \left(\frac{2\sqrt{t}}{1+t}\right)^{2n} C(n) n!, \tag{2.28}$$

where the numbers $C(n)$ are uniformly bounded from above and below. Similar result holds in every $H^{\gamma}(\mathbb{R})$. Thus, (2.22) does not hold, but instead, by (2.28), we have

$$\sum_{\alpha\in\mathcal{J}} \frac{1}{|\alpha|!}\|u_\alpha(t)\|^2_{H^\gamma(\mathbb{R})} < \infty. \tag{2.29}$$

We denote by $(\mathfrak{L})_{0,0}(\mathbb{W}; H^{\gamma}(\mathbb{R}))$ the collection of formal series (2.24) satisfying (2.29); the reason for using $(\mathfrak{L})_{0,0}$ in the notation will become clear later. Note that we had equalities in all computations for Equation (2.1) that lead to (2.29), which suggests that $(\mathfrak{L})_{0,0}(\mathbb{W}; H^{\gamma}(\mathbb{R}))$ is the natural solution space for Equation (2.1). For a more general stochastic parabolic equation of full second order in $\mathbb{R}^d$, the natural solution space turns out to be $(\mathfrak{L})_{p,q}(\mathbb{W}; L_2((0,T); H^{\gamma}(\mathbb{R}^d)))$ for suitable $p, q \leq 0$.

In the next section we address the following questions:

1. How to define the spaces $(\mathfrak{L})_{p,q}(\mathbb{W}; X)$ for $p, q \in \mathbb{R}$ without relying on an orthonormal basis in $L_2((0,T))$?
2. How to construct a solution of a general stochastic parabolic equations of full second order?

3. General constructions and the main result. As before, let $\mathbb{F} = (\Omega, \mathcal{F}, \{\mathcal{F}_t\}_{0\leq t\leq T}, \mathbb{P})$ be a stochastic basis with the usual assumptions and $W = W(t),\ 0 \leq t \leq T$, a standard Wiener process on $\mathbb{F}$. Denote by $H^s = H^s((0,T))$, $s \geq 0$, the Sobolev spaces on $(0,T)$ with norm

$$\|\cdot\|_s = \|\Lambda^{s/2}\cdot\|_{L_2((0,T))}, \tag{3.1}$$

where Λ is the operator

$$1 - \frac{T^2}{\pi^2}\frac{d^2}{dt^2} \tag{3.2}$$

with Neumann boundary conditions. This norm extends to functions of several variables via the tensor product of the spaces H^s.

DEFINITION 3.1. *Given real numbers p, q and a Hilbert space X, $(\mathfrak{L})_{p,q}(\mathbb{W}; X)$ is the closure of the set of X-valued random elements*

$$\eta = \eta_0 + \sum_{k=1}^{N}\int_0^T\int_0^{s_k}\cdots\int_0^{s_2}\eta_k(s_1,\ldots,s_k)dW(s_1)\ldots dW(s_{k-1})dW(s_k), \tag{3.3}$$

$$N \geq 1,$$

with respect to the norm

$$\|\eta\|^2_{p,q;X} = \|\eta_0\|^2_X + \sum_{k=1}^{N}\frac{2^{kp}}{(k!)^2}\|\,\|\eta_k\|_q\,\|^2_X, \tag{3.4}$$

where each η_k, $k \geq 1$, is a smooth symmetric function from $[0,T]^k$ to X.

REMARK 3.1. (a) It is known [2, 7] that, for η of the type (3.3),

$$\mathbb{E}\|\eta\|_X^2 = \|\eta_0\|_X^2 + \sum_{k=1}^{N} \frac{1}{k!}\| \, \|\eta_k\|_0 \, \|_X^2. \tag{3.5}$$

(b) The definition of each individual $(\mathfrak{L})_{p,q}(\mathbb{W};X)$ inevitably involves arbitrary choices, such as the norm in $H^q((0,T))$. Further analysis shows that different choices result in shifts of the indices p,q, and the space $\cup_{p,q}(\mathfrak{L})_{p,q}(\mathbb{W};X)$ does not depend on any arbitrary choices. In the white noise setting, where Ω is the space $\mathcal{S}'(\mathbb{R}^d)$ of the Schwartz distributions and $\mathbb{P}$ is the normalized Gaussian measure on $\mathcal{S}$, the inductive limit $\cup_{p,q}(\mathfrak{L})_{p,q}(\mathbb{W};\mathbb{R})$ is the Kondratiev space $(\mathcal{S})_{-1}$ [2].

REMARK 3.2. If X is the Sobolev space $H^\gamma(\mathbb{R}^d)$, then we denote the norm $\|\cdot\|_{p,q;X}$ by $\|\cdot\|_{p,q;\gamma}$:

$$\|\cdot\|_{p,q;H^\gamma(\mathbb{R}^d)} = \|\cdot\|_{p,q;\gamma}. \tag{3.6}$$

PROPOSITION 3.1. *Let $\eta = f\xi_\alpha$, where $f \in X$ and ξ_α is defined by* (2.20). *Then*

$$\|\eta\|_{p,q;X}^2 = \frac{2^{|\alpha|p}}{|\alpha|!}\mathbb{N}^{2q\alpha}\|f\|_X^2, \tag{3.7}$$

where

$$\mathbb{N}^{2q\alpha} = \prod_{k\geq 1} k^{2q\alpha_k}. \tag{3.8}$$

Proof. Let $|\alpha| = n$. It is known [3] that

$$\xi_\alpha = \frac{1}{\sqrt{\alpha!}}\int_0^T\int_0^{s_n}\cdots\int_0^{s_2} E_\alpha(s_1,\dots,s_n)dW(s_1)\dots dW(s_{n-1})dW(s_n), \tag{3.9}$$

where E_α is the symmetric function

$$E_\alpha(s_1,\dots,s_n) = \sum_{\sigma\in\mathcal{P}_n} m_{i_1}(s_{\sigma(1)})\dots m_{i_n}(s_{\sigma(n)}). \tag{3.10}$$

In (3.10), the summation is over all permutations of $\{1,\dots,n\}$, the functions m_k are defined in (2.12), and the positive integer numbers $i_1 \leq i_2 \leq \dots i_n$ are such that, for every sequence $(b_k, k\geq 1)$ of positive numbers,

$$\prod_{k\geq 1} b_k^{\alpha_k} = b_{i_1}\cdot b_{i_2}\cdot\dots\cdot b_{i_n}. \tag{3.11}$$

For example, if $\alpha = (1,0,2,0,0,4,0,0,\ldots)$, then $|\alpha| = 7$ and $i_1 = 1$, $i_2 = i_3 = 3$, $i_4 = \ldots = i_7 = 6$. Thus, in the notations of (3.4), we have

$$\eta_k = \begin{cases} \dfrac{1}{\sqrt{\alpha!}} E_\alpha f, & \text{if } k = n, \\ 0, & \text{otherwise.} \end{cases} \tag{3.12}$$

Note that

$$\|E_\alpha\|_0 = \sqrt{\alpha!}\sqrt{n!}. \tag{3.13}$$

By definition (3.2) of the operator Λ we have

$$\Lambda^{q/2} m_k = k^q m_k \quad \text{or} \quad \|E_\alpha\|_q^2 = \mathbb{N}^{2q\alpha}\alpha! n!. \tag{3.14}$$

The result now follows. □

COROLLARY 3.1. *A formal series*

$$\eta = \sum_{\alpha \in \mathcal{J}} \eta_\alpha \xi_\alpha, \tag{3.15}$$

with $\eta_\alpha \in X$, *is an element of* $(\mathfrak{L})_{p,q}(\mathbb{W}; X)$ *if and only if*

$$\sum_{\alpha \in \mathcal{J}} \frac{2^{p|\alpha|}\mathbb{N}^{2q\alpha}}{|\alpha|!} \|\eta_\alpha\|_X^2 < \infty. \tag{3.16}$$

Proof. This follows from (3.14) and the equality

$$\|E_\alpha + E_\beta\|_0^2 = \|E_\alpha\|_0^2 + \|E_\beta\|_0^2, \; \alpha \neq \beta. \tag{3.17}$$

□

Denote by $(\mathfrak{L})^{p,q}(\mathbb{W})$ the Hilbert space dual of $(\mathfrak{L})_{p,q}(\mathbb{W}; \mathbb{R})$ relative to the inner product in $L_2(\mathbb{W}; \mathbb{R})$, and by $\langle\langle \cdot, \cdot \rangle\rangle$ the corresponding duality. In the white noise setting, $\cap_{p,q}(\mathfrak{L})^{p,q}(\mathbb{W})$ is the space $(\mathcal{S})_1$ of the Kondratiev test functions [2]. If $\eta \in (\mathfrak{L})_{p,q}(\mathbb{W}; X)$ and $\zeta \in (\mathfrak{L})^{p,q}(\mathbb{W})$, then $\langle\langle \eta, \zeta \rangle\rangle$ is defined and belongs to X.

For $h \in L_2((0,T))$, define

$$\mathcal{E}_h = \mathcal{E}_h(T) = \exp\left(\int_0^T h(s)dW(s) - \frac{1}{2}\int_0^T |h(s)|^2 ds \right). \tag{3.18}$$

PROPOSITION 3.2. *The random variable* $\mathcal{E}_h$ *is an element of* $(\mathfrak{L})^{p,q}(\mathbb{W})$ *if an only if*

$$\|h\|_{-q}^2 < 2^p. \tag{3.19}$$

Proof. Since

$$\mathcal{E}_h(T) = 1 + \int_0^T h(t)\mathcal{E}_h(t)dt, \tag{3.20}$$

it follows that

$$\mathcal{E}_h = 1 + \sum_{k=1}^{\infty} \int_0^T \int_0^{s_k} \dots \int_0^{s_2} h(s_k)\cdots h(s_1)dW(s_1)\cdots dW(s_{k-1})dW(s_k). \tag{3.21}$$

By (3.4) and (3.5), $\mathcal{E}_h \in (\mathfrak{L})^{p,q}(\mathbb{W})$ if and only if

$$\sum_{k=1}^{\infty} \left(2^{-p}\|h\|_{-q}^2\right)^k < \infty, \tag{3.22}$$

that is, $\|h\|_{-q}^2 < 2^p$. □

DEFINITION 3.2. *We say that the function h is sufficiently small if (3.19) holds for sufficiently large (positive) $-p, -q$.*

PROPOSITION 3.3. *If $u \in \bigcup_{p,q}(\mathfrak{L})_{p,q}(\mathbb{W}; X)$ and h is sufficiently small, then*

$$u_h = \langle\!\langle u, \mathcal{E}_h \rangle\!\rangle \tag{3.23}$$

is an X-valued analytic function of h.

Proof. For every $u \in \bigcup_{p,q}(\mathfrak{L})_{p,q}(\mathbb{W}; X)$, there exist p, q such that $u \in (\mathfrak{L})_{p,q}(\mathbb{W}; X)$; by Proposition 3.2, u_h will indeed be defined for sufficiently small h. Similar to (2.17) we have

$$u_h = \sum_{\alpha\in\mathcal{J}} \frac{u_\alpha h^\alpha}{\sqrt{\alpha!}} \tag{3.24}$$

and this power series in h^α converges in some (infinite-dimensional) neighborhood of zero. □

From now on, $D_i = \partial/\partial x_i$, and the summation convention is in force: $c_i d_i = \sum_i c_i d_i$, etc.

Consider the linear equation in $\mathbb{R}^d$

$$du = (a_{ij}D_iD_ju + b_iD_iu + cu + f)dt + (\rho_{ij}D_iD_ju + \sigma_iD_iu + \nu u + g)dW \tag{3.25}$$

with initial condition $u(0, x) = v(x)$, under the following **assumptions:**

B0. All coefficients are non-random.

B1. The functions $a_{ij} = a_{ij}(t, x)$, $\rho_{ij} = \rho_{ij}(t, x)$ are measurable and bounded in (t, x) by a positive number C_0, and

(i) $|a_{ij}(t, x) - a_{ij}(t, y)| + |\rho_{ij}(t, x) - \rho_{ij}(t, y)| \le C_0|x - y|$, $x, y \in \mathbb{R}^d$, $0 \le t \le T$;

(ii) the matrix (a_{ij}) is uniformly positive definite, that is, there exists a $\delta > 0$ so that, for all vectors $y \in \mathbb{R}^d$ and all (t,x), $a_{ij}y_iy_j \geq \delta|y|^2$.

B2. The functions $b_i = b_i(t,x)$, $c = c(t,x)$, $\sigma_i = \sigma_i(t,x)$, and $\nu = \nu(t,x)$ are measurable and bounded in (t,x) by the number C_0.

B3.

$$\begin{aligned} u_0 &\in \bigcup_{p,q} (\mathfrak{L})_{p,q}(\mathbb{W}; L_2(\mathbb{R}^d)), \\ f, g &\in \bigcup_{p,q} (\mathfrak{L})_{p,q}(\mathbb{W}; L_2((0,T); H^{-1}(\mathbb{R}^d))). \end{aligned} \tag{3.26}$$

For simplicity, we introduce the following notations for the differential operators in (3.25):

$$\mathcal{A} = a_{ij}D_iD_j + b_iD_i + c, \quad \mathcal{B} = \rho_{ij}D_iD_j + \sigma_iD_i + \nu. \tag{3.27}$$

DEFINITION 3.3. *A solution u of (3.25) is an element of $\bigcup_{p,q}(\mathfrak{L})_{p,q}(\mathbb{W}; L_2((0,T); H^1(\mathbb{R}^d)))$ such that, for all sufficiently small h and all $t \in [0,T]$, the equality*

$$u_h(t,x) = v_h(x) + \int_0^t (\mathcal{A} + h(s)\mathcal{B})u_h(s,x)ds \tag{3.28}$$

holds in $H^{-1}(\mathbb{R}^d)$.

The following theorem is the main result of this paper.

THEOREM 3.1. *Assume that, for some $p > 0$ and $q > 1$, $u_0 \in (\mathfrak{L})_{p,q}(\mathbb{W}; L_2(\mathbb{R}^d))$ and f, g are elements of the space $(\mathfrak{L})_{p,q}(\mathbb{W}; L_2((0,T); H^{-1}(\mathbb{R}^d)))$. Then there exist $r, \ell < 0$ such that Equation (3.25) has a unique solution $u \in (\mathfrak{L})_{r,\ell}(\mathbb{W}; L_2((0,T); H^1(\mathbb{R}^d)))$ and*

$$\int_0^T \|u(t)\|^2_{r,\ell;1}dt \leq C \cdot \left(\|v\|^2_{p,q;0} + \int_0^T \left(\|f(t)\|^2_{p,q;-1} + \|g(t)\|^2_{p,q;-1} \right) dt \right). \tag{3.29}$$

The number $C > 0$ depends only on $\delta, C_0, p, q, r, \ell$, and T.

Proof. The proof consists of two steps: first, we prove the result for deterministic functions v, f, g and then use linearity to extend the result to the general case.

Step 1. Assume that the functions $v \in L_2(\mathbb{R}^d)$, $f, g \in L_2((0,T); H^{-1}(\mathbb{R}^d))$ are deterministic. Then $v_h = v$, $f_h = f$, $g_h = g$, and classical theory of parabolic equations shows that, for sufficiently small h, Equation (3.28) has a unique solution u_h and the dependence of u_h on h is analytic.

As in the previous section, we write

$$u(t,x) = \sum_{\alpha \in \mathcal{J}} u_\alpha(t,x)\xi_\alpha \tag{3.30}$$

where the coefficients u_α satisfy

$$\begin{aligned} u_{(0)}(t,x) &= v(x) + \int_0^t (\mathcal{A}u_{(0)}(s,x) + f(s,x))ds, \\ u_{\epsilon_k}(t,x) &= \int_0^t \mathcal{A}u_{\epsilon_k}(s,x)ds + \int_0^t (\mathcal{B}u_{(0)}(s,x) + g(s,x))m_k(s)ds, \quad (3.31) \\ u_\alpha(s,x) &= \int_0^t \mathcal{A}u_\alpha(s,x)ds + \sum_k \sqrt{\alpha_k}\int_0^t \mathcal{B}u_{\alpha-\epsilon_k}(s,x)m_k(s)ds, \ |\alpha|>1. \end{aligned}$$

Denote by $\Phi = \Phi_{s,t}$, $t \geq s \geq 0$ the semigroup generated by the operator $\mathcal{A}$. It follows by induction on $|\alpha|$ that

$$\begin{aligned} u_{(0)}(t,x) &= \Phi_{t,0}v(x) + \int_0^t \Phi_{t,s}f(s)ds, \\ u_{\epsilon_k}(t,x) &= \int_0^t \Phi_{t,s}(\mathcal{B}u_{(0)}(s,x) + g(s,x))m_k(s)ds, \\ u_\alpha(t,x) &= \frac{1}{\sqrt{\alpha!}}\int_0^t\int_0^{s_n}\ldots\int_0^{s_2} \Phi_{t,s_n}\mathcal{B}\Phi_{s_n,s_{n-1}}\cdots\mathcal{B}\Phi_{s_2,s_1}(\mathcal{B}u_{(0)}(s_1,x)+g(s_1,x)) \\ &\quad E_\alpha(s_1,\ldots,s_n)ds_1\ldots ds_n, \ |\alpha| = n > 1. \end{aligned} \quad (3.32)$$

Therefore, using the usual parabolic estimates,

$$\begin{aligned} &\int_0^T \|u_\alpha(t)\|^2_{H^1(\mathbb{R}^d)}dt \\ &\leq \frac{C^n n!}{\alpha!}\left(\|v\|^2_{L_2(\mathbb{R}^d)} + \int_0^T \left(\|f(t)\|^2_{H^{-1}(\mathbb{R}^d)} + \|g(t)\|^2_{H^{-1}(\mathbb{R}^d)}\right)dt\right), \end{aligned} \quad (3.33)$$

and then (3.29) follows from (3.16).

Step 2. As in Step 1, existence and uniqueness of solution follows from unique solvability of the parabolic equation (3.28), and it remains to establish (3.29).

Denote by $u(t,x;V,F,G,\gamma)$, $\gamma \in \mathcal{J}$, the solution of (3.25) with $v = V\xi_\gamma$, $f = F\xi_\gamma$, $g = G\xi_\gamma$. If $v = \sum_{\alpha\in\mathcal{J}} v_\alpha\xi_\alpha$, etc., then

$$u(t,x) = \sum_{\gamma\in\mathcal{J}} u(t,x;v_\gamma,f_\gamma,g_\gamma,\gamma). \quad (3.34)$$

It follows from (3.31) that $u_\alpha(t,x;V,F,G,\gamma) = 0$ if $|\alpha| < |\gamma|$ and

$$\frac{u_{\alpha+\gamma}(t,x;V,F,G,\gamma)}{\sqrt{(\alpha+\gamma)!}} = \frac{u_\alpha\left(t,x;\frac{V}{\sqrt{\gamma!}},\frac{F}{\sqrt{\gamma!}},\frac{G}{\sqrt{\gamma!}},(0)\right)}{\sqrt{\alpha!}}. \quad (3.35)$$

Using the results of Step 1,

$$\int_0^T \|u(t,\cdot;v_\gamma,f_\gamma,g_\gamma)\|^2_{r,\ell;1}dt \le \frac{C}{\gamma!}\left(\|v_\gamma\|^2_{L_2(\mathbb{R}^d)} + \int_0^T \left(\|f_\gamma(t)\|^2_{H^{-1}(\mathbb{R}^d)} + \|g_\gamma(t)\|^2_{H^{-1}(\mathbb{R}^d)}\right)dt\right). \tag{3.36}$$

Now (3.29) follows from (3.34) by the triangle inequality. □

REFERENCES

[1] R.H. Cameron and W.T. Martin. The orthogonal development of nonlinear functionals in a series of Fourier-Hermite functions. *Ann. Math.*, **48**(2): 385–392, 1947.

[2] H. Holden, B. Øksendal, J. Ubøe, and T. Zhang. *Stochastic Partial Differential Equations.* Birkhäuser, Boston, 1996.

[3] K. Ito. Multiple Wiener integral. *J. Math. Soc. Japan*, **3**: 157–169, 1951.

[4] S.V. Lototsky and B.L. Rozovskii. Stochastic differential equations: a Wiener chaos approach. In Yu. Kabanov, R. Liptser, and J. Stoyanov, editors, *From stochastic calculus to mathematical finance: the Shiryaev festschrift*, pp. 433–507. Springer, 2006.

[5] S.V. Lototsky and B.L. Rozovskii. Wiener chaos solutions of linear stochastic evolution equations. *Ann. Probab.*, **34**(2): 638–662, 2006.

[6] R. Mikulevicius and B.L. Rozovskii. Linear parabolic stochastic PDE's and Wiener Chaos. *SIAM J. Math. Anal.*, **292**: 452–480, 1998.

[7] D. Nualart. *Malliavin Calculus and Related Topics, 2nd Edition.* Springer, New York, 2006.

1997–1998 Emerging Applications of Dynamical Systems
1998–1999 Mathematics in Biology
1999–2000 Reactive Flows and Transport Phenomena
2000–2001 Mathematics in Multimedia
2001–2002 Mathematics in the Geosciences
2002–2003 Optimization
2003–2004 Probability and Statistics in Complex Systems: Genomics, Networks, and Financial Engineering
2004–2005 Mathematics of Materials and Macromolecules: Multiple Scales, Disorder, and Singularities
2005–2006 Imaging
2006–2007 Applications of Algebraic Geometry
2007–2008 Mathematics of Molecular and Cellular Biology
2008–2009 Mathematics and Chemistry

IMA SUMMER PROGRAMS

1987 Robotics
1988 Signal Processing
1989 Robust Statistics and Diagnostics
1990 Radar and Sonar (June 18–29)
New Directions in Time Series Analysis (July 2–27)
1991 Semiconductors
1992 Environmental Studies: Mathematical, Computational, and Statistical Analysis
1993 Modeling, Mesh Generation, and Adaptive Numerical Methods for Partial Differential Equations
1994 Molecular Biology
1995 Large Scale Optimizations with Applications to Inverse Problems, Optimal Control and Design, and Molecular and Structural Optimization
1996 Emerging Applications of Number Theory (July 15–26)
Theory of Random Sets (August 22–24)
1997 Statistics in the Health Sciences
1998 Coding and Cryptography (July 6–18)
Mathematical Modeling in Industry (July 22–31)
1999 Codes, Systems, and Graphical Models (August 2–13, 1999)
2000 Mathematical Modeling in Industry: A Workshop for Graduate Students (July 19–28)
2001 Geometric Methods in Inverse Problems and PDE Control (July 16–27)
2002 Special Functions in the Digital Age (July 22–August 2)
2003 Probability and Partial Differential Equations in Modern Applied Mathematics (July 21–August 1)
2004 n-Categories: Foundations and Applications (June 7–18)

2005 Wireless Communications (June 22–July 1)
2006 Symmetries and Overdetermined Systems of Partial Differential Equations (July 17–August 4)
2007 Classical and Quantum Approaches in Molecular Modeling (July 23–August 3)
2008 Geometrical Singularities and Singular Geometries (July 14–25)

IMA "HOT TOPICS" WORKSHOPS

- Challenges and Opportunities in Genomics: Production, Storage, Mining and Use, April 24–27, 1999
- Decision Making Under Uncertainty: Energy and Environmental Models, July 20–24, 1999
- Analysis and Modeling of Optical Devices, September 9–10, 1999
- Decision Making under Uncertainty: Assessment of the Reliability of Mathematical Models, September 16–17, 1999
- Scaling Phenomena in Communication Networks, October 22–24, 1999
- Text Mining, April 17–18, 2000
- Mathematical Challenges in Global Positioning Systems (GPS), August 16–18, 2000
- Modeling and Analysis of Noise in Integrated Circuits and Systems, August 29–30, 2000
- Mathematics of the Internet: E-Auction and Markets, December 3–5, 2000
- Analysis and Modeling of Industrial Jetting Processes, January 10–13, 2001
- Special Workshop: Mathematical Opportunities in Large-Scale Network Dynamics, August 6–7, 2001
- Wireless Networks, August 8–10, 2001
- Numerical Relativity, June 24–29, 2002
- Operational Modeling and Biodefense: Problems, Techniques, and Opportunities, September 28, 2002
- Data-driven Control and Optimization, December 4–6, 2002
- Agent Based Modeling and Simulation, November 3–6, 2003
- Enhancing the Search of Mathematics, April 26–27, 2004
- Compatible Spatial Discretizations for Partial Differential Equations, May 11–15, 2004
- Adaptive Sensing and Multimode Data Inversion, June 27–30, 2004
- Mixed Integer Programming, July 25–29, 2005
- New Directions in Probability Theory, August 5–6, 2005
- Negative Index Materials, October 2–4, 2006
- The Evolution of Mathematical Communication in the Age of Digital Libraries, December 8–9, 2006
- Math is Cool! and Who Wants to Be a Mathematician?, November 3, 2006
- Special Workshop: Blackwell-Tapia Conference, November 3–4, 2006
- Stochastic Models for Intracellular Reaction Networks, May 11–13, 2008

SPRINGER LECTURE NOTES FROM THE IMA:

The Mathematics and Physics of Disordered Media
Editors: Barry Hughes and Barry Ninham
(Lecture Notes in Math., Volume 1035, 1983)

Orienting Polymers
Editor: J.L. Ericksen
(Lecture Notes in Math., Volume 1063, 1984)

New Prespectives in Thermodynamics
Editor: James Serrin
(Springer-Verlag, 1986)

Models of Economic Dynamics
Editor: Hugo Sonnenschein
(Lecture Notes in Econ., Volume 264, 1986)

The IMA Volumes in Mathematics and its Applications

1 **Volume 142: Compatible Spatial Discretizations**
Editors: Douglas N. Arnold, Pavel B. Bochev, Richard B. Lehoucq, Roy A. Nicolaides, and Mikhail Shashkov

2 **Volume 143: Wireless Communications**
Editors: Prathima Agrawal, Daniel Matthew Andrews, Philip J. Fleming, George Yin, and Lisa Zhang

3 **Volume 144: Symmetries and Overdetermined Systems of Partial Differential Equations**
Editors: Michael Eastwood and Willard Miller, Jr.

4 **Volume 145: Topics in Stochastic Analysis and Nonparametric Estimation**
Editors: Pao-Liu Chow, Boris Mordukhovich, and George Yin

5 **Volume 146: Algorithms in Algebraic Geometry**
Editors: Alicia Dickenstein, Frank-Olaf Schreyer, and Andrew Sommese

6 **Volume 147: Symmetric Functionals on Random Matrices and Random Matchings Problems**
by Grzegorz A. Rempała and Jacek Wesołowski

The full list of IMA books can be found at the Web site of Institute for Mathematics and its Applications:
http://www.ima.umn.edu/springer/volumes.html

Printed in the United States of America.

GPSR Compliance

The European Union's (EU) General Product Safety Regulation (GPSR) is a set of rules that requires consumer products to be safe and our obligations to ensure this.

If you have any concerns about our products, you can contact us on ProductSafety@springernature.com

In case Publisher is established outside the EU, the EU authorized representative is:

Springer Nature Customer Service Center GmbH
Europaplatz 3
69115 Heidelberg, Germany

Batch number: 10682213

Printed by Printforce, the Netherlands